About Island Press

Island Press is the only nonprofit organization in the United States whose principal purpose is the publication of books on environmental issues and natural resource management. We provide solutions-oriented information to professionals, public officials, business and community leaders, and concerned citizens who are shaping responses to environmental problems.

In 1999, Island Press celebrates its fifteenth anniversary as the leading provider of timely and practical books that take a multidisciplinary approach to critical environmental concerns. Our growing list of titles reflects our commitment to bringing the best of an expanding body of literature to the environmental community throughout North America and the world.

Support for Island Press is provided by The Jenifer Altman Foundation, The Bullitt Foundation, The Mary Flagler Cary Charitable Trust, The Nathan Cummings Foundation, The Geraldine R. Dodge Foundation, The Charles Engelhard Foundation, The Ford Foundation, The Vira I. Heinz Endowment, The W. Alton Jones Foundation, The John D. and Catherine T. MacArthur Foundation, The Andrew W. Mellon Foundation, The Charles Stewart Mott Foundation, The Curtis and Edith Munson Foundation, The National Fish and Wildlife Foundation, The National Science Foundation, The New-Land Foundation, The David and Lucile Packard Foundation, The Pew Charitable Trusts, The Surdna Foundation, The Winslow Foundation, and individual donors.

This publication was made possible by the generous support of the John D. and Catherine T. MacArthur Foundation.

The Business of
Sustainable Forestry

The Business of Sustainable Forestry

Strategies for an Industry in Transition

Michael B. Jenkins
and Emily T. Smith

ISLAND PRESS
Washington, D.C. ● Covelo, California

Library of Congress Cataloging-in-Publication Data
Jenkins, Michael B.
    The business of sustainable forestry : strategies for an industry
in transition / by Michael B. Jenkins and Emily T. Smith
        p.   cm.
    Includes bibliographical references.
    ISBN 1–55963–713–7 (alk. paper : cloth)
    1. Sustainable forestry.   2. Forest products industry.   I. Smith,
Emily T.   II. Title.
SD387.S87J45   1999   99–18533
33.75'15—dc21   CIP

Printed on recycled, acid-free paper ⊛

Manufactured in the United States of America
10 9 8 7 6 5 4 3 2 1

Contents

Foreword   ix
Jonathan Lash and Stephan Schmidheiny

Preface   xiii

Introduction: Sustaining Forests and Profits   1

1   The Pursuit of Innovation   11

2   The Forest Products Industry in Transition   24

3   Emerging Markets for Certified Wood Products   59

4   New Mandates for Technology   88

5   Lessons from a Pioneer   114

6   Harvesting Ecology and Economics   131

7   Opportunities for Private Timber Owners
      in Sustainable Forestry   157

8   Rich Niches   189

9   The Difficulties of Sustainable Forest
      Management in the Tropics   216

10   The Socioeconomics of Brazilian Pulp Plantations   235

11   The Road to Certification   262

12   The Wall of Wood   281

13   Meshing Operations with Strategic Purpose   302

14   Lessons Learned   320

Glossary of Terms   327

The Sustainable Forestry Working Group   337

About the Authors   341

Index   343

Foreword

Jonathan Lash and Stephan Schmidheiny

One of us, Jonathan Lash, is president of the World Resources Institute (WRI), an international environmental policy research group engaged in innovative efforts to slow global forest loss and degradation. Recent analysis by the WRI has shown that logging is the major activity leading to the opening up of remaining forest frontiers all over the world.

The other author of this foreword, Stephan Schmidheiny, is a businessman with several companies in Latin America that turn trees into doors, flooring, and furniture.

The reader might expect that we have a lot to disagree about in thinking about the conservation and use of the world's forests. But the business of forestry is changing fast, and the goals of environmental and business leaders are beginning to line up in ways that few predicted five years ago.

Despite very different vantage points, we are in complete agreement on forest issues. Furthermore, we have been teaching each other a lot. WRI has learned how to work with forestry companies to green investment and to improve their bottom lines. Stephan Schmidheiny's Terranova company, which operates plantations in Chile, is among the first forestry companies to achieve ISO 14,000 accreditation for good environmental practices. Its plantations are planted on degraded farmland.

Despite ongoing efforts from governments and others, rates of deforestation and forest degradation are high and increasing in many regions of the world, especially in the tropics and the boreal zone. In the Brazilian Amazon alone, almost 17,000 square kilometers of forest were lost in the 1997–98 burning season, a 27 percent increase over the previous year. Floods in China, landslides in Honduras and the western United States, and massive fires in the forests of Brazil, Indonesia, and Russia have further accelerated change. Environmental groups are becoming increasingly sophisticated in monitoring the state of forests and threats to their continued survival. They have also successfully engaged the private sector in efforts to remedy some aspects of the problem through independent certification schemes that better inform customers about the production methods for various kinds of wood products.

Those of us in the forest industry are concerned about forest loss. Traditional logging, clear cutting, and associated road building have resulted in deforestation and degradation to the point that almost no natural forest today is economically harvestable. Meanwhile, the public and policy makers have become more aware of these changes, and governments have committed to improving efforts to conserve biological diversity, reduce climate change, and slow forest loss. This has resulted in growing demands from the marketplace and lawmakers for more careful management of dwindling forest resources. These are no longer niche markets or passing fads, as was once thought; instead they constitute multimillion-dollar opportunities for those entrepreneurs prepared to embrace the challenge. Establishment of plantations on degraded lands is a particularly attractive response that restores forest cover, creates jobs, and could reduce the pressure on native forests.

Beyond timber, and fiber commodities, forests have many other values. These include values from the provision of "ecosystem services," such as protection of biological diversity, watershed management, and carbon storage in the biomass of the forest. Local and global markets are emerging for some of these services and have the potential in the near future to further reward forest owners and managers who are conservation oriented.

Some have called for a global convention specifically to address the deforestation crisis. Given the cost and time involved in negotiating such an accord, and the slim likelihood that any convention that most governments would sign would lead to significant improvement, we feel that the private sector and civil society must take the lead.

Innovations in wood processing technology, biotechnology, and plantation management offer business the potential to relieve much of the pressure on the world's last threatened forest frontiers. Profits can grow along with the trees due to increases in efficiency through reduced waste, global green marketing, and reduced costs.

These and other key issues are of great interest to us and the broader business and environmental community. They have brought us together and are covered in depth in this book, which should be of great value to our many colleagues. The message is especially salient for professionals in the forest products industry, private forest owners and managers, environmental campaigners, researchers, and industry executives. From small land owners to industry giants, forest companies are struggling to meet the demands of government, community, employees, shareholders, and beyond. *The Business of Sustainable Forestry* describes what industry leaders are doing to manage their environmental and financial strategies. It reveals the business potential of resource efficiency and waste wood recovery, use of alternative species, and best practice in forest management. The success of these new approaches will help to determine the future of our forests.

The Business of Sustainable Forestry is a snapshot of the dramatic change that is underway in the forest products industry. The book analyzes the global market shifts, technological innovations, and business drivers changing the industry. It profiles some of the most progressive companies and examines what they are doing to make sustainability a competitive advantage. It serves business professionals of any industry, forestry and natural resource students, and environmentalists by presenting an illustrative overview of an industry in flux and of the real challenges of sustainable development. Case studies show how the convergence of ecology and economy creates both limitations and opportunities. Through these cases, executives, managers, and consultants will gain insight into reconciling environmental concerns with their business interests. The take-home message of *The Business of Sustainable Forestry* is that innovative sustainable forest management is not only possible, but potentially very profitable.

Preface

The sustainable forestry case studies that form the foundation of this book grew out of new perspectives among business leaders, work by nonprofit organizations, and creative foundation grant making. Since the early 1990s, conservation of the world's biodiversity has been a priority for the staff of the World Environment and Resources program at The John D. and Catherine T. MacArthur Foundation. Collaboration with a diverse group of organizations to accelerate the adoption of new concepts of sustainable forest management around the world presents perhaps the single most effective strategy to accomplish that goal.

In most countries only a fraction of the forests will be preserved in permanent protected areas. Today at least 85 percent of existing protected areas, particularly those in tropical forests, have human populations estimated to be as high as 500 million people living in and around them. Over the long term, the success of any biodiversity conservation program will depend on improving the livelihoods of the communities that depend most directly on the resources of those forests, while striking a balance between protective and productive activities. The fate of much of the world's biological diversity may be determined not so much by systems of protected areas, but by how the unprotected forest areas surrounding them are managed.

As part of its strategy to preserve biodiversity, the foundation has supported sustainable economic development projects in and around priority conservation areas. Even though projects in places like Peru and Mexico, Papua New Guinea, and India are managing their forests with increasing sophistication, the global market was not evolving in response to those changes. The disconnection between sustainable forest management projects in the field and market forces was nearly universal because those projects proposing sustainability failed to take into account the complexity and realities of the forest products industry. Their founders were unaware of the complex interactions that occur once a tree leaves the forest until it is incorporated into its ultimate end use. Consequently, the projects themselves were not sustainable as business ventures and never achieved significant scale or impact.

In order to address this gap, in 1993 the foundation convened a small group of forestry experts with domestic and international experience in the forest products industry, resource management, the certification movement, green marketing initiatives, the policy and regulatory environment, and donors to devise a more systematic, effective approach to investments in sustainable forest management (SFM). In the process, its members reviewed critical aspects of the SFM concept, including production, harvesting, monitoring, certification, trading, manufacturing, education, and promotion. The effort culminated in the Sustainable Forest products matrix, a poster-size graphic, which illustrates an overall view of the existing forest products system from the resource base to the end user, including the external influences that shape the market. This simplified articulation of the forest products industry's "value chain" captures the sequence of manufacturing and business processes and distribution/marketing mechanisms that convert trees to finished forest products. As the MacArthur Working Group developed the SFM "thinking framework," its members drew a number of conclusions from their discussions, literature review, and conversations with forest products producers, managers, and users.

The group realized that the effort to build a sustainable forest products industry lacked a fundamental element—a body of knowledge that documented that this transformation from a niche activity to a market opportunity was based on a sound business rationale. In order to address this lack of vital knowledge a second Sustainable Forestry Working Group (SFWG) was convened in 1996 to develop a series of business case studies illustrating the financial and market opportunities, as well as the challenges available for converting traditional forest and mill operations into sustainable industries.

The group (more than 30 individuals from major environmental groups, universities, and industry who had broad domestic and international experience in the forest products industry, resource management, business management, field research, and marketing) met five times in the course of the two-year project in 1996/1997. The initial two-day meeting identified companies as candidates for cases and led to the decision to include more representatives from the investment sector who would challenge the process, presumptions, and analysis of the data the group generated. The newcomers included the director of strategic planning at Weyerhaeuser, a consultant to Anderson Windows Company, a senior forest economist for Hancock Timber Resource Group, and the director of investment related to program at MacArthur.

The initial meeting produced case format guidelines for the writers to ensure common discipline and methodology for comparative analysis and identified the companies for cases. The group also settled on the following four studies to put the cases in the broader context of the industry:

1. A definition of sustainable forestry.
2. An analysis of the potential demand or trade in sustainable or "green" timber.
3. An overview of the economic and environmental forces at work on the industry that could affect the adoption of SFM.
4. An analysis of new and emerging technologies that could enhance the adoption of SFM.

At each subsequent meeting, the case writers reported on their progress, and the group discussed and critiqued each draft. At the fifth and final group gathering, members discussed and debated the findings of the case studies to distill the lessons that could be derived from each company's experience, as well as the aggregate findings from the group of cases.

The undertaking stretched the boundaries of conventional grant making, as well as the ability of the foundation to internally manage such a process. The logistics of coordinating the research activities and meetings of over 35 people who were involved with companies on four continents produced an almost overwhelming flow of phone, e-mail, and FedEx traffic. An arduous editing process extended the project another six months. A strategy for disseminating the group's findings, which included not only publishing the cases, but also creating audio-visual materials, presentations for the industry and investors, and an Internet posting, called for adding another layer of expertise. Complications seemed to arise at every step. Several companies in the project underwent management changes or reversals of fortune that extended or complicated the process of completing the cases. More than once Michael Jenkins questioned his judgment in initiating the project. When the cases were completed in late 1997, however, the work justified the resources it had consumed: The cases, over 600 pages, documented the early experiences of companies experimenting with SFM and captured the complex set of business issues involved in the shift toward more sustainable forest management.

The Sustainable Forestry Working Group had decided to produce a book from the case studies as part of its efforts to put them in the hands of industry and forestry professionals and students in forestry, natural resource, and business schools. The task of converting the original cases into a manageable book was hardly straightforward. The experiences of the companies, individuals, and enterprises captured in the cases did not add up to a coherent whole. Instead they were riddled with contradictions, unexpected difficulties, and failures, as well as incomplete discussions of strategic and product initiatives. The "in-process" nature of the cases was understandable, however, since sustainable forest management exists at the margins of the industry. But between the time the SFWG started meeting in early 1996 and the time the original cases were published in late 1997, the momentum of SFM had accelerated markedly, so the situation for some companies also changed. When the group started, for example, one pulp and paper giant, STORA, had announced that it would seek FSC certification. By late 1997, almost half a dozen pulp- and papermakers had jumped on that bandwagon. Similarly, retailers in the United Kingdom, such as J. Sainsbury Plc., went from stocking a handful of certified products to expecting several hundred on their shelves by early 1998.

For those of us who have watched environmental considerations become ever more powerful influences on business operations, the situation in the wood products industry in 1996/1997 with sustainable forest management seemed strikingly similar to that in the chemical industry in the late 1980s. Then, under fire from environmentalists and the government for pollution, the industry began a multibillion-dollar transformation that led to its slashing toxic emissions through technological improvements, new products, and pollution prevention and to its altering irrevocably its relations with suppliers and customers.

In order to capture the new momentum behind SFM and bring the stories of the companies up to date, Emily Smith and Joe Strathmann, the researcher for *The Business of Sustainable Forestry: Strategies for an Industry in Transition,* went back to the management of most of the companies and conducted interviews with the appropriate executives to bring the experiences of the companies current to 1998. They also interviewed suppliers, environmentalists, market analysts, investors, customers, and executives of companies not involved in the cases to revise and augment statistics, as well as integrate a variety of new perspectives on the changes that had occurred in the marketplace for sustainable wood products in 1997 and 1998. In all, we conducted over 100 interviews. Smith then reorganized the cases and overarching analyses into a chapter, integrated the new material into the case, revising and rewriting to integrate the new material.

The industry experiences that the Sustainable Forestry Working Group set out to document in 1996 remain very much a "work in progress." The sustainable forest management activities of each company we examined will evolve, perhaps radically, over the next few years in the face of rapidly changing market dynamics. If several years from now we look at this same group of enterprises, or submit the market and the industry to another round of analysis, we may find that some of the apparent successes documented here will become tomorrow's failures. Or that an as yet not fully executed strategy articulated in these pages will become wildly successful. The forest products industry and the global market for wood products will, no doubt, yield surprises in the next few years. No one may be able to predict the future, however, through the experiences of companies documented here and the analyses of the industry and the market for SFM we will, at least, understand how and why a small number of forward-looking companies decided that sustainable forest management was in their best interests, well ahead of the rest of the industry.

Michael B. Jenkins
Chicago, Illinois

Emily T. Smith
Hastings-On-Hudson, New York
February 1, 1999

Introduction

Sustaining Forests and Profits

On the remote island of Tierra del Fuego in southern Chile, the old-growth forests of southern beech have become a proving ground for sustainable forestry. In that delicate ecosystem, Savia International, Ltd., an affiliate of the Washington-based Trillium Corp. U.S.A., has launched a $260 million timber and milling venture on 370,000 hectares. In the early 1990s, Trillium's Chairman David Syre wanted to expand the company's timber holdings outside the Pacific Northwest, where environmental restrictions were limiting logging, to secure more wood to meet rising global demand for wood products. After examining forest prospects in Canada, Russia, and New Zealand, he decided that Chile, a nation hungry for foreign investments with an expanding wood products industry, was the best bet.

From the start, the Rio Condor project was set up to be ecologically, economically, and socially sustainable. Savia has pledged to forego clear-cuts in favor of selective harvesting practices that leave a forest cover with mature trees, use low-impact harvesting methods, and make sure the forest regenerates. To ensure that its practices will be compatible with the forest's long-term health, the company tapped the knowledge of 100 scientists for a scientific commission, which created a baseline science of the property through 17 studies of everything ranging from ecosystems and soils to aquatic systems and wildlife. The company plans to pursue third-party certification under the rigorous standards of the Forest Stewardship Council (FSC).

1

Savia's management has wagered that ecologically sound forestry—in the long run—will be economically advantageous. The southern beech that grows on Tierra del Fuego can compete with North American cherry and hard and soft maple in high-value applications such as furniture and cabinets, dried hardwood lumber, sliced veneers, and molding. Certification will give the company entrée to northern European markets, which have otherwise shied away from products that come from tropical operations. By producing value-added products and exporting 90 percent of them to Europe, North America, and Asia, Savia management expects the company can eventually earn a 20 percent profit on investment, even though sustainable forestry will cost more and shipping costs will be higher. In the process, Savia will strengthen the local economy by creating some 600 jobs in a region hard hit by unemployment.

As far as Robert Manne, president and CEO of Savia, is concerned a business strategy grounded in sustainable forestry is a "matter of survival" for the wood products industry in the face of disappearing natural forests and society's opposition to its conventional forest management practices. Ironically, local environmentalists, mindful of Chile's history of liquidating its natural forests, stalled the project for several years with legal challenges that went all the way up to Chile's Supreme Court. By the end of 1998, management expected to resolve those issues. Investors, however, recognized that sustainable forestry and environmental certification could bring credibility to a company engaged in the controversial business of logging natural forests in the Southern Hemisphere, and they wanted environmental certification to be a condition for making any loans. "Our bankers are showing a tremendous interest in green certification," commented Sheila Helgath, an environmental forester with the project, in 1998. "They are using certification as a factor in determining our sustainability and environmental credibility."

In the 1980s, a new business venture like Savia based on sustainable forestry would not have existed. In the mid-1990s, though, Savia represented the leading edge of the wood products industry, which, faced with dwindling forests, is being forced to exploit new ecosystems and processes to obtain its wood supplies and must confront the ecological implications of its timber extraction. Savia is just one of a proliferating number of companies, trade groups, and government initiatives surrounding sustainable forestry and environmental certification that took root in the industry between 1995 and 1998. The challenge of implementing sustainable forestry management has set off a wave of global experimentation. Low-impact forestry methods, buyers' groups, plantation forestry, biotechnology and genetics, new technologies that enable producers to make more product with less wood and to use scrap wood in production, sustainable forest certification, and stakeholder involvement—all part of sustainable forest management (SFM)—have emerged in just the past decade. Companies large and small are experimenting with novel ways to supply human wood and fiber needs that do not deplete the resource base or diminish the ecosystem services provided by healthy forests. A few have even tackled the question of poverty in the developing world, which is a leading cause of deforestation.

Together these experiments represent a rich collage of new approaches to competition in the industry. But many of these corporate forays into SFM are still works in progress like Savia. Not all will survive the vicissitudes of global competition. In the future, however, regardless of which prosper or fail, the ability of the wood products industry to implement commercially successful sustainable forestry will be inextricably linked to its prospects to sustain profits.

A Critical Turning Point

The concept of sustainable forest management is not new. But until recently it was relegated to programs conducted by universities and environmental groups and to the practices of a handful of small companies in Europe and the United States as part of tropical forestry projects funded by environmental organizations. That changed in the 1990s when the commercial extinction of native forests, lack of adequate reforestation, conversion of forestlands to other uses, conflicting demands for forest resources, public opposition to logging, more stringent environmental regulations, and diminishing supplies of wood became everyday business realities and forced the industry to reconsider its norms and business strategies.

Environmental pressures are now creating new market niches, shifting the availability of wood, accelerating technology development, and driving new capital investments. At the same time, rising public opposition to conventional practices like clear-cuts and logging in old-growth forests is being translated into sophisticated campaigns in the media, the courts, and the marketplace that are prodding the industry toward sustainable forest practices. Environmental pressures, reacting with a variety of economic and market forces already at work on the industry, will rewrite the competitive ground rules for the global industry, alter world trade, and create new business opportunities over the next 20 years.

The expansion of sustainable forest management and certification marks a critical turning point for the forest products industry. It signals the end of an era when the industry's operating imperatives were based on abundant wood supplies and ushers in an era defined by the realization that forest resources will be increasingly limited in the future—both physically and politically. The industry's broader acceptance of SFM reflects a growing

consensus that sustainable forestry can be a viable alternative to conventional forest management—one that has the potential to reconcile the conflicting demands for economic exploitation that are destroying forests—with
escalating pressures to preserve them. In the process, certification—the verification by an independent third party that a product adheres to specific
standards of forestry that include ecological, social, and economic dimensions—has become both the indicator and the lightning rod for the industry's evolution toward more sustainable forest management.

A Legacy of Destruction

Historically, the forest products industry was oriented almost exclusively
toward the rapid harvesting of standing native forests. The deforestation
underway in the Amazon today is the mirror image of forest exploitation in
the United States during the second half of the nineteenth century. During
that period, the U.S. population tripled and the westward migration peaked,
with farmers clearing three to four acres of forestland for every person
added. The seemingly endless forests allowed policies of laissez faire that,
combined with cheap stumpage, a growing farm economy, and footloose
sawmilling, created a moving lumber front that depleted forests. The trail
of this mobile industry followed the most plentiful resource from the
Eastern United States through the Midwest to the Pacific Northwest and
into the Southeast, "mining" for the most abundant resource. That same
pattern has been repeated throughout the world. Although a variety of causes contribute to global deforestation, major contributors are poverty in the
developing world, logging, and commercial forestry.

The exploitation of the world's forests has taken a drastic toll: rapidly disappearing forests, deteriorating ecosystems, reduced biodiversity, and
increased atmospheric carbon dioxide. Since 1950, 20 percent of the world's
forest cover has been decimated. In the late 1990s, forests were disappearing
at an estimated 35 million acres per year. At those rates, only a tiny remnant
of the native forests that once covered the planet will remain by the middle
of the twenty-first century. The destruction of forest ecosystems, which contain as much as 80 percent of the world's terrestrial biodiversity, has pushed
the current rate of species extinction to 100 times the natural rate. By the late
1990s, tropical deforestation and burning accounted for some 90 percent of
all biotic emissions of carbon into the atmosphere, or 7 to 30 percent of total
atmospheric emissions; thus, deforestation is a major contributor to global
warming.

Global climate change, in turn, negatively influences forest ecosystems.
These complex ecosystems are vulnerable to changes in temperature and
precipitation that are part of climate change and that increase the frequen-
cy and intensity of forest fires and pest infestations. The fires that ravaged

the forests of Kalimantan, the Amazon Basin, central Russia, southern Mexico, and Florida during the mid-1990s were all sharp expressions of the powerful connections among forest health, local land-use patterns, and global climate change.

More Valuable Forests

At the same time, population growth and a better understanding of the connections between the functions of forest ecosystems and essential human needs have made the remaining standing forests far more valuable. On local, regional, and national scales forests provide ecosystem services—clean water, protection from floods, and microclimate effects—all of which have economic value. One estimate tallies the value of these services at an average of US$2,007 per hectare for all tropical forests.[1] In New York City, in Cali, Colombia, and in Quito, Ecuador, municipal governments increasingly recognize the value of the forests to protect the watersheds that provide drinking water. Replacing that function by building plants to purify polluted water can cost an urban area billions.

The 1998 agreement at the Kyoto summit meeting on climate change—to limit global greenhouse gas emissions—will make standing forests even more valuable in the future because forests can sequester carbon. The Kyoto protocol limits developed countries' greenhouse gas emissions but permits them to meet those limits in part by acquiring offsetting emission reductions from other countries, through joint carbon sequestration projects using forests as carbon sinks—a market that some experts predict might reach billions of dollars per year. Ongoing research on ways to quantify the value of forest products and services in economic terms may eventually place biodiversity, carbon sequestration, scenic beauty, and the water resources of forests on an equal footing with timber.

The Consequences of a Dwindling Wood Supply

The loss of access to native forests will squeeze wood supplies in the future. Over the next 25 years, for instance, demand for softwood is expected to leap 25 percent, but supplies will rise just 15 percent. As old-growth or primary forests are exhausted, more wood will come from intensively managed secondary and plantation forests. These forests produce more uniform wood, but the trees are smaller and the wood can be lower quality. Twenty years ago, the average size log from old-growth forests reached 52 inches in some areas; today logs more commonly have diameters from just 8 inches to 20 inches.

Tighter wood suppliers, smaller logs, and lower quality wood will have expensive repercussions for the industry. Already sawmills and processing

plants in many parts of the world—the Pacific Northwest, for instance—
which were originally constructed close to the forest, must now buy
wood from hundreds, even thousands of miles away to supply their
mills. Asian companies, faced with depleted forests in the Pacific Rim,
have accelerated their forest investments in South America and central
Africa. To process smaller logs, mills must go through expensive retooling
and retrain loggers and mill workers who were accustomed to discarding
smaller logs for pulp. They must also develop new products that can use the
smaller sized, lower quality wood. The explosion of engineered wood
products is an example of just one kind of product shift the industry will
make to compensate for tighter supplies and lower quality wood in the
future.

A New Approach to Competition

Over the past decade, as companies have wrestled with the new business
realities created by limited wood supplies and demands for sustainability,
the ranks of corporate innovators in sustainable forestry have swelled. The
21 companies that were part of the research conducted by the Sustainable
Forestry Working Group are representative of the industry's struggle to
meet the challenges of sustainable forestry and to compete in the principal
segments of the forest products industry: solidwood, pulp and paper, panels,
and engineered products. These companies span the spectrum from large,
publicly held industrial corporations like U.S. timber giant Weyerhaeuser
Co. and Sweden's pulp and paper behemoth STORA to small private operations with under $200 million in revenue, such as the Collins Pine Co. The
companies include large and small value-added wood products manufacturers as well as distributors, the retail giant The Home Depot, and the
small private timber owners who supply 49 percent of the U.S. industry's
timber.

This book is intended to describe and assess the practices of these companies to determine which are most likely to succeed or fail. Four overarching
chapters place the SFM activities of these 21 players in the context of the
environmental, social, and economic forces that are transforming the industry. Much about sustainable forest management is controversial and hotly
contested, even the very definition of sustainable forestry. Chapter 1, "The
Pursuit of Innovation," contributed by Jeff Romm, proposes a conceptual
framework for sustainable forestry that goes beyond hair-splitting debates
over technicalities. He defines sustainable forestry as a process of innovation
in forest use and management that arises in response to the social, economic, ecological, and cultural conditions that exist for a given forest at any point
in time. Chapter 2, "The Forest Products Industry in Transition," identifies

the seven major trends that will reshape the global forest products industry in the next 20 year, and then explains how the interaction of business pressures with environmental trends in each segment of the industry will affect the evolution toward SFM.

Chapter 3, "Emerging Markets for Certified Wood Products," unravels the confusion surrounding these markets and identifies market and product opportunities. The momentum behind certification and demand for certified wood products, which is accelerating in European and U.S. markets despite an absence of widespread consumer demand, is documented here. Results may come as a surprise. As the chapter makes clear, pressure from the industry's industrial and commercial customers for environmental accountability in the form of buyers' groups and other associations is a major factor in the wider acceptance of certification schemes among companies in northern Europe and Canada. Chapter 4, "New Mandates for Technology," contrasts the technology requirements for more efficient sustainable forestry with those for conventional technology, explains why it is so difficult for promising SFM technologies to enter the marketplace, and identifies 10 promising candidates.

All but one of the remaining chapters present case studies. Not surprisingly, the companies involved with SFM are taking widely varying approaches to sustainability. Their motives for adopting SFM and their financial performance vary even more widely. Chapter 13, "Meshing Operations with Strategic Purpose," adapted from a paper by Matthew Arnold, Rob Day, and Stuart Hart, presents a framework to capture companies' motivations for pursuing SFM and draws some conclusions about what makes some SFM activities succeed while others fail. Broadly speaking, the motivations that companies have for adopting SFM practices fall into four categories:

1. Franchise Protection: Companies meet certain ethical standards to avoid public and regulatory scrutiny, which include pollution control.

2. Impact Reduction: Companies use pollution prevention, eco-efficiency, or simply risk avoidance to enhance productivity.

3. Product Enhancement: Companies take the extra step and link SFM to product quality by providing "sustainable" wood products, which may yield price premiums over certified wood.

4. Business Redefinition: These few companies are redefining their business as sustainable in process, product, and market. In the process, they are moving toward becoming "fiber service" companies that incorporate other forest values like carbon sequestration, ecotourism, and nontimber forest products into their business strategy.

The researchers conclude that, regardless of which category a company's SFM efforts fit, SFM must reinforce the core business strategy and must have the goal of achieving a permanent competitive advantage or it will fail. The most successful companies in the study recognized sustainable forestry as a business opportunity and integrated sustainability into a broader competitive strategy: It is the fit of R&D, production, marketing, and environmental concepts together that creates the potential for a sustainable competitive advantage.

An early systematic attempt to document the industry's experiments with SFM, the company stories with their wealth of detailed examples and management perspectives provide a baseline of comparable experience that illuminates the issues involved in developing commercially successful SFM. But this book is only a snapshot of a fast-moving industry in transition.

It is premature to draw firm conclusions about the success or failure of many of the SFM experiments discussed here. STORA management, for instance, adopted a certified SFM strategy in 1997 in the belief that FSC certification will give the company a competitive advantage against new lower cost competitors in the European markets, where it does most of its business. It will be several years before the success or failure of that strategic calculation becomes evident. Other companies, like Portico S.A. of Costa Rica, which successfully adopted vertically integrated certified SFM to ensure wood supplies and gain entry to U.S. markets, will face stronger competitive challenges in the future. Moreover, in 1995 less than 0.60 percent of the world's industrial roundwood was certified. The balance of demand and supply for certified products and the broader economic and market trends buffeting the industry will heavily influence how quickly and to what extent sustainable forest management moves from the margins of the industry into the mainstream. Events like the Asian economic crisis of 1998 that reverberated through the forest industry create unanticipated problems and opportunities that may hamper or accelerate the progression of SFM in a given region or market.

Even so, the experience of these 21 companies provides a tantalizing glimpse of the challenges, trade-offs, and rewards involved with the early-stage commercial implementation of SFM. Without doubt, the cost of SFM poses a challenge. Sustainable forestry practices do cost more to implement in the forest than do conventional practices—between 10 and 20 percent—according to the company experience documented here. Lower rates of cutting and the use of multiple species that are required for successful sustainable forestry can have significant implications for milling operations, product development, and marketing. Precious Woods, an SFM start-up in Brazil discussed in Chapter 8, for instance, must develop and market products made from about 30 species, most of them relatively unknown in its target markets, if management is to earn the return that investors expect. The market development for multiple species is both time-consuming and expensive. In the long run, however, under the right conditions, as the companies here demonstrate, a sustainably managed forest can be as productive and produce higher quality timber than a conventionally managed forest.

What constitutes sustainable forestry and certification will no doubt remain contentious for years. To the public, plantation forestry, which is rapidly expanding, is probably the most controversial technological advance included under sustainable forestry. Plantations, typically monocultures of a small number of genetically improved species that are grown like agricultural crops, grow faster and produce more standardized fiber than natural forests. But they are less biologically diverse and more vulnerable to natural or human disturbance than natural forests. Supporters argue that plantations can alleviate stress on natural forests—but that has yet to be demonstrated, as the discussion of Indonesia's plantation strategy in Chapter 10 suggests.

Many industry, government, and nongovernment organization (NGO) initiatives, such as the FSC, continue to try to define principles, criteria, and indicators for sustainable forestry that would be specific enough to be practical and broad enough to apply to the vast range of forest types and specific circumstances around the globe. At a 1998 meeting convened by the World Bank, for instance, industry CEOs and environmental leaders agreed on the fundamental issues—the elements of sustainable forestry operations, the need to have guidelines developed at local/regional and national levels, and the value of some form of verification. Dissension over SFM among industry, government, and NGO groups tends to revolve around the methods or approaches to verification of sustainability and level of public participation. Any prolonged attempts to hammer out an all-encompassing "code of sustainable forest management" and to resolve these disagreements are probably a false target. As Jeff Romm writes in Chapter 1, "The meaning of a 'sustainable forest' is inseparable from the interests, capacities, values, and contexts of those who define it." Natural forests are themselves in constant evolution, just has human demands change rapidly.

In the long run, it will be far more beneficial for the health of forests and the prosperity of the industry to develop institutions—business enterprises, government-based entities, and those in civil society, as well—that can guide society by creating places where the interests of all stakeholders can be heard and the pace of SFM accelerated. Most of the many intergovernmental initiatives created to address forest destruction—the 1992 United Nations Conference on Environment and Development in Rio de Janeiro, Brazil; the 1985 Tropical Forestry Action Plan; and the more contemporary incarnations like the Intergovernmental Panel on Forests (IPF) and the World

Commission on Forests and Sustainable Development—have been remarkably ineffective. Their shortcomings are widely attributed to their inability to engage civil society, both nongovernmental organizations and industry.

Sustainable forestry has the potential to create a new generation of partnerships that brings together the multiple interests—public, private, and nonprofit—that have a stake in the management of the world's forests. Such alliances can encourage appropriate government policy and public education, coordinate regional partnerships, conduct local research and development projects, stimulate investment in improved forestry, and help end nonproductive and expensive confrontations between industry and environmentalists—all necessary elements for a successful sustainable resource management scheme. The experience of the companies in this study underscores the benefits that companies can realize from cooperation and communication with other stakeholder groups—enhanced credibility with the public, environmentalists, and regulators; the ability to influence environmental regulations; better relations with customers; and improved quality control with suppliers.

Better forest management is too important to hinge on the outcomes of international negotiations, environmental regulations, or arguments about what constitutes sustainable management. The global forest products industry—which accounts for some 2 percent of the global gross national product and 3 percent of world trade—and the world's forests are in dire need of more sustainable management, sooner rather than later. The world's forests are precariously balanced between the possibility of a sustainable future and the realities of continued rampant degradation and destruction. The forest products industry will be a determining factor in the fate of the forests.

The Sustainable Forestry Working Group has demonstrated that sustainable forest management and the rising demand for sustainably produced forest products can serve the goals of industry and simultaneously benefit forests. The industry's evolution toward sustainable forestry described here provides a glimmer of hope in an otherwise sorry history of unfulfilled promises and missed opportunities to conserve and enhance the world's forests for the future. For the industry, sustainable forest management represents a new frontier replete with formidable management challenges but rich in new business opportunities.

NOTE

This introduction was prepared by Michael B. Jenkins and Emily T. Smith on the basis of the work of the Sustainable Forestry Working Group, 1997.

1. Costanza et al. (1997). "The value of the world's ecosystem services and natural capital," *Nature* vol. 387, 15 May, 253–260.

Chapter 1

The Pursuit of Innovation

Sustainable forestry is emerging worldwide because the contexts and conditions of forests are changing at an unprecedented rate and in ways that were never before possible. Long viewed as hinterlands valued primarily for meeting the extractive needs of societies, or as preserves of wilderness, forests are now mainstream concerns in the United States and throughout most of the world. Increasingly, forests are recognized as pervasive and crucial features of the social landscape that supply fundamental human needs for wood, paper, water, food, jobs, medicines, minerals, and energy. They form watersheds, agricultural systems, and reservoirs of genes, species, and ecosystems; and they regulate climate. In the process they distribute resources and services among groups, communities, and nations. In this new context, people have come to view forests as critically scarce systems within the bounds of direct human interest rather than as abundant resources beyond those bounds.

The new perceptions of forest scarcity have a biophysical as well as a social basis. Throughout most of the world, the frontier is gone. Few places remain inaccessible or protected from the impulses of global modernization. Population and economic growth explain this fact, in part, while images from space satellites, transmitted through televisions in most villages and farms, graphically demonstrate these impacts. But worldwide democratization of political and cultural access has also given local and indigenous communities voice to stake their claims on forests. And the globalization of forest issues, politics, and organizations has diversified and strengthened once weak or unorganized national and local interests.

Scientific advances have augmented the perceptions of forest scarcity. The ecological sciences are now sophisticated enough to measure and explain the causes and consequences of forest change at the household and farm, at the regional and global scales. They reveal forests as systems that function interdependently as water, climate, and biological infrastructure; as complexes of social rights and responsibilities; and as enduring expressions of the dynamics of human and natural history in which people have always shaped forest conditions everywhere. No longer are forests understood to function simply as large blocks of wood or wilderness "out there," they are now understood to be patterns of trees that permeate every scale and scene of social life.

The time-honored approaches to forest management, which developed in simpler times, assumed that forests were isolated from broader social forces and had singular, unambiguous purposes. These approaches are insufficient in an era when forests are recognized as forceful features of the social fabric. Older approaches to forestry cannot cope with the modern intensity, variation, and complexity of human expectations. The new sense of forest scarcity makes the "cut and run to the next hill or to the next country" mentality toward forests as marginal as the opposing "stop-everything" attitude. It replaces these outmoded views with the urge to manage forest systems as valuable, diverse, and vulnerable assets.

Today's pervasive conflicts over forest uses are symptoms of the disparities among people about what they want and what forests are actually able to do. For the same reasons, forests have become the focus of innovations that promise to achieve regimes of action consistent with the scale and complexity of forests' new mainstream role. Sustainable forestry, the pursuit of the means to govern, finance, and manage forests as central features of social life, in the largest sense, signals that forestry has come of age.

The Sustainable Forest

Forestry is a regime of actions, but the forest is a concrete biophysical system. The sustainable forest is an aggregation of trees that people preserve in a dynamic social and natural environment for the ecological qualities, services, and yields that they want. Active effort is required to prevent unwanted changes that human activities and natural phenomena would otherwise produce. Preservation requires investment to control forest uses and the social forces that determine them; to conserve and replace necessary ecological capacities; and to develop the organizations, policies, and technologies that forest replacement, maintenance, and enhancement are likely to need.

The sustainable forest strikes a dynamic balance among economic, environmental, and social forces that people control to prevent the loss of whatever forest state they prefer.

While this definition explains the central role of sufficient investment to ensure forest sustainability, it does not explain the ubiquitous conflicts that exist among people who fully agree on the need to sustain forests. Conflicts over forest use arise because people have multiple, different interests in forests. No one forest structure and composition can satisfy all of those interests and their infinite combinations. The meaning of "sustainable forest" is inseparable from the interests, capacities, values, and contexts of those who define it. A short-rotation monocultural plantation or a wild preserve can be sustainable forests through the eyes and actions of those who champion them as such. They are not sustainable from the perspectives of those with different values. And these forests change at different rates and for different reasons over time.

Scale Matters

Nor can all the diverse interests in forests be attained on the same spatial scales. Some interests reflect global perspectives, as is the case with stabilizing climate, preserving biodiversity, and sustaining the international wood products industry. Most interests, however, fit along a continuum of possible scales from the farm to a public forest to a region of nation-states. When it comes to sustaining forests, scale matters. A forest that is sustainable at one scale, such as a national forest, will not be so at another. In this respect, an intensive forest plantation may relieve pressures to harvest wood from natural forests, thereby increasing their sustainability, even if the plantation itself is unsustainable from the perspective of natural diversity. Yet restrictions on the harvesting of wood from public forests may shunt consumption pressures to privately held lands, which may reduce the sustainability of the forest region as a whole. The paradox of scale in forest management often scuttles constructive debate on sustainability.

Moreover, the forms and scales of desired forests change readily with shifts in market forces and credit institutions, demographics, social and political trends, developments in science and technology, communications networks, and natural catastrophes. For any specific interest, the sustainable forest in one period will differ from that in the next. If all interests were taken together on their own terms, the sustainable forest would be a swarm of interacting scales, structures, and compositions that change for different reasons over time.

Although sustainable forests are as different as the people who define them, they share common needs. The foremost of these is protection from and adaptation to changes that would otherwise disrupt the desired pattern, whatever that may be. The sustainable forest is a concept of stability in an unstable world. It is a longer-term vision in the midst of short-term forces that shape forests differently than people prefer. Fire, hunger, disease, financial crisis, market booms and busts, discoveries, conflicts, and political necessities are the normal instabilities that change any forest structure, composition, and spatial pattern if allowed to do so. The sustainable forest is protected from such forces, by excluding or modifying them, and is able to adapt to those forces that penetrate it in a manner that maintains its desired attributes.

Even though most people may agree about the need to sustain forests, few agree about which kind of forest should be sustained, or about what the rates, forms, and sources of forest innovation and investment should be. Ironically, such disagreements too often erode the climate for innovation and investment from which all visions of the forest would gain, and a "tragedy of the commons" arises in the realm of the social capacity to resolve forest interests.

Sustainable Forestry

Forestry is the social process through which people organize the effort to perpetuate a forest's desired attributes. It consists of the regime of actions that shapes the forest's attributes for specific purposes. It is supposed to maintain the desired balances within the forest and between the forest and the external conditions that shape it. Its actions regulate the structure and processes within the forest by strategies of harvest, conservation, enhancement, and exclusion. Its actions also protect the forest from external disruption through systems of ownership, cooperation and coordination, fire protection, and market organization, to name a few. These actions are designed to use public policy, education, and science, for example, to reduce the pressure from sources of disruption. They are accountable to the social purposes that apply in the particular case.

The essential difference between "forestry" and "sustainable forestry" lies in the complexity and scale of relations between people and trees that are now recognized to define forests and to motivate the actions that affect them. Regimes of action that were suitable in an era of abundant frontiers are insufficient in a world in which the population has quadrupled in this century, most people have been freed from the more obvious forms of political and economic tyranny, economies and states are urbanized and integrated on a global scale, and the social and scientific basis for valuing and acting on forest conditions has been revolutionized. Forests are increasingly becoming diverse overlays of different systems of social interests that interact uniquely in any one place. The actions that predominate are those that affect the pressures in that context and the opportunities that happen to converge in that place.

Pursuit of Innovation

As yet, sustainable forestry does not provide a regime of actions to respond to these emerging circumstances and shifts of priority. Rather, it is the pursuit of innovation in science and technology, the organization of forest ownership and jurisdiction, the development of business, financial, and market institutions, and the formation of political interests and modes of popular engagement that conventional forestry did not address. As a social process, sustainable forestry is characterized by the expansion of ideas, instruments, and applications that diversify experience and the social capacities to learn from it. Sustainable forestry challenges us to appreciate change, to respect differences, to encourage innovative endeavor, and to grow means to capture and use the lessons it will yield.

Opportunities

Sustainable forestry has already spawned a variety of innovations that are demonstrating practical worth. These innovations are taking place in several key arenas that open up new opportunities and pose new challenges.

From Silviculture toward Ecoculture

Silviculture, an elegant field of applied ecology, has historically focused on cultivating timber products. In sustainable forestry, however, it encompasses a much wider range of goods and services that forests can provide and addresses ways to manipulate the entire forest ecosystem to diversify, finance, organize, and sustain its aggregate productive capacities. Foods, spices, medicines, ornamental plants, feeds and fuels, as well as "waste" tree species and wood are developing as marketable sources of income. Environmental services, such as water yields and water quality protection, recreational opportunity, and wilderness and open space, increasingly represent values that can yield economic returns through new forms of exchange and asset valuation. The values of nontimber products from a forest are now competing with those of timber products in an increasingly wide range of conditions. In these circumstances, the purposeful manipulation of forest ecosystems requires a much more diversified scientific program than conventional silviculture has provided. Researchers and practitioners are just beginning to respond to the challenges of "ecoculture."

From Volume toward Quality

Previously, products of the forest typically were considered in terms of volume rather than quality. Market and industrial structures have favored the production of bulk wood products that satisfy a minimum quality and economies of scale associated with highly capital-intensive technologies rather than ways to value and exploit differences in the quality of products. As these strategies have displaced workers, it is not surprising that quality differentiation is related to efforts in sustainable forestry to create knowledge- and labor-intensive job opportunities in forest ecosystem management.

This is occurring through, for example, the development of specialty products and markets, the restoration of stream systems, the regeneration of forests, the reduction of hazards to forest health, and the development of systems of finance and insurance that suit these activities. Although household enterprises are the pervasive ones in the world's forests, the potential organization, scales, and technologies to produce, process, market, and finance a diversified and quality-oriented forest economy are only beginning to be explored.

From Stands toward Landscapes

Until recently, forests were treated primarily as collections of timber stands—discrete homogeneous units of timber and timber potential. Today, they are increasingly considered patterned aggregations of trees in landscapes of functionally interdependent units. The distinction between treating forests as collections of separate timber stands or as unified landscapes of interdependent forest elements takes on qualities of a religious war in some arenas, in large part because of the strong value content embedded in any such classification. The concept of landscape itself contains similar sources of difference because people want those particular landscapes that express the interactions they think are important. The tension between stand and landscape perspectives, and among perspectives of landscapes, has provoked innovations that are accelerating the transition to a landscape view of forests and enriching the concepts and techniques involved. "Ecosystem management," watershed agreements, agroforestry, community forestry, riparian forests, urban forestry, carbon forests, and bioregional councils merely characterize a broad field of trials that have come to dominate public understanding of what the forest is and can be.

These innovations increase the economic value of all the forest's attributes by establishing quantifiable comparative relations among them. The transition depends on how well these relative weights among attributes, meaning their values, can be transformed into financial returns that will sustain reinvestment in the forest. The growth of knowledge, organizations that dilute institutional barriers, and proven techniques are critical factors in the transition. Advances in information science, environmental survey, simulation and monitoring, and geographic information systems have brought these possibilities within view but not yet into normal practice.

From Ownerships toward Councils and Communities

Forest management has typically been carried out within the lines, authorities, and interests of specific forest ownerships and jurisdictions. Sustainable forestry diversifies the concepts, patterns, and scales of forests to such a degree that no unilateral form of control can encompass them. Indeed, many conflicts in forestry persist because of the absence of mediating interest in circumstances where formal controls and de facto interests have grown increasingly inconsistent with one another. The response of sustainable forestry is to diversify the sources of authority. Including local communities in forest management is particularly significant in disputes that have previously been waged, and stalemated, primarily among national interest groups. Although local communities depend on the forest and possess human resources critical for its viability, they have long been disenfranchised by the characteristic patterns of absentee control of forests, both public and private. The legitimization of their interests is helping to build a "third force" means to resolve conflicts and to increase the resiliency of dispute resolution generally.

Engaging communities in forest management is a central force in sustainable forestry. It usually involves ratifying local entitlements to share in forest governance and in creating motives for local investments in forest stewardship. Experimentation in types of partnerships, cooperative management, community management, enterprise and job creation, and the like is increasing. Multijurisdictional approaches to watershed management and community-centered systems of forestry are but two examples of these new approaches. They are part of a broad search for the mixes of entitlement that will mesh the interests and capacities necessary to achieve forests that, even from diverse points of view, are sustainably productive. To sustain such processes, knowledge of forest ecosystems must be sufficiently predictive, and the benefits to communities sufficiently large and equitable, to justify the effort involved.

From the Forest as Product toward the Forest as Capital

From an economic perspective, sustainable forestry is intended to enhance forest assets, to essentially increase nature's capacities for the growth of economic value. But few economic systems yet have the capacity to capitalize the full benefits that forests provide, to demonstrate the true capital gains that investments in forests yield, to distribute the returns to those who make the effort to create them, and thereby to motivate the levels and combinations of investment that forest benefits justify.

Despite the obvious economic contributions of forests—as water storage and flow, to energy supply and conservation; as infrastructure that is exceedingly expensive to replace, to local subsistence and enterprise; as a source of foreign exchange through sale of materials and tourism; in housing construction and landscape organization of settlements—a small fraction of these contributions is assessed to forest value. And a much smaller fraction is invested to sustain them. Those who benefit from those contributions typically have little obligation and no mechanism to invest in the source of the benefits. The problem of assessment is even more severe for less obvious contributions—wildlife, climate, open space, biodiversity—or potential future contributions, such as specialty foods, medicines, and woods.

Sustainable forestry is stimulating efforts to close the loop between those who benefit from the forest and the forest sources of those benefits. Efforts to incorporate natural assets in national capital accounts are strengthening the capacity to treat forest investment as an aspect of national economic strategy. Institutional reforms are breaking down tenurial and jurisdictional barriers to exchanges between sources and recipients of forest benefits, thus increasing opportunities for return flows of investment from forest beneficiaries. The concepts underlying such efforts are emerging in public discourse to complement and enrich long-standing debates about preferred forest conditions.

From Current Income toward Natural Capital and Green Finance

Forests are exceedingly vulnerable to the force of financial markets. Landowners typically accrue large debts to purchase timberlands and develop processing facilities, then convert timber stocks in accordance with the pressures to repay their loans. Governments exploit forests to transform "free" natural capital into money rather then run up debts to expand their infrastructure or industrial base. Throughout the world, forest composition and structures are historic expressions of financial forces rather than ecological and silvicultural judgment. A critical aspect of sustainable forestry is the pursuit of means to regulate relations between the dynamics of forest ecosystems and the financial markets that dictate patterns and rates of forest exploitation.

Innovations are occurring on a number of fronts. New institutions are translating ecological and financial values into comparable terms that permit monetary payment for retained ecological attributes. Some institutions, such as ecosystem management organizations, create arenas of exchange that force ecological, social, and financial interests to come to terms with one another. Others, such as differentiated markets for forest qualities and attributes, strengthen financial incentives for forest preservation or enhancement. Still others, including "green" brokerage and mutual funds, debt-for-equity swaps, and forest trusts, are reducing the costs of holding and buying forest assets for long-term conservation interests. As knowledge of long-term forest values improves, enterprises oriented toward asset growth, such as insurance companies and pension funds, are increasing their shares of forest ownership, a trend that is countering the tendencies of the financial market to value forest assets primarily for current income. Such changes are expanding the investment in natural capital.

From Blind Consumption toward Consumer Awareness

Forest products embody "gifts of nature" this generation did not create and as yet does not replace. Forest nutrients, moisture, and ecological structures and functions consumed in the production of forest goods are treated as "free." In a competitive economy, their unpaid costs are excluded from the prices people pay for the goods that forests produce. Consumers are rarely aware of the "costs to nature" of the depletion of this production capacity, which their purchases embody. Sustainable forestry includes pursuing the means to close this loop.

The increase in forest certification efforts in the past several years represents an effort to inform consumers about the conditions of the specific forest operations from which the products they buy are derived. These efforts distinguish products in the market by the sustainability of their source operations. Some of these efforts are made by neutral bodies that certify forestry operations as sustainable only if they meet stringent standards. Certification entitles producers to label their products as certified, which distinguishes them from others that come from operations that are not validated as replacing the gifts from nature that they use. Certification enables consumers to vote for forest management in the marketplace. In the past several years, a number of wholesale and retail outlets have begun to carry certified products and advertise the differences in content among forest goods that may otherwise appear to be identical.

Challenges

Sustainable forestry is proving to be a catalyst for innovations in forest management. But it is a pursuit, not a prescription. Many unresolved issues will demand new ideas in the future. Several are indicative of the challenges that remain in transforming sustainable forestry from concept to reality.

Unified Concept versus Reality of Conflicting Practices

Although unified within the process of sustainable forestry, the emerging visions of the sustainable forest will prove to be competitive or inconsistent with one another when put to the practical test in the same place. Economic, community, watershed, and biodiversity visions of the sustainable forest, for example, are unlikely to shape the same sets of choices and actions or to yield the same ecosystem structures and functions. Community visions include sustainable and growing forest employment opportunities, while biodiversity visions tend to limit or exclude human activity. The process used to resolve these diverse views will determine how effective sustainable forestry will be as a practical pursuit. As yet, these processes have not faced a serious test.

One-Scale Visions versus Many-Scale Realities

Sustainable timbersheds, watersheds, community forests, habitat ranges, biodiversity regions, and metropolitan forests exemplify the differences of scale at which different sustainable forests function. A forest that is sustainable when viewed at one scale is not necessarily sustainable when viewed at another. The diversification of forest concepts involves equivalent diversification of scales at which forests are managed. Relations among actions at these various scales become fundamental to the sustainability of the parts and the whole.

Because relations among scales are poorly recognized and therefore often counterintuitive, they tend to seed conflict rather than constructive strategy. An intensively managed monoculture plantation, for example, can be sustainable on a small scale but unsustainable for biodiversity at a regional scale. Yet it may satisfy sufficient economic need at the smaller scale to reduce production pressures on larger-scale forests that then can be managed more easily for biodiversity and other natural attributes. A regional mosaic of intensively diverse, small-scale monocultures can be more sustainable than a natural forest at the same scale, even if many of the monocultures are not individually managed in a sustainable way. Moreover, a forest that seems sustainable on any scale may not be sustainable on the global scale of financial forces that will determine its long-term viability.

Forest resilience and sustainability on a global scale require a reasonable complementarity among actions at different forest scales. Achieving that, however, requires a readiness among different interests to withdraw from absolute forest prescriptions that may suit one scale or circumstance but that adversely affect the whole if applied generally. Better explanation of the relationships among scales should facilitate the process by demonstrating the potential complementarities at current edges of conflict.

General Government Authorities versus Flexible Forest Institutions

The pervasive spontaneous experimentation fueled by sustainable forestry is rapidly transforming forestry from a rigidly controlled to a remarkably inventive field of action. If the past is a guide, such creative flexibility will eventually run into the more general and less resilient authorities of formal governments at all levels. Governments have broader concerns than forest conditions and have strong needs to apply uniform rules. These tendencies tend to confine institutional innovations within limits that are relatively insensitive to the requirements of the problem and the motives for its resolution. Confinement may be avoidable to the extent that the configurations of governmental policy begin to integrate forest conditions as mainstream rather than specialized and marginal public concerns. In other words, the framework that governs all public actions would change in ways that strengthen the context of sustainable forestry.

Sustainable Forestry as Adaptive Learning

Forestry has existed as a marginal, specialized, and narrowly vested interest segment of the environment, economy, and society for too long and at great cost. Sustainable management of timber could be accomplished within conventional ownerships and jurisdictions. Sustainable management of forest ecosystems cannot. The knowledge, techniques, and institutions it requires cannot exist without general public interest that has expanded the systems and scales of the forest, its governance, and its finance. The public increasingly weighs forest choices against the impact those choices may have, and it is increasingly aware of how the broader forces of society affect the forest. This recognition started the processes of innovation, investment, and institutional reform that constitute sustainable forestry.

Are these processes necessarily spontaneous pursuits, or can method accelerate their progress? The answer to both questions is "yes." In complex and uncertain situations, nothing substitutes for creativity, initiative, and enterprise. Imaginative action has social virtues that reflex caution can never claim. But method can help to show why different endeavors are relatively more or less successful, how they are likely to benefit from various changes in approach and context, and what changes are likely to be feasible and effective.

Methodical approaches to sustainable forestry might begin with fundamental requirements for long-term viability. Sustainable forestry must be sufficiently profitable to sustain the necessary levels of investment, sufficiently suitable ecologically to avoid depletion of nature, sufficiently responsive socially to avoid human harm and conflict, and sufficiently dynamic to learn rapidly from experience over time. As the operational conditions under which any sustainable forestry enterprise exists are relatively more or less forgiving within the envelope of these thresholds of sufficiency, a viable strategy positions sustainable forestry activities to maintain an effective balance within their contextual envelope, utilizing opportunities, and managing risks. Since operational conditions can be characterized, different strategies can be assessed or defined for different ranges of conditions, including those that do not offer reasonable prospects.

Assessing Strategies

Such contextual conditions might be assessed along three planes. The first, the horizontal, includes relations over space that affect the advantages and disadvantages—financial, ecological, social—of conducting activities in a particular place. An isolated enterprise, for instance, may face labor scarcities but be relatively free of social pressures. Or it may be in a natural environment that is relatively fragile or resilient, and may enjoy broad managerial discretion but limited scope, for its use. A metropolitan business will face a different configuration of issues and will have better opportunities for cooperation with others who confront similar disadvantages.

The second plane, the vertical, includes the relationships on which a business depends. Sustainable forestry enterprises may be particularly affected by (1) the quality, resilience, and security of supply and sales chains; (2) the strength, suitability, and cost of access to finance; and (3) the links to responsive sources of science and technology. The type, scale, and history of an enterprise would affect the relative influence these conditions have on its viability and, thus, the strategy it may need to adopt to compensate for the risks it confronts.

The third plane of context, policy, includes explicit influences such as interest and inflation rates, trade and investment mechanisms, taxation, regulation, subsidies, and the distribution of functions among central, provincial, and local governments. It also includes implicit expectations regarding how the enterprise operates. These expectations are difficult to characterize—their significance may not be apparent until problems arise—but they can be sufficiently powerful to dominate long-term considerations. They include cultural and political notions of corporate responsibility and the ensuing plausible streams of future liabilities and opportunities.

Conclusion

Sustainable forestry is a process of adaptive learning that depends on spontaneous innovation, investment, and institutional reform. It needs enterprise to progress, organizations that reward it, and organizations that embrace creativity, critique, and change as sources of success. It is a process of learning and adapting for entrepreneurial organizations that are attuned to their circumstances and committed to the long term. The business cases in this project present such organizations. They provide examples of innovation and leadership and collectively demonstrate the value of methodical efforts to capture and explain the experience and to improve the approaches and conditions that shape it. The sustainability of forests depends, in large measure, on how well these activities serve one another over time.

NOTE

This chapter was prepared by Jeff Romm for the Sustainable Forestry Working Group, 1997.

Chapter 2

The Forest Products Industry in Transition

A trio of corporate actions in 1997 provided a glimpse into the future of the forest products industry. That year International Paper Co. acquired Federal Paper Board Co., Inc., in a $3.6 billion deal that transformed International Paper into a $22 billion company. The combined mill operations will increase International's paper manufacturing capacity and efficiency. As part of the agreement, International gained over 300,000 hectares of forest close to its mills in the southeastern United States. STORA, Sweden's $6 billion pulp and paper giant, committed the first portion of its 2.3 million hectares of forestland to independent environmental certification to strengthen the company's position in its all-important European markets. Half a world away in Guyana, South America, Barama Ltd., a joint venture of Samling Strategic Corp. of Malaysia and The Sunkyong Group of Korea on 1.6 million hectares of tropical forest, was producing plywood and other wood products for export to the Caribbean and U.S. markets. The venture is part of an expansion by Samling, one of Malaysia's largest timber and manufacturing companies, into new markets as logging in Malaysia's forests is declining.

At first glance, consolidation in the pulp and paper business, a European paper titan embracing environmental certification, and an Asian timber producer migrating to South America may appear to have little in common.

But each event represents a reaction to economic, environmental, and economic pressures that will, over the next 25 years, restructure the global forest products industry. As part of that transformation, sustainable forestry will move from an idea that existed at the margins of the industry to a force that will shape the business decisions of forest products companies.

In the 1990s, the industry confronted a rapidly changing business environment. Population growth and economic expansion pitted a rising demand for wood against diminishing supplies. In search of new sources of wood, the industry is expanding into new regions, especially South America and South Asia, and increasingly turning to plantations in the Southern Hemisphere to supply its fiber. Rising wood production in the Southern Hemisphere promises to make the region a global manufacturing center. Innovation in technology and product development is accelerating as companies scramble to stretch wood supplies and cut costs through efficiency gains and standardization. Over time, the industry will invest more money to make less valuable—but available—species of trees, like rubber wood, look and perform like more expensive—and scarce—species such as oak or cherry. Finally, globalization is picking up momentum. In the future, the successful company will routinely source its wood around the globe, manufacture in equally diverse locations, and ship products to far-flung markets.

Environmental trends and pressures will influence every facet of change. In the next two decades, already strained wood supplies will be stretched further as public scrutiny and government oversight of the industry's forest management practices intensify. New regulations and market initiatives will further curtail the industry's access to government-owned forests and dictate management practices even on privately held lands. "Certification" is defining a market for wood products grown in an environmentally sound fashion. Just as environmental trends will shape the economic and market forces acting on the industry, restructuring in the industry will have its own environmental consequences—both negative and positive—for the world's forests.

The ability of environmental trends to shape the industry's everyday business realities has moved environmental concerns higher on the list of industry priorities. Ten years ago the industry considered environmental issues "tangential to the core business of making profits," recalled Carlton N. Owen, vice president of forest policy at Champion International Corp. "We've now come to the point that in most forest products companies, environment is core to the business." At Champion, environmental concerns are part of corporate decision making and forest management and are a strategic consideration in the company's efforts to adapt to the industry's changing business conditions. (See Box 2.1.)

Box 2.1. Champion: Adapting to a Changing Business Environment

In the mid-1990s, management at Champion International Corp. recognized that the future of the forest products industry was not necessarily going to mirror the past. They began positioning the $5.8 billion company to compete in a global business where the cost and availability of wood would be strategic considerations. The company took a variety of steps in its timber, pulp, plywood, and lumber operations to improve efficiencies, ensure wood supplies, and integrate environmental considerations into timber production.

In 1996 Champion made two acquisitions to protect itself from wood shortages and price increases. The Lake Superior Land Company with its 288,000 acres of hardwood forest lands in Michigan and Wisconsin will provide wood to Champion's Quinnesec, Michigan, pulp and paper mill. Champion also expanded its 40-year-old Brazilian plantation operations by buying an additional 438,000 acres, which brought its Brazilian land holdings up to about 1 million acres.

Efficiencies in all operations took on a new urgency to lower costs and stretch wood supplies. Between 1986 and 1996 productivity in pulp and paper manufacturing, measured by the ton of product produced per employee, leaped 47 percent in response to a variety of changes. The Mogi Guacu mill in Brazil, for instance, stretched the interval between regular paper machine maintenance from 15 to 45 days.

New practices and technology enhanced efficiencies in the lumber mills. In 1996, the mill in Whitehouse, Florida, installed equipment that reduced the average gap between logs on the processing lines by 18 inches, which allowed the mill to handle 5 percent more logs. It also invested in an optical scanning system that analyzes the size and shape of 40- to 50-foot trees and indicates what length logs to cut to optimize lumber recovery; this enabled the mill to get 4 percent more lumber from the same amount of wood.

Champion advocated SFM early to improve its long-term ability to supply wood to its mills and because SFM reflects the values and concerns of multiple groups with interests in forests. To conform with SFM guidelines, Champion implemented a series of management changes on its 5 million acres of U.S. forestland. A land classification scheme identifies forests sites by those that must be protected, or those where logging is restricted for social or environmental reasons, from those that can be managed primarily for timber production and high-yield forestry. The company created protected stream zones that in many instances exceeded state and federal requirements, as well as programs to protect endangered species. The mills increased their wood inventories to minimize the need to log during unfavorable conditions when harvesting can rut wet roads and degrade soil and water quality. Champion also helped develop training programs to improve the practices of the independent logging contractors that harvest the company's lands and those of private landowners. The independent loggers supply 70 percent of Champion's wood. Loggers who pass and exceed minimum standards are eligible for a preferred supplier program.

Some practices—carrying larger wood inventories at mills, for instance—have added substantial costs. Others were of questionable value. Carlton T. Owen, vice president for forest policy, was confident that those SFM issues would eventually be resolved, even though the goal of sustainability remains elusive. "Sustainability is a journey, not a destination," he commented. "You will never get to a point where you say this is a sustainable forest." (See Box 2.2.)

The Pressures Transforming the Industry

Overall, seven major trends in the industry—(1) a gap between supplies of wood and demand for wood products; (2) a rise of plantation forestry in the Southern Hemisphere; (3) the globalization of trade in wood products; (4) greater product standardization; (5) greater efficiencies; (6) demand for environmental sustainability and the emergence of certification; and (7) more government regulation—will act differently on the pulp and paper, panels and engineered products, and solidwood businesses. The interaction of business pressures with environmental trends in each segment will determine how SFM and sustainable forest products are integrated into the industry. In some parts of the business, conditions are considerably more favorable for sustainable forestry business opportunities than others. In those arenas, environmental considerations will separate which markets and products flourish from those that will languish.

The Rift in Supply and Demand

For the next 25 years, a widening rift between rising demand for wood and constrained supplies will be the defining business reality for the industry. Population growth and economic development, which continue to boost the numbers of consumers for forest products, caused consumption of roundwood to soar 63 percent from 2.056 billion cubic meters in 1961 to 3.354 billion cubic meters in 1995.[1] That hunger for wood is rapidly consuming the natural resources on which the industry depends. Over 180 million hectares of forest, an area the size of Mexico, were destroyed between 1980 and 1995, largely through clearing for agriculture.[2] The poor who use fuelwood for cooking and heating will continue to contribute to deforestation in developing nations, while economic expansion and population growth will exacerbate the pressures on forests worldwide.

As forests diminish they become more valuable. In the past forests were prized primarily for timber production, now they are coveted for the biodiversity they shelter, the scenic beauty they provide, and the water resources they harbor. Other stakeholders are making competing claims on forest services and values. Just as the industry faces dwindling resources, it also confronts unprecedented pressure from environmentalists, scrutiny from government over forest management issues, and competing demands for forest use.

Under these conditions, even though the industry continuously improves its efficiency, world supplies of fiber are expected to fall short of demand. The worldwide consumption of industrial roundwood rose at about 1.3 percent per year between 1983 and 1993, according to industry supply expert Robert W. Hagler.[3] This relatively modest growth rate, however, cloaks wide regional variations. In developed countries, consumption of forest products showed little or no growth; but countries of the Pacific Rim experienced huge spurts in demand. More significant, the growth in demand for individual sectors of the industry generally rose faster than increases in roundwood consumption. The industry's ability to convert roundwood into products more efficiently accounted for the difference between demand and consumption. But in the future several factors, most related to environmental pressures, will constrain softwood and hardwood production, creating local, and even regional, supply shortfalls. For primary and secondary processing facilities, which are typically located near the forests that supply their wood, supply issues will become critical when nearby forests are exhausted or are taken out of production.

Forest products fit into one of two basic categories: softwoods, which come from coniferous forests; and hardwoods, which are harvested from nonconiferous trees. Both types of trees feed the principal forest products industries of pulp and paper, panels, and sawnwood.

Most of the world's wood, 54 percent, is consumed as fuelwood for cooking and heating (Figure 2.1). Lumber and sawnwood products represent the

	Millions of Cubic Meters	Percent
Fuelwood Demand	1,971	54%
Industrial Roundwood	1,680	46%
Lumber and Sawnwood Products	949	26%
Plywood, Panels, Indus. Wood Products	329	9%
Pulp and Paper Products	402	11%
Total Demand	**3,651**	**100%**

Figure 2.1.

A Snapshot of Global Fiber Demand—1995.
Source: Robert W. Hagler, 1995

second largest category at 26 percent, followed by pulp and paper products at 11 percent. Panels of all types consume 9 percent of industrial roundwood.

Uncertainties and contingencies make predictions about the future supply and demand for wood problematical. Historically, when faced with supply challenges, the industry bred new trees, developed new technologies, substituted products, and took advantage of lesser known species. But recent experience demonstrates that robust economies and rising individual incomes lead most societies to value their forest resources for services other than wood production. The extent to which newly enriched middle classes in some parts of the world may demand that harvests be curtailed in the future is unknown. Analysts also disagree about the rate of logging in Russia, rates of reforestation in South America and Southeast Asia, and how price increases in forest products will affect demand. In the mid and late 1990s, a variety of views on supply and demand were expressed in analyses from Jaakko Pöyry, Roger Sedjo, and Aspey and Reed.[4] The reports had their differences, but all expressed a common theme—that wood supplies will be tight for the next two decades.

A Shortfall in Softwood Production

Softwood, which accounted for 32.7 percent of global fiber production in 1996, grows largely in the temperate forests of Canada, Scandinavia, the United States, and Russia and supplies construction lumber and the pulp used to make long-fiber paper, such as newsprint. Over a 25-year period the total production of softwood is estimated to rise less than 15 percent to 1,085 million cubic meters. Demand, which is concentrated in the Northern Hemisphere, is projected to grow 25 percent. The shortfall between supply and demand is expected to be some 315 million cubic meters by 2020 (Figure 2.2).

Regional developments will be chiefly responsible for curbing the expansion in softwood supplies over the next 20 years. In Russia, which holds 50 to 60 percent of the world's coniferous forests, production is stagnant and well below cutting potential. Even though production should rise slowly, a lack of infrastructure, an uncertain investment and political climate, and corruption are expected to hamper full-scale development of those forests.[5] In 1994, Weyerhaeuser canceled plans to invest in a Siberian joint venture with Koppensky Kombinant, a Russian forest products group, citing just those reasons.[6]

Environmental restrictions on logging in government-owned coniferous forests will put the damper on softwood production in North America. During the 1980s and 1990s, between 10 million and 12 million acres of public forestland were pulled out of production in the Pacific Northwest to protect the spotted owl, according to the U.S. Forest Service. Such restrictions

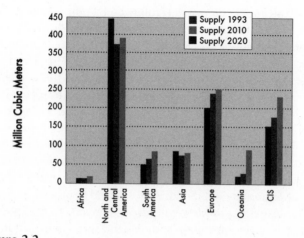

Figure 2.2.

Softwood Industrial Roundwood Supply Projections 1993–2010.
Source: Apsey and Reed, Council of Forest Industries (Canada) 1995

will continue to constrain production. Even though softwood production in the southern United States will increase marginally, it will not be enough to offset the drop in sales from U.S. public lands. Canada faces constraints, as well. Canada's productivity is already suffering from the lack of reforestation in earlier decades and will continue to do so. Provincial governments are also expected to enact stricter logging controls, in response to environmental concerns, during the late 1990s and early next century.

In Europe, softwood production will rise, but not enough to keep up with demand, so Europe will continue to import much of the softwood fiber that it needs. In South America and Oceania, softwood production will rise as plantations come on line over the next 25 years. Because the growing cycle for these plantations is as little as 15 years, supplies from these areas will be more elastic than elsewhere.

Scarce Hardwood

Tropical and temperate regions both harbor large stocks of hardwoods, which are traditionally used for solidwood, plywood, and paper. Even so, hardwood, too, will be scarce relative to demand. The total worldwide production of hardwood industrial roundwood is projected to grow just 6 percent between 1993 and 2020 to some 599 million cubic meters, an increase that will be concentrated in the United States (Figure 2.3). Product and environmental trends are likely to prevent any significant new sources of hardwoods from becoming available in the near future.

Most tropical forests are hardwood, so the availability of tropical hard-

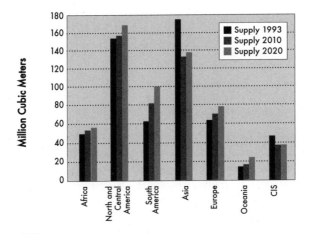

Figure 2.3.

Hardwood Industrial Roundwood Supply Projections 1993–2020.
Source: Apsey and Reed, 1995

woods is closely linked to the controversial issue of rainforest conservation. Tropical regions are expected to show a modest growth in hardwood production, primarily from eucalyptus plantations in Brazil and acacia farms in Indonesia. But environmental pressures will curtail the production of tropical hardwoods from natural forests. Malaysia and Indonesia, Asia's top hardwood producers, rely almost exclusively on their natural forests. For decades both countries have logged forests at an unsustainable rate. In the late 1990s and early in the next century, both governments are expected to slow down the liquidation of those forests. In Indonesia, Asia's second largest producer, hardwood production from natural forests is projected to plunge 50 percent from 40 million cubic meters in 1985 to 20 million cubic meters by the turn of the century.[7] Both countries are trying to develop hardwood plantations that should produce harvests over the next 20 years, although the total area of these plantations is open to question. Neither country, however, has initiated plantations in most of the previously logged areas, so those areas will continue to supply local paper mills with mixed second growth for low-quality pulp.

The large U.S. hardwood reserves will not produce up to their cutting potential either, largely for environmental reasons. Most hardwood forests are privately owned, which makes U.S. hardwood production less dependent than softwood on government lands. Nevertheless, more stringent regulation of wetlands, biodiversity guidelines, and other government-mandated management practices will prevent hefty increases in production from private lands.

The constraints on hardwood and softwood production will exacerbate other pressures already working on the industry. As we will see, the way companies and the industry try to fill the gap between rising demand and limited supplies will define corporate strategies, market opportunities, world timber trade, and opportunities for SFM over the next generation.

The Rise of Southern Plantations

In a world where wood is getting scarcer and natural forests are increasingly off-limits, plantations in the Southern Hemisphere have become the industry's solution of choice for a dependable supply of wood. Over the next 20 years, plantation forestry will make the Southern Hemisphere a major world supplier of fiber (Figures 2.1 and 2.2). In the mid-1990s, less than 10 percent of the world's industrial roundwood was produced on plantations, but production was growing at a double digit rate—faster than any other source of wood. According to some estimates, production from South American plantations was expected to spurt 226 percent from 95 million cubic meters in the mid-1990s to 215 million cubic meters by 2005.[8] During the same period, harvests from plantations in Southeast Asia are expected to rise almost 9 percent to 14 million cubic meters.[9] Chile, Brazil, and New Zealand, which have had government policies favorable to plantations, are receiving the largest infusions of investment capital.

Plantation forestry offers the industry distinct advantages over natural forests. Because plantations produce more standardized fiber than natural forests they are extremely attractive to the pulp and paper industry, whose expensive equipment runs better on the predictable wood of plantations. Trees grow faster on plantations, so they produce a quicker return on investment. Historically, plantations have also been less subject to government regulation or intervention by environmentalists. In the long run these characteristics are likely to make capital markets more comfortable with investments in plantations than in natural forests.

The Southern Hemisphere is the prime locale for plantations because it costs less to grow fiber there than anywhere else. The mild climate creates favorable growing conditions for the highly developed genetics and intensive cultivation and harvesting involved in modern plantations, so trees produce higher yields per hectare than other regions. In Brazil just 50,000 hectares of land can produce 500,000 tons of fiber each year for a pulp mill. In British Columbia it takes 1.6 million hectares to produce the same amount of fiber (Figure 2.4). Labor costs are also lower than in other regions, and degraded land appropriate for plantations is available at a relatively low cost. Finally, the sophistication of modern communications, transportation, and capital movement make the international expansion of fiber production more feasible today than it was in the past. These conditions give producers in the Southern Hemisphere a significant cost advantage (Figure 2.5).

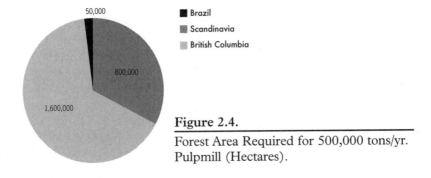

Legend:
- Brazil
- Scandinavia
- British Columbia

50,000
800,000
1,600,000

Figure 2.4.

Forest Area Required for 500,000 tons/yr. Pulpmill (Hectares).

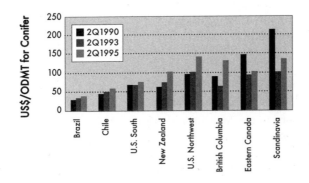

Legend:
- 2Q1990
- 2Q1993
- 2Q1995

Y-axis: US$/ODMT for Conifer (0, 50, 100, 150, 200, 250)

X-axis: Brazil, Chile, U.S. South, New Zealand, U.S. Northwest, British Columbia, Eastern Canada, Scandinavia

Figure 2.5.

Cost Advantage of Fiber Production in Southern Hemisphere.
Source: Robert W. Hagler

The rise of plantation forestry in the Southern Hemisphere will have global implications. Eventually, the industry may develop a dual supply situation: Limited areas of natural forest may be preserved or minimally harvested, while most of the world's fiber would be produced on plantations or other "fiber farms." Industry analysts, such as David Price of the United Kingdom, suggest that all of the world's pulp needs could be met from a massive tree farm no larger than 2 percent of the land area of Brazil. As supply becomes concentrated on plantations, producer nations may be able to wield more influence over world markets. Indonesia and Malaysia, for instance, have moved from merely banning the export of unprocessed logs to trying to set plywood prices. In late 1995, the two countries agreed on minimum plywood prices for the Chinese, South Korean, and Japanese markets.[10] Malaysia also hopes to ban the export of sawn timber by the year 2000 to help encourage primary and secondary manufacturing.

As the Southern Hemisphere produces more wood from plantations, primary and secondary processing industries will migrate there, for financial reasons and because producer nations will exert pressure and offer incentives to do so. The shift will open up new business opportunities in wood and fiber processing. The evolution of The Rio Iata Group in Chile is one destined to be repeated in the Southern Hemisphere. In the 1960s, the Chilean government began to encourage the development of radiata pine plantations. In the mid-1990s, the Chilean industry was ready to leverage those plantations by building processing industries. At that time The Rio Iata Group, according to management, went into particleboard and veneer manufacturing in anticipation of a global wood shortage. The company added a sawmill, particleboard, and veneer mills to its plantation operations and started to export all of its value-added products, such as moldings and fingerjoint stock—most of it to Japan. By the year 2000, the plantations should supply 70 percent of the manufacturing operations' raw materials.[11]

A Reversal of Trade and Globalization

The constraints in hardwood and softwood production in tandem with the rise of plantations in the Southern Hemisphere will foment a revolution in the global timber trade. The combination of Asia Pacific's growing population and rapid economic growth with the depletion of its own natural forests will create a looming fiber deficit in the region. As a result, in 20 years the Asia Pacific will reverse its status from a major wood exporter to a major importer: In 1990, the region had an almost even balance between wood exports and imports; by 2010, it will become a net importer.[12] Japan, the region's largest importer, will continue to absorb the bulk of the imports. The shift in trade will open up major exporting opportunities for wood-producing countries to the Asia Pacific region.

The shifts in the sources of wood supply and centers of consumption and the rise in the global timber trade will accelerate the globalization of the industry. Today, it is not uncommon for an international paper company to sell its equity in Tokyo, grow its fiber in Brazil, purchase papermaking equipment in Europe, and sell its products worldwide. Similarly, companies that use sawnwood often source around the world to service equally dispersed markets.

High economic growth rates in some regions and the relaxation of trade barriers under new trade agreements such as the General Agreement on Trades and Tariffs (GATT), the North American Free Trade Agreement (NAFTA), and other regional accords will further stimulate international trade in forest products. Over the next 15 years exports from North America and Asia will plunge. Rising exports from Russia and Oceania will compensate for some of that decline. More wood will flow from Africa to Asia, as well as from South America to Europe. As the industry optimizes production costs across borders in the wake of globalization, freight and logistics will play an increasingly important role in the search for international efficiencies. Industry leaders have already adapted to these global realities with sophisticated systems to handle logistics at lower costs and to gain competitive advantage.

Even More Standardization

Consistency, as the saying goes, may be the hobgoblin of little minds, but in the forest products industry consistency and standardization will increasingly contribute to success. Just as consumers want predictable quality in the food they buy and the cars they drive, they also look for paper and wood products with no surprises. Greater standardization and consistency in production and products are inevitable in all except small niche markets. The paper industry, for example, is increasingly using genetically engineered plantation species that yield fibers that behave predictably inside a paper mill. New technologies allow medium density fiberboard (MDF) to look like dyed mahogany, even though the MDF's raw material may be a fast-growing secondary growth species. Oriented strand board (OSB) and other engineered wood products are successful because their performance qualities do not vary.

The trends toward more consistency and standardization will alter the way the industry allocates investments in forests, plants, and equipment. In a world where consistency is king, with other considerations being equal, products that directly incorporate solid wood will be undervalued relative to their engineered peers. The former will always look less perfect than the latter. And with downstream processing creating "wood" products out of a wide range of species, less and less of the final value of forest products will be determined by the forest itself. As a result, future processing will attract disproportionately more investment than forests.

Increased consistency and standardization will also contribute to the development of plantation forestry. To lower costs and produce more standardized products, the paper industry, which owns and manages huge tracts of forests worldwide, will increasingly prefer pulp made from genetically consistent trees grown on plantations. In other segments of the industry in the past, the quality of the forest determined the quality of final output: Old and well-managed forests were better than new and unmanaged. The emergence of OSB, MDF, and plantation-cloned eucalyptus means that the quality of the forest will dictate the quality of products in fewer segments of the industry. The separation of forest quality from the quality of the final product makes generating the largest amount of raw fiber the priority and favors plantation forestry.

The Continued Quest for Efficiency

Historically, technological improvements have enabled the industry to continually produce more forest products from the same amount of wood. This is why growth in industrial roundwood consumption has risen only about 1.3 percent per year over the past decade, while consumption of many categories of forest products such as paper and panels increased more rapidly. In the paper industry, alkaline pulping technologies and improved papermaking machinery have yielded impressive efficiency gains. Similarly, the panel industry has flourished in its ability to use scrap materials that would otherwise be used as boiler fuel.

The industry's emphasis on efficiency will be as important in the future as it has been in the past to bridge the gap between rising demand and constrained supplies. As resource efficiencies become more important and natural forests scarcer, more efficient systems and technologies should take root in the Southern Hemisphere. In the late 1990s, harvesting and milling operations in those regions were wasteful: They converted just 25–30 percent of the wood processed to product, compared to rates of 45–50 percent in more developed regions.

Environmental Certification

Systems for forest management certification, which emerged in the early 1990s, are the leading edge of pressures on the industry to improve forest practices through sustainable forest management. The degree of impact, or the speed at which certification will move through the industry, is difficult to predict and will vary greatly from country to country and among different market sectors, depending on business conditions. By the end of the 1990s, nearly half a dozen different certification systems existed. They ranged from the third-party independent Forest Stewardship Council (FSC), which certifies groups that audit forest management practices, to an international system under the auspices of the International Standards Organization (ISO) and an industry-backed effort in the United States under the American Forest and Paper Association.

Certification, which emerged from the nongovernmental sector as a voluntary, nonregulatory, market-based mechanism to improve forest management, is controversial with the industry, and few companies have publicly endorsed it. Nevertheless, some experts and analysts have predicted that certification will reshape the forest products trade within the next 10 years and become as essential for market access as the Underwriters' Laboratory (UL) seal of approval on electrical appliances.[13] By the late 1990s, certification as represented by the FSC had become a catalyst for raising forest management standards and spreading SFM, particularly in Europe, the United States, and among producing nations that competed in the European market. (Chapter 3 discusses certification.)

More Government Oversight

Government oversight will remain a spur on the industry to improve its environmental performance. International, national, provincial, and local forestry management regulations are increasingly defining the industry's access to raw material, the terms of international trade opportunities, and forest management practices. Forestry regulations are already fairly well established in developed countries. In developing nations, regulation ranges from practically nonexistent to significantly structuring trade and opportunity. But every country with a forestry industry is engaged in a debate over how forest assets should be managed, and more regulations are sure to come. New rules generally increase harvesting costs, by either forbidding certain types of cuts or prohibiting logging of certain areas such as watersheds and habitats. They are also reducing the overall harvest of wood from natural forests.

In the international arena, forest management took its place high on the environmental agenda in 1992 at the United Nations Conference on Environment and Development (UNCED) in Rio de Janeiro. While disagreements between developed and developing countries scuttled any binding international agreement on forests, the meeting did adopt a statement of principles for sustainable development of all types of forests. Then in 1994 The International Tropical Timber Agreement set an objective for all trade in tropical timber to come from sustainably managed forests by the year 2000. These agreements, along with several other regional accords and an increase in annual development assistance for forest management from $400 million a year in 1985 to $1.5 billion in 1993, have helped nations and governments reach an understanding of the issues and the need for action. Yet these international agreements, which are important frameworks for consensus and action, do not contain standards, penalties, or means of enforcement strong enough to materially affect forest management practices.

In the future, effective international forestry accords may become more feasible. The certain failure of the International Tropical Timber Organization's (ITTO) attempt to achieve forest management sustainability by the year 2000 is likely to lead to another international agreement. With one failure already in hand, ITTO or some other group of tropical producers are more likely to draw up a more effective accord the second time around. Continued United Nations activities, which grew out of the 1992 UNCED meeting, and the Commission on Sustainable Development charged with reviewing progress on global forest issues will put modest pressure on countries to improve their management in the future, as well. Finally, as governments gear up to take action to reduce carbon dioxide emissions as agreed at a 1997 conference in Tokyo, the forest products industry is likely to be a target. Government actions taken to mitigate global warming may well include more restrictions on harvesting natural forests, which would diminish wood supplies in the short run. But those actions are also sure to involve enacting incentives for reforestation to create "carbon sinks" that, long-term, would increase wood supplies.

Environmental Risks from an Industry in Transition

Economic, market, and environmental pressures behind the industry's restructuring will have their own environmental repercussions. In some cases those impacts have the potential to improve the prospects for natural forests; in others they threaten to accelerate their destruction. Simultaneously, these forces will also have both beneficial and negative effects on the prospects for business opportunities in sustainable forest management.

The shift in the sources of wood supply to the Southern Hemisphere, globalization, and the rise of imports in Asia Pacific could be an unhealthy combination for natural forests in the Southern Hemisphere. Most of the growth in the forest products industry lies either in the nondeveloped world or other areas of the Pacific Rim, markets that have demonstrated little interest in evaluating the environmental performance of their wood suppliers. As companies roam the globe seeking wood supplies, pressures to harvest natural forests in more remote, fragile, or marginally profitable areas will mount—for example, in Guyana, Suriname, areas of the Amazon, and in the Golden Triangle of Southeast Asia. The vertically integrated giants that dominate logging in Asia, particularly in Malaysia and Indonesia, are already migrating to South America and Africa in search of wood as the natural forests in their home countries have become exhausted. If those companies continue the logging practices they used at home, the prospects for greater long-term damage to tropical hardwood forests in those countries are real.

Some of these new producing nations do not have the regulatory infrastructure to respond appropriately to the escalating demand for wood from well-organized companies, which increases the risk of unsustainable harvests—a risk that was realized in Gabon during the late 1980s and 1990s. Massive buying from the Asian Pacific markets drove log exports from just 75,000 cubic meters in 1990 to 700,000 cubic meters in 1995 and threatened the country's forests with destruction. Only by halting log exports did the government and the industry return harvesting to a more even keel.[14]

Environmental restrictions that limit production in developed nations could also have a detrimental impact on forests in the Southern Hemisphere. Any drop in exports from the United States or Canada to importers like Japan, for instance, will open an export opportunity that is likely to be filled by unsustainable Asian producers.

The increase in plantations will also have environmental fallout. Plantations require less land for wood production than a natural forest, which gives them the potential to take pressure off natural forests. But tree farms also have less capacity to supply other valuable services, such as wildlife habitat, water supply, and recreation than do natural forests. As the industry plows more capital into plantations because they are attractive investments on technical, financial, and environmental grounds, companies may invest less in natural forest management. If that happens, it could hinder the development of a sustainable natural forest industry and may, under some conditions, leave natural forests more vulnerable to exploitation.

New Business Opportunities for Sustainable Forest Management

Restructuring will also create new business opportunities built around sustainable forest management. Shifts in wood supplies, for instance, will create opportunities for a variety of new business ventures. Improved logistics and transportation will make it easier for processors to buy wood from a wide variety of sources. Capital will be more mobile, so processing assets will more easily follow wood supplies. Long term, companies that can manage natural forests in an environmentally beneficial way will find new opportunities in the Southern Hemisphere. In Suriname, Chile, Brazil, and Peru companies that practice SFM already get preferential access to superior forest sites and financial resources and more cooperation from local governments. Savia International has adopted SFM as a business strategy on 400,000 hectares of natural forest in Tierra del Fuego for just those reasons. Similarly, forward-looking nations that can enforce regulations may be able to exploit opportunities created by the industry's restructuring if they invest in infrastructure, marketing, and appropriate controls and if they position themselves with buyers of natural forests and investors in plantation forestry.

Globalization will also create conditions more favorable for SFM business opportunities. If rising demand raises hardwood prices, the enhanced margins might make SFM, which is more expensive, more commercially viable in countries with sophisticated regulatory systems. Worldwide information systems, enhanced telecommunications, and more efficient distribution mechanisms that go along with globalization will also be favorable for niche markets. Much of the "certified" wood products industry originated from these niches, and further globalization should generate additional opportunities. With information on markets and management techniques readily disseminated globally, new trends like certification and sustainability will spread through the industry more quickly, just as large trade flows will help spread sustainability into other markets. Improvements in information transmission, shipping, and distribution should also make it more feasible to introduce species previously not used commercially, particularly from the tropics. Creating markets for underexploited species of trees is essential for sustainable tropical producers. These producers need to maximize the value of all natural forest species to make sustainable forestry economically viable, instead of harvesting just a few commercially valuable species as is typically done.

The priorities of the forest products industry are likely to continue to dominate forest management for many years. In the future, however, the value of the nonwood "products and services" that forests provide may determine how forest economics are structured in some areas. In Indonesia and Malaysia, for instance, the combination of still significant natural forest reserves, relatively effective governments, important biodiversity issues, and a mature national industry could force the industry to balance wood production with other competing interests—such as exploiting the commercial potential of the region's rich biodiversity. If that happens, new opportunities may open for companies that own forestland.

Difficult Business Conditions for the Pulp and Paper Industry

Sustainable forestry may take place in the forest, but whether it takes hold in the industry depends on the use of wood in final products. The industry's three major segments—pulp and paper, panels, and sawnwood—operate under different business conditions. The pressures at work on the industry will react with the unique business conditions in each segment to create both opportunities and obstacles for sustainable forestry.

The global pulp and paper industry, which in the late 1990s had estimated annual sales of between $500 billion and $600 billion, produces thousands of wood-based products ranging from newsprint and printing papers to cardboard and other packaging. U.S. pulp and paper sales totaled $170 billion in 1997. The U.S. industry accounts for roughly 2 percent of GDP, while the Canadian industry represents about 6 percent of GDP.[15] In Asia, pulp and paper products generate roughly 3 percent of regional GDP. The global pulp and paper industries consumed about 600 million cubic meters of industrial roundwood and produced 157 million metric tons of pulp in

1996. Four business conditions will dominate the segment in the coming years:

- Relatively Stable Demand: Global demand for paper products is expected to rise between 2 percent and 3 percent annually over the next five years, driven principally by economic and population growth.[16] The United States wins the per capita consumption award at 330 kg per year, followed by Finland (266 kg) and Japan (231 kg). Developing nations, such as China and India, lag well behind with per capita consumption of less than 30 kg each year. In the 1990s, the fast-growing economies of Asia, Latin America, and even Eastern Europe registered the highest growth in paper consumption—a trend that is expected to continue. The long-term impact of the electronics revolution and the "paperless" office is the unknown variable in paper demand. The internal business market consumes 70 percent of all printing and writing paper. As corporations more fully automate their payment, invoicing, and "paperwork" systems, the business sector will use less and less paper; but most observers predict that the rising appetites for paper in developing countries will compensate for any decline in the internal business market.

- Fragmented Supply: The industry is entering a period of volatile supplies. Ownership of supplies and manufacturing facilities is fragmented in most world markets, although consolidation appears to be accelerating. Between 1980 and 1990, the number of paper and board mills in key markets plunged 21 percent in the United States, 25 percent in Europe, and 23 percent in Japan.[17] Further consolidation shook the U.S. market in 1995, when the high-profile merger of Kimberly-Clark and Scott Paper created an $11 billion combined company and again in 1997 with International Paper's $3.6 billion acquisition of Federal Paper Co. Some analysts expect supply issues to trigger a round of mergers and acquisitions as companies try to lock up sources of stable wood supplies over the next decade.

- Rising Capital Intensity: High capital intensity is a distinguishing characteristic of the industry (Figure 2.6). In the United States, capital expenditures of pulp and paper manufacturers average 10 to 11 percent of sales, or roughly double that of other highly capital-intensive industries like chemicals, primary metals, and manufacturing; and the industry's capital intensity is rising. From 1980 to 1990, the percentage of all mills in the United States producing more than 450,000 tons per year (tpy) of pulp rose from 40 percent to almost 60 percent, while the number of mills producing more than 500,000 tpy doubled.[18] The cost of a larger mill has grown proportionately: In

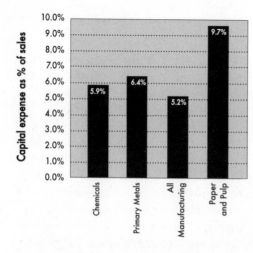

Figure 2.6.

Capital Intensity of Pulp.
Source: AFPA, 1995

the late 1990s, a new pulp mill with the standard capacity of 500,000–700,000 tons cost roughly $1 billion. The huge scale of production needed to produce pulp economically is the chief reason for the industry's high capital requirements. With few exceptions, pulp and paper products are commodities—manufacturers compete primarily on price. Price drives most purchasing decisions, which gives buyers great power over producers. This intense price sensitivity requires producers to continually lower production costs through new technology and efficiencies in growing, harvesting, and processing trees.

• Variable Profits: The industry's fragmentation, capital intensity, commodity nature, and tendency to overspend on plant and equipment contribute to a cycle of steep swings in pulp pricing with proportionate variations in corporate profits. Over the past 30 years, the U.S. industry's net profit margin has averaged roughly 4.5 percent. The industry's return on net worth has averaged roughly 9 percent since 1960.[19] Excessive debt has also eroded the industry's return on equity, which is frequently lower than the average cost of capital. For these reasons, long-term returns in the paper industry underperformed other global financial securities by more than 35 percent from 1970 to 1996.[20]

A New Cycle of Environmental Adaptation

Over the past 30 years, pulp- and papermakers have been forced to make costly and disruptive changes to adapt to environmental pressures. During the 1970s, the industry was hit with pollution and pollution-abatement regulations. In most years since 1973, the U.S. industry has spent more than 10 percent of its capital expenditures to cover the cost of complying with pollution control regulations; costs have ranged from a high of 35 percent in 1973 to lows of between 5 and 10 percent in the 1980s.[21] The U.S. Environmental Protection Agency's Cluster Rules that combine air and water regulations are expected to cost the industry an estimated $5 billion to $15 billion by 2001.[22] In the 1980s, recycling hit the paper segment, propelled by a public perception that landfill space was disappearing. U.S. paper mills increased their use of recovered papers by 94 percent between 1985 and 1995 to more than 32 million tons. Recycled fiber, less than 25 percent of fiber input in the U.S. paper industry as late as 1986, made up roughly 35 percent of total production by 1998, according to the U.S. Forest Service.[23] During the same period, recycled fiber capacity grew more than 100 percent in the United States. The American Forest and Paper Association has projected continued but more moderate growth.[24]

The pressures on the pulp and paper industry for more sustainable forest management will be so great over the next 20 years, according to some industry observers, that they will provoke radical changes. Adaptation is, in fact, underway. The size and cost of a modern mill have made a steady, reliable fiber supply far more important than it was in the past for a mill to operate efficiently and cost-effectively. To ensure those supplies, pulp producers have become heavy investors in plantations in the Southern Hemisphere. In 1993 original old-growth forests supplied 16 percent of the industry's pulp versus just 11 percent for plantation.[25] In the future those exotic plantations will supply a greater proportion of the industry's fiber. Under pressure from large customers, pulp and paper producers are moving toward more sustainable forest management and the adoption of environmental certification. By the late 1990s major Scandinavian paper producers had begun to certify their forestlands, and by the turn of the century markets for certified pulp and paper products are expected to emerge in Europe. (Chapter 3 discusses market opportunities for certified paper products.) The enthusiasm in some markets for environmentally differentiated products should create a variety of opportunities for early movers, which a handful of companies has begun to exploit in the late 1990s (Box 2.2).

The Age of Panels and Engineered Wood Products

Until recently, most engineered wood products were panels, basically wood fibers reconstituted into flat, thin sheets. In the 1980s and 1990s, however, producers leveraged panel technologies to create new products that substitute for solid lumber principally in the construction and furniture industries. Overall in 1995, panels consumed 9 percent of industrial roundwood.

Box 2.2. Environmental Opportunities in Paper

A handful of companies, taking a cue from public opinion and lobbying from customers and regulators, have already organized their strategies around sustainable forest management issues. By acting early they have reaped, or expect to reap, a competitive advantage:

- Three top Swedish paper producers—SCA, STORA, and Assi-Doman— participated in the Forest Stewardship Council (FSC) working group, which drew up national Swedish certification standards. By 1997 all had certified portions of their land under the FSC system. In early 1998 the first certified products from those lands were shipped to European customers. Assi-Doman, for instance, delivered certified timber to the Homebase stores in the United Kingdom, owned by Sainsbury Plc., and certified pulp, which enabled Sainsbury's to introduce FSC hygienic tissue.
- The Rock-Tenn Co., a major U.S. producer of recycled paperboard and folding cartons, has successfully implemented a strategy of vertical integration, high technology, and extensive use of waste fiber. The company's emphasis on recycled paper has given it flexibility in managing costs and helped keep earnings variability lower than that of most of its competitors.
- Fort Howard Corp., the world's leading maker of commercial and industrial tissues, pioneered recycling technology that has enabled the company to use low-cost mixed office wastes as a raw material and maintain the highest profit margins in the tissue business. Most of the company's products are made from 100 percent recycled waste paper and are processed with a proprietary de-inking technology.
- Brazil-based Aracruz Celulose S.A., the world's largest producer of bleached hardwood market pulp, has established itself as the market's lowest-cost producer of a premium product. In the process, Aracruz has forged ahead of competitors in eucalyptus forestry practices and has made environmental performance a cornerstone of its strategy.

In the next 15 years, however, demand will soar for a new class of engineered panels, which includes MDF (medium density fiberboard), OSB (oriented strand board), and particleboard. (See Figure 2.7.)

Environmental considerations, rapidly advancing technology, and attractive economics will make these engineered products increasingly more popular substitutes for solid wood and plywood:

- Oriented Strand Board: Until OSB was developed in the 1970s, "structural panel" was synonymous with plywood. In the late 1990s, OSB was rapidly replacing plywood in North America and Europe

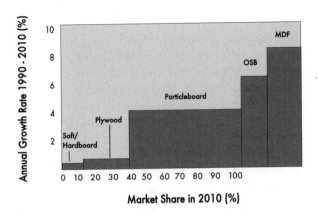

Figure 2.7.

Global Demand for Wood Panels—2010.
Source: Jaakko Pöyry, 1996

in many applications and will gain ground in other markets. By 2010, it is expected to hold a respectable share of the global panel market. OSB is made by aligning "strands" of wood in a multilayered mat with the layers of strands aligned perpendicularly. The layers are then glued together under heat and pressure, producing a board of uniform, consistent strength. The panels can be made from aspen, birch, and other "weed" species, common in secondary forests, that have few alternative uses and are inexpensive. The cost of OSB plants is dropping faster than the cost for plywood mills. Between 1990 and 2010, the North American OSB capacity will surge 362 percent to 23.9 million cubic meters. Meanwhile, plywood capacity is expected to drop about 29 percent to 15.63 million cubic meters. The same trends are evident elsewhere.

- Particleboard: This panel, which is less expensive than OSB, dominates the production of engineered panels worldwide. The panels are made by bonding together small flakes of wood fiber—often sawdust or plywood shavings—with an adhesive under heat and pressure to create an economical board that is used in furniture making, cabinetry, floor underlayment, and door cores. The explosive growth of the ready-to-assemble (RTA) furniture market, which depends heavily on particleboard, caused demand to soar in the late 1980s and 1990s. In the future, technical innovations promise to improve the product's structural characteristics and lower its cost. Particleboard will continue to dominate the worldwide panels market. Industrial uses, including furniture, will account for most of the industry's growth in the United States and other regions. Like other segments of the industry, the particleboard sector is evolving toward manufacturing cut-to-order pieces and other differentiated products, which enables manufacturers to operate more efficiently and tailor their products to fit customers' needs. No substitutes are in sight that can match the performance and cost characteristics of particleboard. It is, however, a low-value product relative to its bulkiness and weight, which limits its potential for international trade. The industry will continue to be chiefly regional.

- MDF Panels: In the 1990s, MDF panels became one of the most successful new product categories in industry history. The panels, produced by reducing individual wood fibers—typically from wood chips—in the manufacturing process, then forming them into boards with pressure, heat, and urea formaldehyde, can be made from a variety of fibers, including rubberwood and eucalyptus, as well as annual crops such as cotton stocks. Between 1994 and 1997, U.S. MDF capacity almost doubled, while output increased fivefold in Europe and capacity was expected to triple Asia.[26] MDF, a more expensive panel alternative than others, is sought after in the high end of the panels market because it is easy to machine, easy to coat, is natural looking, has smooth surfaces, and is homogeneous. In the 1990s, MDF was used primarily for furniture and cabinetry but was also making inroads into the thin plywood market, door skins, and millwork applications. As MDF with special characteristics—water resistance and thin sheets, for instance—enters the market, it will gain ground in new applications. As a result, demand for MDF panels is expected to grow at an average annual clip of just over 8 percent each of the next 15 years.

The Importance of the Environment for Engineered Wood Products

Producers of engineered panels will benefit from an era of environmental constraints. OSB, particleboard, and MDF are filling a void created by the increasingly scarce supplies of peeler logs that are used to make plywood and by the decline in the quality of available wood, both of which are linked to deteriorating forests. Because OSB, MDF, and particleboard all use non-traditional trees, low-grade material, or by-products, and do not depend on high-quality raw material, producers will be able to readily adapt to any difficulties or shifts in wood supply—unlike producers that depend on natural forests for their raw material. Waste wood, annual crops, and other fiber sources can fill any gaps in traditional fiber supplies. In Asia, producers are already using annual crops as the fiber to produce particleboard. And OSB is replacing plywood primarily because the peeler logs that provide the raw material for plywood are more difficult to find, especially in Asia, as the natural forests that contain such high-quality timber are cut down.

MDF producers, particularly in Asia, are in a favorable position to market the panels as a "green" solution in furniture and construction markets. In the 1990s, MDF capacity in Asia exploded as processing capacity grew around wood sources, and the region began to produce furniture for export (Box 2.3). Much of the furniture is exported to Europe, where concern over forest management is greatest. Positioning products as "green," or certifying them, might be prudent for MDF manufacturers that export to Europe, if only as a defensive measure. The dynamics of the office furniture market might also favor certified MDF. A handful of companies dominates the office segment of the furniture market, including Knoll, Herman Miller, and Steelcase, while Fortune 500 companies account for most office furniture consumption. The desire of large corporations to be environmentally correct may eventually lead those customers to ask their large suppliers to offer sustainably produced furniture products.

Even under favorable conditions, however, the characteristics of the panels business will make it difficult for a viable market in "certified" products

---

Box 2.3. An Advantageous MDF Partnership

IKEA, the Swedish manufacturer renowned for its trendy, economical ready-to-assemble furniture, and Golden Hope Plantations Bhd of Malaysia have found profit potential in the forces reshaping the industry. The two companies have forged a joint venture to develop manufactured products based on the Malaysian company's extensive rubberwood plantations. Golden Hope will build a 100,000 cubic meter plant that will make furniture components and manufacture MDF furniture to IKEA designs. IKEA, the leader in RTA particleboard furniture, will handle the MDF furniture for all markets outside Malaysia. Initially, the furniture will not be marketed as a "green" product, but the emerging demand for such furniture and IKEA's leadership in the market positions the venture to do so in the future.

The venture takes Golden Hope from mere fiber production into value-added manufacturing. IKEA is catching the currents of globalization by designing products in Europe that will be produced near the new sources of wood from plantations in Southeast Asia, then distributed worldwide. The engineered wood products that the two companies produce will have competitive advantages over traditional wood products that depend on natural forests because they are produced from formerly unattractive—but available—species.

to develop. Manufacturers would have trouble meeting the chain of custody requirements involved in certified products since much of the raw material in engineered panels is a by-product of other wood processing industries. The industry is also fragmented, produces relatively undifferentiated products, and has no dominant players exercising market leadership—all obstacles to the development of markets for certified products. In addition, the power of buyers is widely dispersed among tens of thousands of builders, remodelers, furniture manufacturers, and others who use panels in their final products.

Consumers are also at least one link away from the decision on what particular type or brand of panel will be purchased, so they exert less influence. For these reasons, large-scale markets for certified panels will probably develop later than certified markets for other forest products. Despite these obstacles, however, opportunities will exist for engineered wood and panel producers to market certified panels to the furniture sector, which has been more receptive to "green" and certified products than the construction industry.

A Global Veneer Market

Environmental trends will exert both positive and negative effects on the veneer sector of the industry. Worldwide production of the veneers that are used as decorative covers to appear like solidwood for furniture, wall paneling, and other premium applications totaled 6.1 million cubic meters in 1996.[27] The U.S. veneer industry, which produced 80,000 cubic meters of veneers in 1996, is relatively fragmented and dominated by family-owned and run businesses that have operated for decades. The capital required to build a plant is small relative to the rest of the industry—in the low seven figures. In recent years, though, requirement for higher capital investment for new equipment and technology has forced some consolidation. Larger firms with good access to roundwood supplies and the capital to invest in closer tolerance and highly efficient equipment will increasingly dominate the market.

Veneer is among the most "globalized" of forest products. The product's relatively high value, light weight, the industry's ability to customize and differentiate products at a reasonable cost, and wide knowledge about the product among furniture designers make it a good candidate for international trade. Exports account for nearly 50 percent of U.S. production and are expected to remain strong between 1997 and 2002 as European consumers continue to avoid tropical veneer.[28] Worldwide veneer manufacturers are evolving toward greater uniformity in manufacturing processes to satisfy customers, who increasingly want veneers that are as "identical" as possible. This preference gives a market advantage to lighter colored veneers from temperate forests that can be stained to a uniform color over darker, tropical veneers. In the next few years, MDF poses the greatest threat to veneer products because it is expected to become available in high-quality overlays or prints that look just like hardwood plywood.[29]

Long term, however, the lack of quality hardwood roundwood will pose the most significant challenge for veneer makers. Demand for veneers will rise substantially over the medium and long term. But to produce a high-quality veneer you need a high-quality forest. The precious few hardwood trees that can provide furniture and panels with a rich, natural look and feel are becoming scarce and, when available, more expensive. Veneer, which represents less than 1 percent of total global roundwood production, has extremely high value. A top-quality veneer log can sell at a 600 percent premium over alternative uses. As larger-dimensional wood becomes more expensive, manufacturers will substitute engineered wood products and use veneers to cover them, thus meeting customers' aesthetic requirements at a reasonable cost.

Many of the characteristics of the veneer sector are favorable for the production of certified veneer products. Much of the veneer export market is in Europe, where demand for certified products is strongest. Veneer producers operate more like "job shops" than most other branches of the industry, so the custom orders and chain-of-custody issues typically involved in producing certified products should be less burdensome for them than for other types of manufacturers. Some producers, such as Curry-Miller Veneers, Inc., in Indiana, already use bar-coding systems to track logs and their products from log yard to customer because veneer buyers often want to buy entire sliced logs. Veneer producers could readily adapt this type of tracking system to third-party certification schemes that require monitoring of certified wood from logging through manufacturing to final retail destination. Finally, certified forests produce higher-value logs that can yield top profits when directed toward the right market, which should make them prime candidates for veneer production.

The Prospects for Sawnwood

Worldwide, the same pressures at work on other sectors of the industry will shape the production trends of sawnwood from softwood forests and temperate and tropical hardwood forests.

Stagnant Consumption of Softwood Sawnwood

Softwood accounts for just over 300 million cubic meters of lumber per year, most of it harvested from native forests in temperate countries (Figure 2.8).

Softwood lumber, the most important traded solidwood product, accounted for about 26.6 percent of world softwood consumption in 1995, or about 1.145 billion cubic meters. The trade in softwood lumber flows primarily from Canada to the United States, from the Pacific Coast of North America to Japan, and from Canada to Europe. With the exception of a rise in trade from Oceania and South America to the Pacific Rim, these patterns are expected to remain relatively unchanged until 2005.[30]

Softwood lumber production will be constrained by the same forces that are limiting the production of softwood generally in Canada and the United States, the world's largest producers. Traditionally, the Canadian industry has depended on government-owned lands for its softwood. In British Columbia, 90 percent of the harvest comes from public lands. The United States has become less dependent on government lands for softwood in recent years than it has been in the past (Figure 2.9). Nevertheless, the large public ownership of softwood forests in both countries makes the industry more vulnerable to shifts in public priorities and policies toward forest management than would be the case if most softwood came from privately held land.

The construction industry consumes most softwood lumber, primarily for home building. The United States dwarfs all other countries in its use of softwood lumber. U.S. consumption of softwood lumber, however, is expected to remain flat between 1995 and 2010.[31] In Japan, the second largest softwood lumber market, new housing dominates consumption because the Japanese tend to tear down older houses and rebuild rather than repair and remodel them.

Several factors will contribute to stagnant softwood lumber consumption. As public forests are taken out of production for environmental reasons, and

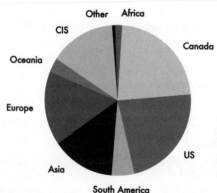

Figure 2.8.

World Softwood Sawnwood Production, 1995.
Source: FAO, 1995

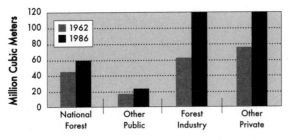

Figure 2.9.

U.S. Harvest Trends—Softwood 1962 and 1986.
Source: William J. Lange, U.S. Forest Service, 1992

second growth and plantation fiber replace old-growth trees, the quality of softwood is declining. This has made higher quality grades more expensive and more difficult to find. In response, contractors are turning to substitutes. Softwood lumber has lost market share to steel, engineered wood products such as composite I-beams, and other substitutes. In the future, these trends will be exacerbated, as softwood lumber users choose substitutes more often to protect themselves against price swings caused by supply difficulties. Substitutes tend to make inroads only when prices are high; when prices drop off, substituting technologies tend to keep the market share they gained.

Greater efficiencies will also help keep consumption flat. Building construction is becoming much more efficient in its use of wood. New technologies that enable sawmills to "scan" lumber and grade it according to strength are expected to have a significant impact on efficiency. Currently most lumber is graded by dimension, so that all lumber is assumed to have the lowest common denominator in terms of strength. These new technologies will permit more efficient use of wood in selected applications.

Mixed Prospects for Certified Softwood

The softwood lumber market would seem to be a prime candidate for certified products because it is so susceptible to environmental pressures. But several factors will hinder any market for certified softwood lumber. Historically, the construction industry has not proactively sought solutions to environmental issues. Furthermore, there are multiple steps between final consumers and forest managers and the market consists of highly fragmented end users. Thus, it is difficult for consumer demand for certified products to work back to suppliers. It is also more difficult to develop markets for certified products in a commodity business, such as softwood lumber, than it would be in high-value market segments, such as veneers.

In the long run, circumstances may mitigate some of these difficulties. The large Scandinavian pulp producers that have begun to adopt environmental certification also produce large volumes of softwood lumber and sawnwood products. The Nordic producers will undoubtedly seek to distinguish their products to create a competitive advantage, which would pressure U.S. and Canadian producers to follow. In addition, the large flows of softwood lumber from environmentally important regions like British Columbia into environmentally sensitive markets like California could be an arena where certified softwood lumber products might find a receptive market.

The Status Quo for U.S. Hardwoods

Hardwood lumber production worldwide totaled 120.3 million cubic meters in 1996, down 8.6 percent from production levels in 1990 of 131.7 million cubic meters. Most hardwood lumber is consumed in the country that produces it, so trade accounts for a small percentage of the world hardwood lumber consumption. Major trade in hardwood lumber flows from the United States, Southeast Asia, and Brazil to Europe and from Southeast Asia and the United States to Japan. These flows are expected to increase over the next decade, particularly to Japan. West Africa will become a more important player in the hardwood trade with the Pacific Rim as native production declines in those markets.

During the 1990s, the United States produced as much as 40 percent of the world's temperate hardwoods, largely from privately held land, according to estimates from officials at the U.S. Forest Service. During this period, U.S. production rose, while that in Europe and Russia fell. Most U.S. hardwood forest is owned in relatively small parcels that were inherited, sold, or somehow passed along from an agricultural past. Little consolidation is expected among those privately owned lands, so small and medium nonindustrial land holdings will probably continue to dominate the production of U.S. hardwood lumber.

Between 1995 and 2010, little change is projected for either U.S. hardwood lumber production or consumption patterns. In the past 40 years, U.S. hardwood growth has exceeded harvest. But a combination of passive management practices—withdrawals for environmental reasons such as wetlands preservation and poor stocking—will probably keep hardwood lumber production fairly steady at about 29 million cubic meters, according to several estimates.[32] The pallet and crating industries consume the most hardwood lumber by volume, about 35 percent or roughly 10.5 million cubic meters. The furniture industry, the second largest consumer, accounted for about 8 million cubic meters of hardwood in 1995.

Consolidation of Hardwood Sawmills

The sawmills that process temperate hardwoods are undergoing consolidation. These sawmills, typically small- to medium-sized businesses that serve a nearby geographic area, depend on that area for their timber supply. The leading hardwood-producing state of Pennsylvania, for example, has 578 sawmills. Increasingly, hardwood mills need to become more efficient and produce value-added products to improve their margins and stay profitable. But midsized mills, those with sales between $1 million and $10 million, typically do not have the technology, such as high-tech CNC milling machines and scanners, to maximize milling yields. Nor do many smaller mills have the organizational skills, technologies, and links with markets needed for successful value-added production. Medium to large mills are better able to meet the requirements for technology, investments, marketing know-how, and product development demanded by value-added production than smaller mills. These pressures will accelerate consolidation among hardwood mills.

Environmental Opportunities for U.S. Hardwoods

During the 1990s, concerns in Europe over the destruction of tropical hardwood forests were a boon for U.S. producers. U.S. exports of hardwoods spurted from $400 million in 1988 to $2 billion in 1996, according to the American Hardwood Export Council. U.S. producers also profited from the growing popularity of light-colored woods, many of which come from the United States. The pace of growth in U.S. exports will slow in the future. But continued concerns about tropical hardwoods and diminishing supplies coupled with robust U.S. forests, and a growing familiarity among European customers with U.S. hardwoods, should position U.S. exports for additional growth in the next decade.

Conditions in the U.S. hardwood business would seem to be conducive for certified products. Several U.S. companies such as Kane Hardwood, MTE, and Seven Islands already offer certified products; this has helped them enter European markets. New market opportunities will undoubtedly arise as markets for certified products evolve in Europe. Private U.S. hardwood forest owners also exhibit a number of characteristics that might predispose them to adopt SFM practices and certification under the right conditions. Typically these owners hold their land as a long-term investment. They do not necessarily seek an annual return or try to maximize returns. Many private landowners manage for multiple objectives, such as hunting and hiking, not just timber production—they already husband their forestland for several objectives, which is part of SFM. Selective harvesting practices, also used in SFM, are common among private owners of hardwood forests. Although the fragmentation of small owners would make chain-of-custody certification difficult, several certification schemes aimed at small landowners have surfaced that should make it easier for them to certify land and lower the costs of the process, as well. (See Chapter 7.)

Continued Controversy over Tropical Hardwoods

Tropical hardwoods are—and will continue to be—the most controversial segment of the industry. Since the early 1990s, Western-developed nations have pressed tropical countries to conserve their hardwood forests because they are the planet's richest reservoir of biodiversity and one of its most significant "carbon sinks." Developing nations, which consider their tropical hardwood forests a resource ripe for economic development, have for the most part resisted outside influences over forest management.

Little of the production of tropical sawlogs and sawnwood enters the world market—between 4 and 6 percent, according to the International Tropical Timber Organization (ITTO). But trade in plywood, which is important to the world trade of tropical hardwoods, accounts for over 80 percent of plywood production in some years. Although the total volumes of plywood represent a relatively low percentage of total tropical hardwood production—slightly more than 1 percent of production in 1993—the extraction of high-quality peeler logs used to make plywood is a controversial tropical forest management issue. Asia dominates tropical hardwood harvests, with production concentrated in Indonesia, Malaysia, and Brazil. According to the ITTO, Asia and Brazil account for 70 percent of logs, 63 percent of sawnwood, 84 percent of veneer, and 90 percent of plywood produced from tropical hardwoods.

The unease over tropical hardwood products in developed countries, particularly European markets, has had an impact on producing nations. Imports of tropical hardwoods, particularly those from Malaysia and Africa, to Europe plunged during the 1990s in response to bans on tropical imports, certification efforts, and general skittishness. Malaysian imports to Europe dropped as much as 30 percent, according to some estimates.[33] Tropical producer nations have instituted several initiatives to improve the image of their hardwood exports, ranging from public relations tactics to serious action to ensure the long-term viability of the industry.

Meanwhile, African exports have shifted from Europe to Pacific Rim nations, which import the majority of tropical hardwoods to feed their thriving economies and compensate for a lack of wood at home. Japan, the largest importer, buys wood from tropical and temperate sources to meet 70 percent of its wood needs; in 1994, Japan imported some 12.688 million cubic meters of tropical hardwood logs, sawnwood, plywood, and veneer.[34] In Japan, Taiwan, Thailand, and other Pacific Rim importing nations environmental concerns are not priorities. As long as this is the case, tropical hardwood products will find an enthusiastic market there.

Europe is likely to offer new opportunities for sustainably produced tropical hardwoods. European buyers still yearn for tropical hardwoods because they have unmatched characteristics. But they want environmentally correct tropical hardwoods, which is why demand for certified tropical hardwoods far exceeds the supply. Most of the few certified tropical forest product companies that exist are small. Medium term to long term, as the market for certified wood develops, tropical producers should find new business opportunities in sustainably produced species such as mahogany and teak that are in demand in Europe.

A Future Molded by Environmental Forces

In the late 1980s, environmental concerns for the wood products industry were largely confined to manufacturing processes, chiefly pollution in pulp mills and the recycled content of paper products. A decade later, environmental issues and pressures were strategic considerations for all three major segments of the industry—pulp and paper, panels and engineered wood products, and sawnwood—and for all types of companies whether timber producers, manufacturers, or retailers. Although environmental pressures are complex and their effects diffuse, they are affecting wood supplies, creating market opportunities, dictating processing technologies, and shaping the regulatory atmosphere of the wood products industry.

Environmental pressures have become strategic considerations for the industry in three key arenas: to mitigate the risk of public censure and government intervention, to secure wood supplies, and to create new business opportunities. Sustainable forestry is proving to be an insurance policy for companies against criticism of their environmental performance from the public and environmentalists, pressure from customers, and even the loss of market share in environmentally sensitive markets like Europe. Malaysia's exports of tropical hardwoods to Europe plunged an estimated 30 percent in the 1990s—a graphic illustration of the risk that questionable environmental performance will pose in the future. Adaptation to the public's escalating demands for better forest management through sustainable forestry may provide the industry with an often effective means to stave off more stringent government regulations of forest management in North America, as well as in some nations in the Southern Hemisphere.

For all segments of the industry, securing dependable sources of fiber in a world of disappearing natural forests will be a critical issue. Here, too, environmental concerns will be central to decision making. As one industry executive has pointed out, "In the next 20 years the company that has the wood, will win."

Superior forest management will have its own rewards in the competition for fiber supplies. Governments in some Southern Hemisphere nations are increasingly likely to grant preferential access to natural forests to those companies with sound forest management. Hardwood producers that, through sustainable forestry, can produce high-quality hardwood products such as veneers may well gain an advantage in a global market where demand for high-quality hardwoods will outstrip supplies. The industry's growing investments in Southern Hemisphere plantations are, and will be, driven in part as a way to avoid the environmental risks increasingly posed by logging in natural forests.

Environmental pressures will create new markets, give some products competitive advantages over others, and upset global trade, opening up business opportunities for those companies that are positioned to exploit them. The rising tide of imports to Asia Pacific as those countries exhaust their forests at home will create trading opportunities for companies from producing countries. The increase in wood production from southern plantations offers opportunities for a variety of manufacturing ventures, which companies outside the region can exploit—as Sweden's IKEA did in the mid-1990s. In Europe, the strong preference for environmentally correct wood products should give sustainable producers new business opportunities, especially those who produce certified tropical hardwood products.

North American and European companies that can market environmentally differentiated products, including paper products, will also find opportunities in Northern European markets. Producers of MDF, particleboard, and OSB are poised to gain market share at the expense of plywood and softwood lumber producers because their engineered panels can be produced from an eclectic mix of readily available and less expensive fiber. For the wood products industry, already in a hurry to keep pace with a changing world, environmental pressures have become a primary strategic consideration that must be integrated into the business of producing wood products.

NOTES

This chapter is adapted from "Sustainable Forestry within an Industry Context," prepared by Tony Lent, Diana Propper de Callejon, Michael Skelly, and Charles Webster, Environmental Advantage, Inc., 1997.

1. Price, David (1996). *World Fibre Supplies, Pira International Survey,* United Kingdom.

2. FAO (1997). *State of the World's Forests* 36.

3. Hagler, Robert W. (1995). "The Global Wood Fiber Balance: What it is, what it means," from 1995 TAPPI Global Fiber Supply Symposium Proceedings, Atlanta.

4. A number of studies of world fiber supply and demand were carried out in the mid-1990s. See Jaakko Pöyry (1995), *Solid Wood Products Competitiveness Study,* American Forest & Paper Association, Washington, D.C.; Sedjo, Roger A., and Kenneth S. Lyon (1995), "A global pulpwood supply model and some implications;" Final Report to IIED: *Resources for the Future,* Washington, D.C.; Aspey, Mike, and Les Reed (1995), "World Timber Resources outlook, current perceptions: A discussion paper," second edition. Council of Forest Industries, Vancouver, B.C., Canada; Solberg, B. et al. (1996), *Long-term trends and prospects in world supply and demand for wood and implications for sustainable forest management: A synthesis,* European Forest Institute and Norwegian Forest Research Institute.

5. Jaakko Pöyry Consulting, Inc., Tarrytown, N.Y. 1996.

6. "Weyco Kills Plans for Russian Project," *Wood Technology* July/August 1994, 7.

7. Ministry of Forestry (1997). Welcoming Remarks by the Minister of Forestry at the opening of the 1997 National Ministry of Forestry Workshop, April 14, 1997 1A: 14–15.

8. Palmer, W. J. Bayard (1995). "A Timber Supply Outlook for South America." Proceedings, TAPPI Global Fiber Supply Symposium, Oct. 1995. Chicago.

9. McKenzie, Colin R. (1995). "Global Supply Outlook: Southeast Asia/China." Proceedings, TAPPI Global Fiber Supply Symposium, Oct. 1995. Chicago.

10. "Malaysia and Indonesia Attempt Plywood Pact" *Tropical Timbers,* December 1995, Surrey, U.K.

11. Blackman, Ted. "Adding Value: Chilean Mills Get Full Value from Forests." *Wood Technology* January/February 1995, 34–35.

12. Jaakko Pöyry Consulting, Inc. (1996). *Solid Wood Products Competitive Study.* Tarrytown, N.Y.

13. Cassells, David (1996), personal communication.

14. "A Big Opportunity—But Easy to Miss." *Tropical Timbers* June 1996, 1.

15. *The 1996 North American Pulp and Paper Factbook,* 2, based on 1993 data sources from the FAO.

16. *Pulp and Paper International* July 1996, 118, cited from an AF&PA capacity study.

17. Bonifant, Ben (1994). "Competitive Implications of Environmental Regulation in the Pulp and Paper Industry." *Management Institute for Environment and Business* 18.

18. *Pulp and Paper 1992 Fact Book,* 4–6, cited in Ben Bonifant (note 17, above), 19.

19. See David Null, "The Challenge of Capital Intensity," *Pulp and Paper* December 1995, 67. *The 1996 North American Pulp and Paper Factbook,* 19, provides a slightly higher estimate of 10 percent, citing information from the American Forest and Paper Association.

20. Morgan Stanley International Investment Research, *Paper and Forest Products—Midyear Outlook,* August 15, 1996, cover page.

21. See *The 1996 North American Pulp and Paper Factbook,* 20 (note 19, above).

22. This figure is variously estimated at between $5 billion and $15 billion, the lower figure cited by the EPA and the larger by industry. Citation from "Cluster Rule Would Place Unrealistic Demand," in *Pulp and Paper* December 1995, 103.

23. Jince, Peter. *Recycling of Wood and Paper Products in the United States.* U.S. Forest Service, January 1996, 3.

24. Jince, *Recycling Wood and Paper.*

25. Price, David (1996). *World Fibre Supplies,* citation from IIED, Robert W. Halger, Wood Resources International Ltd., Pira International: Surrey.

26. Haylock, Owen (1995). "The Development of the Medium Density Fiberboard Industry in the Tropical Areas of the World," in Proceedings of the 29th International Particleboard/Composite Materials Symposium, Washington State University.

27. Estimate from the United Nations Food and Agricultural Organization (FAO).

28. Personal communication with William Altman, president of the hardwood Plywood and Veneer Association, September 1996.

29. Maloney, Thomas, quoted in *Panels: Products, Applications and Production Trends,* Miller Freeman, 1994, 171.

30. Pöyry, Jaakko (1996). *Solid Wood Competitiveness Report,* Tarrytown, N.Y., 8.

31. Resource Information Systems, Inc. (RISI). *Wood Products Review,* 1994.

32. Several groups support this estimate—officials of the U.S. Forest Service, Jaakko Pöyry, and others.

33. Personal communication with Rachel Crosley, Environmental Advantage, Inc., October 1996.

34. *Annual Review and Assessment of the World Tropical Timber Situation 1993–1994,* ITTO, 1996.

Chapter 3

Emerging Markets for Certified Wood Products

Ultimately, it is the global marketplace that will measure whether sustainable forest management is a success. In the late 1980s, optimists predicted that a parade of "green" forest products would march onto the shelves in the early 1990s, responding to demand by consumers concerned over the destruction of the world's forests. When that demand never materialized, skeptics argued that any potential market for sustainably based wood products was an illusion.

By the late 1990s, however, the marketplace reflected a different reality. The consumer was just one actor among an ensemble of players that was slowly shifting the industry toward sustainable forest management (SFM). These other factors promise to be more powerful influences initially than consumer demand in fueling a market for sustainable wood products. These forces are converging on the supply and demand sides of the forest products industry to drive environmental concerns from retail stores to pulp and sawmills back down to the forest floor. The issue is not whether demand for sustainable forest products will develop but rather how fast and in which markets sustainable forestry will become an important variable.

In the late 1990s, a variety of signals indicated that SFM was gaining momentum. Certified products—those that carry an "ecolabel" awarded after an independent third party has evaluated forest management—represent the leading edge of sustainability. In 1997 certification and certified wood supplies were poised for exponential growth. By year-end 1998 at least 10 million hectares of land worldwide were expected to win approval under standards set up by the Forest Stewardship Council (FSC), up from 3.8 million hectares in October 1997, according to the organization. The World Bank announced a joint commitment with the World Wildlife Fund to ensure the certification of 200 million hectares by 2005—a target that was two orders of magnitude larger than all of the certified hectares in 1997. In 1997 the FSC also agreed to allow manufactured products with as little as 70 percent certified wood to carry FSC labels, a change that is expected to entice more producers and manufacturers to adopt certification. And in early 1998, executives from 20 companies that produced certified wood products reported that orders were picking up. R. Wade Mosby, vice president of marketing at Collins Pine Co., characterized the market for certified wood products as "a snowball rolling down hill."

Buyers' groups formed by commercial customers of wood products to stimulate supplies of certified products, which started in the United Kingdom in 1991, had proliferated to 12 by 1997 in Europe and the United States. Voluntary certification programs had also gained industry support in Canada, Finland, Sweden, and the United States. The Canadian industry in 1996 adopted its own set of national standards for SFM based on the International Organization for Standardization (ISO) environmental management systems. In 1998 Sweden became the first country to have an FSC-approved set of national standards. Certification, which initially focused on roundwood, spread to the paper products segment of the industry in 1996–1997 when a triumvirate of top Scandinavian pulp producers—STORA, Assi-Doman, and Kornas—embraced certification. And certification in North America took a big step in 1998: MacMillan Bloedel, Canada's largest producer, announced that it would no longer clear-cut old-growth forests and that it would seek certification—a turnabout that came in response to pressure from environmentalists and a successful boycott in Europe.

How quickly—and to what extent—sustainable forest management becomes an integral part of the industry will depend on the ebb and flow of economic and market trends. The market for sustainable forest products is in its infancy. Using environmental certification to distinguish products is a phenomenon that has existed only since 1992 and is currently concentrated in niche markets. Any demand for certified products is restricted by the small supply of certified wood products—less than 0.60 percent of the world's supply of industrial roundwood was certified in 1995, according to a survey by Environmental Advantage Inc. The speed with which the industry responds to pressures for SFM will vary depending on the structure and conditions in different geographic markets. Europe, where conditions are most favorable, will be the hub of demand for sustainable wood products during the late 1990s and early in the next century.

The Push and Pull for Sustainability

Several forces in the marketplace promise to simultaneously "push" and "pull" the industry through a transition to sustainably produced wood products. Pressure from environmental groups on industry to minimize the environmental damage of forestry operations, the rise of government regulations forcing timber companies to manage their operations more sustainably, environmental certification systems that distinguish sustainable companies and products in the marketplace, and early adopters are all forces propelling the industry toward more sustainable practices.

Self-appointed surrogates for consumers are also providing powerful incentives for the industry to adopt more sustainable practices and offer certified products. Buyers' groups, industrial customers, and professionals such as architects have become a chorus demanding and cajoling the industry to supply products with less environmental risk. These intermediaries between the consumer and the industry have emerged as perhaps the most powerful catalysts generating demand for sustainable wood products.

The Pressure to Adopt Sustainable Practices

During the 1990s, outsiders—regulators, the public, and environmentalists—and a handful of pioneering companies began to exert pressure on the industry to move toward more sustainable practices. Environmental groups with conservation agendas, trade associations in defense of their membership, and companies with certified products to sell became agents for change in forest management.

The Clout of Environmental Activists

Throughout the century, environmental groups have repeatedly tangled with forest products companies and successfully forced changes in the industry. During the 1960s and 1970s, activists successfully pressured paper mills to reduce effluent discharge. During the 1980s, they were instrumental in prodding the industry to increase the content of recycled fiber in paper products and to develop less-polluting pulp-making technologies. Greenpeace International led the campaign in Europe that forced many paper companies to produce totally chlorine free (TCF) paper and to make elemental chlorine free (ECF) the minimum threshold of environmental acceptability. In the 1990s, environmentalists turned their energies toward pressuring the forest products industry to improve forest management.

Environmental groups possess clout in the political process because they have credibility. Polls conducted in the 1990s repeatedly showed that they held more respect and trust among the general public and educated elites than did industry or government. This standing with the public often enables environmental groups to set the agenda of the forest sustainability debate and influence outcomes even before governments get involved. Respect from the public also allows environmental groups to shape consumer demand and corporate behavior. The groups wield their influence, according to the head of the German magazine publishers' association, Wolfgang Furstner, "because they represent the guilty conscience of an industrial society which knows it has to pay the price for its prosperity with the often ruthless exploitation of nature and its resources."[1]

During the 1990s, environmental activists in Europe and North America scored a number of victories that either helped to change forestry practices or impinged on the way companies operate. In 1991 Greenpeace started chastising European magazine and newspaper publishers for their high levels of paper consumption and pulp producers for their destructive forestry. The campaign of demonstrations and other activities paid off in 1996 when Enso Oyj, a large Finnish logging firm that had been a Greenpeace target, agreed to a one year moratorium on cutting and selling timber from old-growth forests in Russia. In the United Kingdom, the World Wildlife Fund (WWF) successfully pressured major retailers and forest products buyers during the early 1990s to form a buyers' group that pledged to buy paper and wood from third-party certified forests.

In North America environmental groups won comparable victories. A decade long campaign against logging old-growth forests in the Pacific Northwest resulted in court victories over the federal government, which shut down the federal timber-sale program in the Cascade Region between 1991 and 1994.[2] Actions by environmentalists also led to other management changes in the region's public forests that have curtailed logging. In British Columbia 10 years of protests against clear-cutting culminated in changes to the province's forestry code in 1995 that restricted the size of clear-cuts and increased the amount of protected forest.

Outside North America and Europe environmental groups do not possess the political influence to force significant changes in forest management practices. But intensified globalization of the industry will expose forest products companies to greater pressure from environmental groups and will bolster the groups' ability to influence individual and corporate consumers. That, in turn, will exert additional pressures on the industry to improve forestry practices and will stimulate demand for sustainably produced wood products.

The Rise of Voluntary Sustainability Initiatives and Third-Party Certification

Voluntary initiatives to improve forest practices, which have gained credibility and adherents in industry since 1992, are also indicative of the rising influence of SFM. The FSC, the International Organization for Standardization (ISO), and the Sustainable Forestry Initiative of the American Forest and Paper Association (AF&PA) represent different approaches to setting standards and improving forest management. All three, however, are based on the assumption that skeptical corporate buyers and consumers will not accept at face value producers' claims that products are sustainably produced. In fact, bogus claims were probably made in the past. In 1991, the WWF surveyed British retail outlets and found more than 360 wood products that claimed to come from sustainable forests. When the group asked for proof, all but four claims were withdrawn.

Underlying these initiatives is a strong philosophical conflict over the relative desirability of second- or third-party certification of forestry practices. Large segments of the industry maintain that second-party, systems-based programs run by trade associations or other industry bodies based on landowner reporting can provide credible assurance of sustainability claims. Proponents of third-party certification, represented by the FSC, argue that independent auditing of companies' forest management practices, similar to the way accountants audit a company's financial transactions, is essential to make claims of sustainability credible.

The Forest Stewardship Council (FSC), founded in 1993, is the only independent, not-for-profit membership organization that certifies forest management and allows manufacturers to label their products as those produced from certified wood to distinguish them in the marketplace. The FSC international principles of forest management are designed to not only foster sustainable timber production but also to protect ecosystems, water quality, wildlife, and further sustainable economic development. The standards are adapted at the national and regional level by environmental groups, the timber industry, economic development organizations, and the general public. The FSC itself evaluates, accredits, and monitors the independent bodies that conduct audits for certification. Costs for FSC certification vary. According to a 1996 study by the German government, certification of tropical timber accounted for less than one-half of 1 percent of the timber's value. FSC certifying bodies have reported that costs for initial certification can run from 7 cents to 21 cents/hectare for initial certification and 1.6 cents to 2.5 cents/hectare for an annual audit.

FSC-accredited certifiers conduct two types of certification audits—for forest management and for chain of custody. The forest management audit, which reviews management plans and on-the-ground practices, determines whether FSC standards are being met. The chain-of-custody audit for production operations reviews procedures for tracking certified wood from the forest to the final consumer. Eventually, the group hopes to establish one globally accepted set of standards for certifiers of forests and wood products. By mid-1997, three countries—the United States, the United Kingdom, and The Netherlands—had accredited certifying bodies; groups from seven countries, including Brazil, Canada, and Sweden, had applied to become certifiers; and initiatives to develop national certification standards were underway in 16 nations.

The International Organization for Standardization (ISO), a global standards body which in 1993 created its "14000" series of environmental management standards (EMS), takes a different approach. Unlike the FSC, which sets performance and product standards, the ISO has adopted a systems approach based on general environmental management tools that are designed to improve environmental performance. The Canadian Standards Association (CSA), an industry-backed initiative, has developed SFM standards that it hopes to certify under the ISO. If ISO standards, which have attracted the interest of large producers in the United States, New Zealand, and Europe, are widely adopted, they could become an alternative to the FSC.

Industry associations and leading producer nations have also set up their own forest management initiatives that rely on company-reported information. The International Tropical Timber Organization (ITTO), an association of tropical timber producing and consumer nations, has set a goal to have its member nations manage their forests sustainably by 2000. The American Forest and Paper Association (AF&PA), which represents 90 percent of U.S. industrial timber companies, required members to adopt its Sustainable Forestry Initiative (SFI) in 1996. Members must follow forestry guidelines that include reforestation, protection of water quality, enhancement of wildlife habitat, and minimizing of the visual impact of logging. The goal is a gradual, ongoing progress toward higher forestry standards. Members report their progress to the association annually, which then publishes a report based on aggregate reporting. The SFI initiative requires no verification. But the guidelines, particularly those that limit the size of clearcuts, were sufficiently onerous that the AF&PA dismissed 15 companies from its ranks between 1996 and 1997, for refusing to comply, and 10 others voluntarily resigned. Between 1995 and 1997, membership dropped over 10 percent. By 1997, however, SFI covered 53 million acres of industrial forestland.

Each certification system has its proponents and detractors. Not surprisingly, environmentalists favor the FSC with its performance standards and independent audit. Predictably, the FSC is controversial with industry. Industry critics, generally, object to outside oversight of their activities and argue that forests are already managed sustainably, which makes certification unnecessary. They also maintain that FSC standards are unreasonably high; that certification is too expensive for private forest owners; that chain-of-custody requirements are too cumbersome; and that the proven benefits of certification are insufficient to justify the costs.

Environmental groups fault initiatives like those of AF&PA and ITTO because they are self-policed. Whatever their merits, industry-sponsored efforts are unlikely to satisfy a number of the actors pressing the forest products industry for change, primarily because the "public won't accept the fox guarding the henhouse," commented Bruce Cabarle, senior associate of the World Resources Institute. Even executives from companies that subscribe to the AF&PA's SFI have indicated they are aware of the SFI's limitations and question whether SFI will serve their long-term interests.

It is impossible to predict which, if any, certification initiative will eventually win the most adherents or predominate in which markets. In the initial stages, no one system will provide the silver bullet to validate sustainability because circumstances vary too greatly from company to company and market to market, so multiple standards will coexist. But FSC certification is raising the profile of SFM, acting as a catalyst for other voluntary initiatives, and raising the standards for SFM in the industry as it gains momentum. "FSC certification means that the big players have to modify the way they do business not to seem offensive," said Paul H. Fuge, president of Plaza Hardwood, Inc., which markets certified products.

In 1997 the AF&PA began to evaluate the feasibility of voluntary third-party auditing for its SFI initiative and internally discussed shifting from SFM guidelines to performance standards. That the group, which initially opposed FSC certification and outside auditing, even considered such change is a testament to the growing influence of the FSC. Overall, the proliferation of sustainability schemes and the involvement of industry associations, environmental groups, and timber companies provide ample evidence that industry and government recognize the need to improve forest management—with or without certification. The more companies participate in sustainability initiatives, the more nonparticipants will feel the pressure to improve their practices and the more supplies of sustainable product will increase.

The Influence of Early Movers

Those companies that certified their forestlands, mills, and manufacturing operations early have become high-profile missionaries for SFM in the industry, using their status to stimulate demand for certified products and increase certified wood supplies. In the process, they have raised the standards for what is acceptable practice in the industry and even compelled other players in their niche markets to follow their lead. The pioneers cite several business rationales for their decision to become certified suppliers:

- Company Positioning: For some companies certification is a natural outgrowth of exemplary forest stewardship that predated third-party certification. Collins Pine Co. and Menominee Tribal Enterprises in the United States, for instance, have long traditions of forest stewardship, which readily lent themselves to third-party certification. Certification validates their commitment to sustainable forestry and communicates its value to regulators, customers, and neighbors. Increasingly, though, executives cite certification as a way to position their companies as "green" to stave off criticism or the threat of action that might limit their ability to operate. As pressure mounts on companies to improve forest management, "we'll look a lot better because we're certified," said David Ayer, chairman of Keweenaw Land Association.

- Access to Resources: Under some circumstances companies need certification to simply guarantee access to forestland or gain entrée to other resources. Portico, FUNDECOR, and Ston Forestal, which operate in Costa Rica, went through FSC certification for their forestlands because it helped them secure timberland and operate under Costa Rica's strict forestry laws. Ston Forestal, for instance, found that certification helped diffuse opposition from environmentalists to its proposed gmelina plantations. Through certification FUNDECOR not only obtained land, but the company also became eligible for international sources of debt financing for reforestation and natural forest management.

- Access to Markets and Customers: Certification can be a ticket for entry into new markets and a drawing card for new customers. Colonial Craft, a midsized U.S. manufacturer of window and door grills, picture frames, and molding, attracted European customers because it offered a line of certified products. Later one of the same customers awarded the company a far larger order for noncertified products. In Europe where public concern over the destruction of tropical forests has caused a sharp decline in tropical hardwood imports to those countries since the early 1990s, companies with certified tropical products stand a better chance of entering European markets or regaining lost market share. According to Daniel Heuer, a board member at Precious Woods Ltd., which produces certified tropical sawnwood and semi-finished wood products from forests in

Brazil and Costa Rica, "without the FSC label exports to Central and Northern European markets would simply not be possible."

- Product Differentiation: Still other companies use certification to distinguish their commodity products from a sea full of competitors. Collins Pine has created brand name recognition for its FSC certified products as "CollinsWood." The distinction has helped the company stand out among hundreds of otherwise similar milling operations in the Pacific Northwest.

Typically the decision to certify involves a complex set of circumstances, as it did for Seven Islands Land Co. The company, which manages close to a million acres of forestland in Maine, pursued FSC certification in 1993 because, according to John W. McNulty, vice president, management saw a market for green wood products emerging. They decided that certification would enable the company to capitalize on its existing system of SFM and provide a "self-check," as McNulty described it, for its forest management systems. But as one of the largest landowners in Maine, Seven Islands also wanted to distinguish its forestland from the rest of the commercially held lands in the state, enhance the company's public image, and insulate itself from critics. At the time, certification seemed prudent because commercial forest owners in the Northeast were under fire from environmentalists. The company preferred certification as a type of self-regulation to the prospect of externally imposed regulations. Certification, completed in 1994, gave the company's line of certified flooring "almost instant recognition in a dog-eat-dog market," said McNulty.

The Incentives for Sustainable Management

By the late 1990s, the attitudes of consumers toward forest management, niche markets created by certified producers, industrial demand, and demand orchestrated by buyers' groups had emerged as the most potent forces stimulating demand for certified products.

More Thought Than Action from the Green Consumer

Just how consumers influence the development of markets for sustainable products is often misunderstood. Numerous studies indicate that European and North American consumers prefer forest products produced in an environmentally sound way and that they will pay more for them. The European Commission's Eurobarometer study of consumer attitudes on a wide range of issues in 1995 found that 58 percent of those polled in the United Kingdom and 50 percent of Germans were willing to pay more for green products. Studies in the United States have also found that a significant percentage of consumers express a similar willingness. One 1996 study

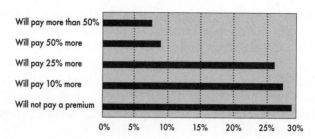

Figure 3.1

Respondent Willingness to Pay for Certified Wood Products.
Source: Richard P. Vlosky, "Willingness to Pay for Environmentally
Certified Wood Products"

conducted by Richard P. Vlosky of the Louisiana State University Agricultural Center found that 72 percent of consumers said they were willing to pay premiums between 10 percent and more than 50 percent (Figure 3.1).[3] The problem such studies raise, however, is how to account for the difference between what consumers say they will do and what they actually do.

The research on consumers' "willingness to pay" appears to be overly optimistic about the likelihood of people actually paying premium prices for environmentally differentiated products. Many in the forest products industry share the view that the consumer, when faced with two similar products whose only differentiating characteristics are price and a green label, will opt for the less expensive item. A number of companies have conducted their own unpublished studies, which show that most consumers are not interested in certified or sustainably produced wood and do not yet know what certified forest products are. The Home Depot's lackluster success with a handful of certified sawnwood products in the mid-1990s would seem to support the industry position.

The more meaningful finding of the various "willingness to pay" studies is that they consistently find discomfort among consumers about their consumption of forest products. A study of German consumers conducted by FINNPAP, the Finnish Paper Mills Association, in 1994 found that the majority of those surveyed thought that paper production caused "big" or "very big" environmental problems. Another private survey of the Swedish forest industry completed in the mid-1990s found that 75 percent of Britons worried about how much paper they used and tried to use as little as possible, and 62 percent disagreed with the statement that the world's forests are easily able to meet the world's needs. Consumers are also less likely to believe industry claims about environmental issues than those made by environmental groups. According to a study by the Angus Reid Group, a Canadian market research group, 79 percent of adults believe all or some of what environmental groups tell them about the environment, and 78 percent trust scientific evidence. Only 37 percent believe business or industry claims.

Surrogates for Consumers

Consumers may not yet have acted on their environmental concerns, but consumers' concerns do find their way back to the marketplace. Most consumers do not directly buy the industrial roundwood that they consume each year. They consume it as paper, part of a house or a remodeling job, hidden in a piece of furniture, or in some other form. That is why industrial demand for sustainable forest products—at least initially—will be the determining factor in the creation of markets for sustainable forest products. Industrial players have great leverage in the marketplace because most forest products flow through industrial channels. In the late 1990s engineers, retailers, architects, and other intermediaries became the primary catalysts for demand of certified wood products.

These intermediaries, according to studies, believe consumers are interested in environmentally sound products. In one study by The Western Wood Products Association of individuals who make the buying decisions on behalf of consumers, more than 30 percent of architects thought their clients believed they might be harming the environment by using wood products (Figure 3.2). An even greater percentage of specifiers (people who specify wood over other products for a given application) and a significant number of engineers said the same about their clients. Over 70 percent of those who said they believed that wood consumption harmed the environment also said that the industry was "not taking care of national forests" (Figure 3.2.). And both engineers and architects reported that their customers were more concerned about forest management practices than they had been in the past.

These stand-ins for consumers also seem to prefer certified products. In the same Western Wood Products Association survey, 84 percent of architects, 75 percent of engineers, and over 90 percent of specifiers said their clients would be interested in wood products endorsed by a third-party certification organization. Another 1996 study by Vlosky of the Louisiana State University Agricultural Center examined what kind of entity architects, building contractors, and home center retailers would trust most to certify forest management and harvesting practices. From the perspective of what might be considered a relatively pro-business group, third-party certifiers emerged as the most trustworthy group, ahead of government, environmental groups, and the forest products industry (Figure 3.3).[4]

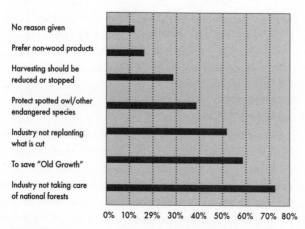

Figure 3.2.

Reasons Given for Respondents' Belief That
Using Wood Products Harms the Environment.
Source: Western Woods Products Association

	Architects	Building Contractors	Home Center Retailers	Weighted Average
Third-Party Certification Entity	2.0	1.6	1.7	1.7
Forest Products Industry	2.6	2.4	1.9	2.2
Federal Government	2.8	3.1	3.0	3.0
Non-Governmental Environmental Groups	2.5	3.0	3.6	3.1

Figure 3.3.

Business Customer Perspectives on Forest Management Certification.
Level of Trust to Certify Forest Management and Harvesting (1 = Trust
Most to 4 = Trust Least).
Source: Richard Vlosky

The Engines of Demand

Until recently, architects, cabinetmakers, furniture manufacturers, flooring
companies, and contractors were primarily responsible for pulling small volumes of certified products into the marketplace; these niche markets have
been the most important outlets for certified wood products. In the future,
buyers' groups will be the more significant engines generating demand for
certified wood products. Between 1991 and 1997, 12 buyers' groups were

created, primarily in Europe and the United States. The groups, founded because members had such great difficulty finding certified forest products, have begun to wield their collective and individual influence to facilitate the trade in certified forest products. The U.S. buyers' group, the Certified Forest Products Council launched in 1997, was set up to help major buyers in the world's largest wood products market promote SFM by buying and using independently certified forest products. The group ended the year with 140 members (including The Turner Corp., a leading U.S. commercial construction company; Donghia Furniture and Textile; and Habitat for Humanity, one of the largest homebuilders in the United States) and a roster of activities to match certified suppliers with buyers.

The original buyers' group, the 1995+ Group in the United Kingdom, has existed long enough to illustrate how much potential these groups have to influence demand for sustainable wood products. The 1995+ Group grew from about a dozen members, notably leading do-it-yourself (DIY) chains that were under attack from environmental groups in 1991, to 79 members in 1997. Members have goals to stock increasing numbers of certified products. DIY members, which include J. Sainsbury Plc., have a combined 80 percent share of the DIY market. Other members—several dozen supply and distribution companies, two large supermarkets, the leading news agent, and the top pharmacy chain—represent the top players in their respective sectors. Together members account for 15 percent of all U.K. industrial roundwood consumption.

But supplies of certified forest products are insufficient even for the relatively small U.K. market. In 1996 the group made stimulating supply a top priority for products made from roundwood and paper. Producers in Sweden and Finland were the prime targets since they are the largest sources of wood for members. Members were instrumental in persuading STORA and Assi-Doman to adopt FSC certification, and in the Swedish industry's decision to develop national FSC standards, according to executives at both companies. Brian McCloy, a marketing specialist with the Canadian Pulp and Paper Association (CPPA), also credited efforts by the group's members for being the catalyst for Canada's ISO certification initiative.

The Motivations Behind Industrial Demand

Large buyers in the United States and Europe, whether members of a buyers' group or not, increasingly want reassurance that the wood and paper products that they use or stock come from suppliers that practice responsible forestry. One internal study by Weyerhaeuser Co. in the mid-1990s found that forestry practices topped the list of issues identified as areas of concern by their industrial and retail customers. Large buyers often simply wish to avoid the risk of public embarrassment or censure over products that might carry an environmental stigma. So they ask suppliers to guarantee that the forest products they buy are produced in an environmentally sound fashion. Similarly, those companies that sell name-brand products want to prevent environmentally tainted materials from entering their products, even in packaging. In the words of one executive at Procter & Gamble Co., one of the largest customers of fiber for packaging and diapers, "sustainable forestry should be a given."

Retailers that have curried trust with their customers to ensure long-term patronage are particularly sensitive to environmental issues that may tarnish their reputations. In the United Kingdom during the early 1990s, the largest DIY home improvement retailers confronted environmentalists wielding inflatable chain saws in their parking lots to protest the sale of products made from tropical hardwoods. The action drove home the risks the chains faced if they were ignorant of the environmental implications of the products on their shelves. They quickly committed to buying certified forest products because certification is a tool they can use to answer queries about the nature of their products and to avert the risk of damage to their reputations by selling environmentally suspect products.

The Power of Government Regulation

Governments, too, are pressuring the industry to improve forest practices. In recent years, governments in far-flung reaches of the world have adopted regulations that dictate where companies log, how much they log, what species they harvest, and the harvesting techniques they use. Some of these actions are designed to encourage SFM, because governments recognize that it can be a vehicle to conserve forests while also reaping economic benefits from them. Brazilian regulators in 1996, for instance, imposed a two-year moratorium on harvesting mahogany because of overcutting. Several nations, including Indonesia, the Philippines, Brazil, and Argentina, have adopted restrictive log export policies—high royalties, taxes, and even outright bans—to capture a share of value-added processing. California and Oregon have enacted regulations based on ecosystem management, which make timber production only one of several variables that the management system seeks to optimize. The system has added more oversight and complexity to the region's forest management.

Governmental entities, primarily local and municipal, in the United States and in Europe, are also offering incentives for SFM. A number have passed ordinances that make sustainability a criteria in their own purchasing decisions and in the markets over which they hold jurisdiction. The General Agreement on Tariffs and Trade (GATT) and other international trade agreements make most regulations that restrict trade on the basis of environmental considerations illegal, a constraint that has impeded the development of many trade-related initiatives but by no means all. As early as 1988, for instance, the European Parliament passed a nonbinding measure proposing that the European Community (EC) import only tropical hardwood products certified to be produced under forest management and protection programs. Local and city governments in Britain, Germany, and the United States that had already banned the use of tropical timber in public buildings and projects began to change those policies to favor certified products in the late 1990s. And the U.S. government began to specify certified wood in building projects. In 1997 the U.S. Department of Defense, for instance, submitted a bid package for a large-scale renovation of the Pentagon that sought sources of certified forest products and lumber.

The Slow Convergence of Supply and Demand

Despite pressures on both the supply and demand side of the forest products industry, the quantities of sustainably produced and certified products in the marketplace, as measured by the supply of FSC-certified industrial roundwood, are almost negligible. As noted earlier, certified wood made up less than 0.60 percent of the world's supply of industrial roundwood in 1995. Moreover, by far the majority of that certified wood goes directly into the mainstream supply as undifferentiated product. The companies that produce certified wood, mostly players in niche markets, produce a broad spectrum of products, from charcoal and plywood to gmelina chips, but few have successfully marketed their products as certified. If major intermediate and final customers want to buy certified products and have organized to do so, why are certified suppliers unable to sell their products as certified? The apparent contradiction accounts for much of the skepticism about the development of a market for certified wood products.

The many steps between a tree in a forest and a finished product have contributed to the slow evolution of demand for sustainable wood products. Forest products can flow from landowner to logging company to sawmill to broker to secondary processor to wholesale distributor to retailer before reaching the final consumer. The industry is built around its ability to pull specific species of trees and specific products through the value chain efficiently. Orders for forest products are made for specific times and quantities, which adds to the complexity. In this environment, certified producers face the task of finding customers for their otherwise undifferentiated products in the sea of possible purchasers, while prospective buyers must identify, from the limited pool of certified producers, those that can fill orders for the species, grade, dates, and volumes they want. Primary and secondary processors and distributors and current chain-of-custody requirements further

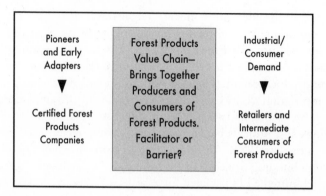

Figure 3.4.

Two Disparate Ends of the Value Chain for Certified Forest Products.

complicate matters. Most current certified producers do not own secondary processing facilities and are, at best, familiar with companies in the value chain immediately ahead or just below them. To create a vibrant market, sustainable forest owners and environmental consumers at opposite ends of the value chain need to meet across a considerable distance (see Figure 3.4).

Europe—The Trend-Setting Market

Europe, where circumstances are most favorable, will lead the world in demand for certified wood products and will shape many of the SFM trends that will ripple out to other regions. Europeans to a greater extent than Americans, or people in other regions, accept the idea that overconsumption of finite resources is the root cause of many environmental problems. Historically, Northern Europeans have also been more concerned about environmental issues than their counterparts in other developed countries. That concern supported the rise of "green" parties in Germany and The Netherlands in the 1980s and 1990s. That history and attitude toward environmental issues has helped to make Europe "the most informed (environmental) market," according to William Martin, senior technical manager for Sainsbury Plc., a retailing giant that has committed to sell 100 percent certified wood and paper products. Martin described the European market as one where "relationships between producer, chain of supply, and customer are growing closer and more transparent. Customer expectations are rising all the time. And environmental pressure groups are effective and find a ready and influential audience in the public at large."

Industrial and commercial wood products' customers in Europe, in general, have a different attitude toward environmental issues and certified products than their U.S. counterparts. U.K. retailers like B&Q Plc. are not waiting for consumer markets to develop—they are creating them (Box 3.1).

Box 3.1. B&Q: Making the Market for Certified Products

Customers of Britain's B&Q will soon have little choice but to buy certified wood and paper products. The retailer intends to carry 100 percent FSC certified wood and paper products in its 281 stores by the year 2000. At the end of 1997, about 9 percent of the chain's wood products were made from certified material. By year-end 1998, an estimated 20 to 25 percent are expected to fit that category. After that the chain's certified product offerings will grow "exponentially," according to Allen Knight, Environmental Policy Controller for B&Q.

B&Q, which had sales of $2.9 billion in 1997, set its course in certified wood products "even though there is no indication of any increase in customer demand for certified products," acknowledged Knight. In Knight's opinion, companies that wait for consumer demand before stocking certified products have misread the issues of sustainability and the dynamics of new markets. Customers do not ask for certified products because they are unaware of them: Raising awareness and creating markets are the retailer's role. Knight was confident that if the products are there, the demand will come: "How many customers would prefer unsustainable timber?" he asks.

The task of putting certified products on B&Q shelves was—and will continue to be—daunting, even though certified wood and pulp from Scandinavian producers were expected on the market in 1998 and U.K. producers, which supply 35 percent of the chain's wood, were moving toward certification. B&Q, for instance, sold 30 percent of all wallpaper in the United Kingdom, but in 1997 it still did not have a source of certified material. The company had asked all of its suppliers to draw up plans to show how they would meet B&Q's target. Suppliers reluctant to make the shift were told that "we . . . will look for another supply if needed," according to Knight.

B&Q demonstrated that it would keep its word. The chain, the largest retailer of garden furniture in the United Kingdom, had purchased its product line—about 600 containers-worth a year—from suppliers in Southeast Asia, primarily Vietnam and Indonesia. The Asian suppliers, however, showed little interest in certification. But a small cooperative of certified producers in Bolivia, IMR/CIMAL, saw an opportunity and approached B&Q about supplying the furniture. In 1998 the Bolivian venture planned to send B&Q about 50 containers of outdoor furniture, or 10 percent of the chain's needs. In 1999 IMR/CIMAL planned to quadruple its production for B&Q to supply close to 50 percent of the 600 container loads. For the Bolivian cooperative, "B&Q represents a huge opportunity," according to Robert J. Simeone, president of Sylvania Forestry, who brokered the arrangement.

From their perspective it is retailers that make markets, not consumers, by anticipating consumer preferences, developing new products, and adapting production systems in response to them. Historically, retailers have launched a variety of new products and markets in advance of any consumer demand, including the personal computer from International Business Machines Corp. and the Walkman from Sony Corp.

From Solid Wood Products to Paper

FSC-certified forest products originated with solid wood products, and they still dominated the market in 1997. In Europe, however, certification has moved beyond solid wood products to pulp and paper producers, largely under the influence of buyers' groups and the publishing industry. According to a 1994 study by FINNPAPP, German purchasing managers and other individuals involved in buying paper supplies considered "environmentally sound production" one of the most important criteria for selecting suppliers—after price and quality.

Since publishers rarely own forestland or production facilities and are acutely conscious of public opinion, they are susceptible to such pressure and are predisposed to pressure paper producers on sustainability issues. European pulp and paper producers are as reactive to lobbying by their customers as publishers are to public opinion. The pulp and paper industry's value chain is relatively short (Figure 3.5). In Europe, publishers typically buy directly from suppliers, not through brokers, as is common in the North American market. Paper prices are determined by the normal "paper cycle" that dominates the industry, which periodically drives prices up then sends them plunging when new paper mills come on line. In a downturn, large orders can mean the difference between running a paper mill at capacity or shutting it down—capacity utilization is the key to profitability. During the periodic downturns, discounting often becomes the norm as suppliers try to move product in saturated markets. Producers figure that if they are certified during a downturn, they might maintain access to premium markets while undifferentiated producers might be forced to accept discounts. Publishers not only possess the buying power in a commodity market but they also dominate the most desirable top end of this market.

Embracing Certification in Scandinavia

Under prodding from their major customers, Scandinavian pulp and paper producers dropped their earlier opposition and embraced FSC certification. Sweden, which produced approximately 56.1 million cubic meters of industrial roundwood in 1995, adopted national standards for forest management that the FSC endorsed in 1998. By then, Sweden's three top producers—

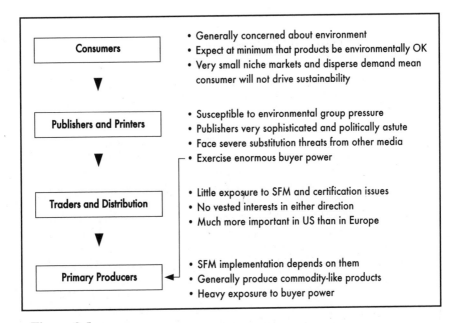

Figure 3.5.

Sustainable Forest Management Along the Value Chain Pulp and Paper Industry.

STORA, Assi-Doman, and Kornas—had committed to full certification of their holdings in anticipation of meeting demand from their large customers (see Box 3.2). Finland, which had constituted a working group to implement a national Finnish certification system compatible with both FSC and ISO systems, was expected to begin large-scale certification in 1998. The addition of even a modest percentage of Finland's annual production of 46.1 million cubic meters of industrial roundwood plus that of Sweden's would significantly boost the world's volume of certified wood supplies. Large-scale certification in both countries would probably produce more certified wood than those two markets could absorb. Scandinavian producers would then probably try to differentiate their green products from those of competitors' to gain a more favorable market position, which would further stimulate demand for certified products.

The Scandinavian forest products industry, by virtue of history and structure, is primed for large-scale certification. The export-oriented industry is historically already sensitive to customer and market demands. Implementing FSC certification promises to be easier and less costly in Scandinavia than in some other regions. Scandinavian companies do not log old-

Box 3.2. Assi-Doman: FSC Certification for Credibility

Certification took several steps toward the mainstream when Sweden's $2.5 billion Assi-Doman AB committed to the FSC and ISO certification systems. The timber, paper, and packaging board giant, which is the world's largest publicly held forest owner with 3.2 million hectares of forestland, was a member of the group that drew up Sweden's national forestry standards and sits on the FSC board. By year-end 1997, 50 percent of its holdings had the FSC stamp of approval; the rest were expected to follow in 1998.

The FSC imprimatur is the only one that satisfied Assi-Doman's criteria for credibility. According to Olaf Johansson, senior ecologist at Assi-Doman Skog & Tra, on issues of forest management "credibility is a key issue for the forest industry" and certification provides the vehicle to demonstrate that the company can satisfy expectations of its customers for environmental improvements. The FSC is the only system that offers a common platform for environmental, social, and economic interests by involving industry, environmentalists, government, and academics. "Any system that lacks the cooperation among those interest groups will also lack the necessary credibility," said Johansson.

The costs of meeting FSC standards were not a major consideration in the company's decision. According to estimates, Swedish companies will lose 12 to 13 percent of potential production if they adopt Sweden's FSC standards. But Assi-Doman changed its forest management during the 1990s to comply with other forestry regulations and adapt to rising environmental standards. Those actions have already reduced production levels by nearly that much. Instead, management expects certification to be commercially favorable for the company. "We know that there is a genuine and increasing demand for certified goods" in segments of the European market where the company competes, said Johansson. "We see certification as strengthening our competitiveness."

growth forests—they were harvested long ago—which eliminates controversy, and their forestry practices are among the most advanced in the world. Much harvesting is done during cold winter months when erosion problems are not as severe and harvesting equipment can be mobilized with relatively low impact on the frozen ground. Sweden's IKEA Group, the world's largest furniture maker, which has a reputation for environmentally sensitive policies and has positioned its products as "clean, green, and Swedish," is expected to be a ready customer for certified wood products.

Finally, the Scandinavian industry, which prides itself on world leadership in forestry, considers certification an opportunity to exercise this leadership. Some in the industry have argued that the Scandinavian producers are using certification as a nontariff barrier to protect their share of the European market against lower-cost competitors. But one company's barrier is another's competitive advantage, and some top Scandinavian producers have indicated they welcome the opportunity to compete on environmental grounds.

A U.S. Market Poised for Change

The U.S. wood products industry lags well behind Europe in support for third-party certification. Conditions in the United States make the industry less receptive to outside oversight in forest management than companies in Europe. The AF&PA SFI initiative has undercut the FSC. The initiative has muddied the issue of third-party certification with industrial customers, who need to be informed to identify environmentally differentiated products. The U.S. industry also depends less on exports than do European and Canadian producers. Weaker links to the European markets where environmental concerns are strongest have enabled U.S. producers to avoid the close scrutiny of their forest management practices that suppliers to the European market face. U.S. imports of tropical forest products have been traditionally low and, as a result, draw less attention from environmental groups than have tropical imports to Europe. U.S. producers are also less dependent on public lands than Canadian producers and have traditionally defended the prerogatives of private property owners.

Influential U.S. commercial and industrial customers have not felt compelled to press suppliers for certified products. U.S. environmental activists are generally less confrontational than those in Europe and, in general, public awareness of environmental issues is lower. For most U.S. commercial customers, particularly retailers, certified products are a niche play directed at the "green" consumer: They will carry the products if consumers will buy them. The Home Depot, for instance, has experimented with certified products since the mid-1990s, but the products' have not flown off the shelves. In late 1997, the retail giant carried hardly more than a handful nationally (Box 3.3).

Nevertheless, resistance to outside monitoring within the U.S. industry softened in the late 1990s. U.S. companies with Canadian operations were involved in that industry's forest certification initiatives. According to a senior executive at Weldwood, a subsidiary of Champion International Corp., the company was positioning itself to register a one million hectare tract with an annual allowable cut of two million cubic meters under the CSA standard. "The CSA process clearly benefits all of our stakeholders," he said. Weyerhaeuser Canada also participated in the CSA process, along

Box 3.3. Home Depot: A Contingent Commitment[5]

The Home Depot's commitment to certified wood products is far more contingent on consumer behavior than its DIY counterparts in Europe. The world's largest home improvement retailer, which rang up sales of $24.156 billion in 1997, advocates third-party certification, has a goal to offer certified products, and a policy to not do business with suppliers that are "not doing the right things." Buyers work with suppliers to motivate them to improve forest practices. They also work to screen out those with egregious practices. In 1994 Home Depot became the first home center in the world to offer temperate and tropical region wood products from certified forests—shelving from Collins Pine Co. and doors from Portico S.A.

But Home Depot's efforts are not part of its business strategy. In 1996 the chain quit stocking the pine shelving produced by Collins Pine and in late 1996 carried no forest products with the FSC logo. Home Depot executive vice president Larry Mercer stated that everything else being equal HD would always choose certified products. "What really limits the success of the program is the fact that it's not seen as something that is really ringing the register," commented Mark Eisen, the company's former director of environmental marketing. Nor do Home Depot managers feel that they have the resources or ability to unilaterally develop the market for certified products.

The Home Depot's posture toward certified wood products is shaped by the business conditions under which it operates, which differ dramatically from those in the United Kingdom. The U.S. market is six times larger in sales volume than that of the U.K. Its size and diversity make decentralized buying practical for Home Deport. But decentralized buying also makes it more difficult for the chain to buy and carry certified products on a national basis. Moreover, The Home Depot has no other comparable companies in the U.S. retail home improvement market—or other markets—committed to certified products, so unlike U.K. chains, thus far it has borne the entire burden of moving suppliers toward forest certification and cultivating awareness among consumers.

Home Depot's strategy dictates how it implements its certified products program. "What is at stake is the ability to be on the cutting edge of environmental concerns and market sustainably produced products while remaining profitable," according to Mercer. The DIY giant is positioned in the discount end of the market. Efficiency, economies of scale, preferential access to supplies, and low overhead are crucial for success. The circumstances make it more difficult for The Home Depot to move ahead of consumer demand than for a retailer like B&Q, which has the support of the 1995+ Buyers' Group. The Home Depot may intend to stock sustainably produced products and expose them to the marketplace, but whether consumers buy them will ultimately determine whether the program succeeds and whether the chain puts more certified products on the shelves.

with several other U.S.-based companies. At the time Weyerhaeuser and Champion had also begun to explore the prospects of FSC certification.

Certification of public lands, which began in 1997 with pilot projects on state lands, could also accelerate the acceptance of SFM in the United States. In 1997 a total of 1.8 million acres of public woodlands in Minnesota and Pennsylvania received FSC certification. Pennsylvania planned to eventually certify all of its 2.1 million acres and make SFM the centerpiece for its next 15-year management plan. New York, Michigan, and Wisconsin were expected to undertake similar initiatives. Minnesota implemented the test because "the public wants to be reassured" that public forest lands are being properly managed, according to John Krantz of the Minnesota Department of Natural Resources. The U.S. Forest Service also initiated an internal review to study the feasibility of certification. If certification were widely implemented on public lands, and if the industry and environmentalists reached a credible compromise over forest use, some analysts predict that the industry could regain access to public forests that have been off-limits to logging in recent years.

Large U.S. customers have shown subtle signs that they, too, want to know more about the origins of the wood and pulp they buy and thus avoid environmental risk. Procter & Gamble Co. has slowly backed out of the pulp and paper business and ownership of forest land to reduce the risks related to forest management and pulp production for the company and its flagship brand name products, according to observers. More often, U.S. companies with brand-equity or a base of environmentally concerned consumers try to insulate themselves from controversy over forest management issues by requiring suppliers to give them evidence that they practice SFM or that they avoid certain practices such as using tropical woods.

In niche markets where certified products already have a beachhead—in furniture, cabinetry, residential construction—requests for certified wood picked up in 1997 and 1998. Many of these requests came from small manufacturers like the Loft Bed Store in Washington, D.C., a $3 million dollar furniture maker that introduced a line of "Eco-Furniture" in 1997. More producers had started to pay attention to these requests. In mid-1997, for instance, States Industries Inc. introduced a line of 100 percent certified plywood veneer, which by year-end totaled some 2 percent of the company's annual sales. The company rolled out the new line, according to William Powell, marketing manager, because corporate customers that use environmental responsibility as part of their marketing strategy, including The Gap Ltd., The Body Shop, and The Rain Forest Café, wanted the material. With the prospect of rising volumes of certified wood coming on the market in the next several years, Powell predicted that "the demand will be there."

The U.S. construction market was poised to become a promising new venue for certified wood products. In 1997 Turner Construction Co., which did $3.3 billion business in 1996, joined the U.S. buyers' group. At the time Ian Campbell, Turner's director of sustainable construction, considered certified wood "the next frontier in environmentally responsible design and construction." Campbell predicted that the "green" segment of the construction market, which totaled about 2 percent of the general construction market in 1994, or about $200 million in contracts, would reach $1.2 billion by year-end 1998.

The Lesson of Recycling

Historically, markets in both Europe and the United States are subject to pressures from buyers, concerned consumers, and activist environmental groups. The industry's shift to the production of paper with a higher content of recycled paper in the 1980s may be a harbinger of the evolutionary path that SFM will take. The recycled fiber content of paper in the United States and Europe rose from 16.5 percent in 1985 to about 37.5 percent in 1998. This evolution required the industry to invest several billion dollars in plants and equipment and to assemble complex collection and distribution systems; forced producers to reposition paper products to seize new opportunities; and changed consumers' behavior.

Recycling gained momentum from both ends of the product life cycle, as is happening with SFM. In the mid-1980s, the notion that the United States was rapidly running out of landfill space to house paper garbage took hold with the public. Underlying that concern was a broad preoccupation about the ability of forests to produce enough fiber to satisfy demand for paper. The public perceived recycling as an antidote to the scourges of deforestation, clogged municipal landfills, and a lack of resource stewardship. Industry fought recycling content initiatives initially. Gradually, however, it succumbed to regulations, pressure from buyers pushing for more recycled content, and consumers' concerns, as well as the realization that under the right conditions it costs less to produce paper from recycled fiber than from virgin. By 1997 millions of consumers routinely packaged their paper, glass, metal, and plastic to be recycled: Fifteen years earlier, observers doubted that such a rapid change either in industry practice or in consumer behavior was feasible or possible.

The wholesale adaptation by the U.S. marketplace to sustainable forestry arguably presents an even greater challenge. But the rapid innovations in recycling and the industry dynamics that lent velocity to recycling—and exist with SFM—have primed the industry for change and suggest that a pronounced shift to SFM is not only possible but perhaps more likely. Even though the green consumer is not yet a market-maker, environmental issues consistently top the polls on any list of social concerns. U.S. wood products customers have begun to react to those latent concerns over forest management, and European markets are moving toward SFM. U.S. industry will find it increasingly difficult to ignore those realities.

ISO-14000 Standards for Canada

The market dynamics that are fueling a demand for sustainable forest products in Europe are also behind a certification initiative in Canada. The Canadian forest products industry ships over 50 percent of world exports of newsprint, 34 percent of wood pulp exports, and contributes nearly 15 percent of the world trade in printing and writing papers, according to the Canadian Pulp and Paper Association's annual statistical review. The Canadian industry's international exposure, environmental pressures, and the need for sustainable production at home where trees grow slowly in the harsh climate prompted Canadian forest products companies to develop a set of national forestry standards in 1996. The industry hoped that the standards, an application of the ISO-14000 system, would become internationally acceptable. By 1998 more than 20 timber producers that accounted for over 20 million hectares of forest reported interest in certification. The first supplies of Canadian certified wood were expected on the market by 1999.

The Canadian industry, however, did not adopt chain-of-custody protocols or create an ecolabel for its certified products. In designing their certification system, the Canadians have assumed that sustainability will become a general market condition, something that customers expect from suppliers. The CSA standards were drawn up to provide those assurances, not create market opportunities for niche players. FSC certification, with its logo and procedures for chain of custody, is better suited for the creation of differentiated products.

Little Interest in Japan and Asia

By contrast, SFM and certification have caused barely a ripple in Asia and Japan. The large Asian producers, in fact, take great umbrage at efforts by European countries to boycott Asian tropical hardwoods. In 1996 Malaysian producers even considered organizing a counterboycott of German equipment. European concerns have undoubtedly cost Asian producers market share in Europe, though it is difficult to quantify the loss. According to Aw Beng Peck of Asian Timber, the demand within Asia for paper products is so great that the pulp and paper industries are, so far, unaffected by market access issues in Europe. As a defensive maneuver most Asian countries have taken up some sustainability initiative—often intended to fulfill the ITTO campaign for SFM. These initiatives fall far short of actions in Scandinavia and Canada, but attention focused on SFM may eventually lead to improvements on-the-ground.

In Japan, a bellwether Asian market, the prospects for a market for certified forest products any time soon are remote. In 1996 no certified wood products were sold in Japan. That year influential companies and environmental groups—including Canon, Sanyo, Sony, and Matsushita Electric—formed the "Green Purchasing Network," a coalition designed to kindle a market for "green products." The group's goals, however, are modest and not focused on wood products. Japan, the largest importer of tropical forest products, has the power to influence tropical suppliers to improve their forest management, but it has never taken such action and appears unlikely to do so.

The Globalization Factor

The forces of globalization that are reshaping the industry promise to exert a catalytic effect on SFM. It will become easier for global customers to implement policies on sustainability for all supplies rather than just those for specific markets. Few companies care to explain that sustainability is only a condition for certain markets; most prefer to attribute their interest in forest sustainability to progressive environmental policies. Until recently, though, a global commitment would have been risky because the supplies of sustainable products were unavailable and the potential to overstate commitments high. But as more suppliers enter the arena, large, multinational customers will be emboldened to demand sustainable supplies from all of their suppliers worldwide.

In a similar fashion, the same factors that have kept global players out of the sustainable products market in the past will encourage them to participate in the future. That was the case with Manadnok Paper Mills of New Hampshire in the late 1990s. Management was concerned that its line of wallpaper products might be excluded from the U.K. market if it could not deliver certified paper. If it was unable to supply certified material, the company expected that Scandinavian producers would quickly fill the gap. The prospect of losing that business prompted the company to look for certified suppliers.

Producers, too, will be subject to the same catalytic effects. The modern, top tier, fiber producer is a multinational player with global operations. When a company sees its competitors move toward certification, it will react. The forces converging on the industry are beginning to spark a chain reaction that will fuel a market for sustainable forest products. Suppliers, in response to pressure from environmental groups, governments, consumer and industrial demand, and wood products industry leaders, aware of the market's burgeoning interest in forest sustainability, are looking for the appropriate vehicle to communicate their efforts and concerns over forest management. Over time, the cumulative effects of these forces will make SFM an integral part of the industry's overall value equation. Quality, price, reliability, and service will continue to dominate purchasing decisions, but sustainability will also take its place as an essential variable. And as sustainability is integrated into production and purchasing decisions, the standards that suppliers must meet will rise.

Geography and industry structure will moderate the rate of change across the industry. But by the opening years of the next century, third-party certification is almost certain to emerge as a major market force in Europe. In premium markets any producer interested in competing will probably need to be certified, which will make certification a general market condition for suppliers to the upper ends of the market. As certification takes hold in Europe, those countries that want to compete in the European market will be forced to adopt certification to compete, as Canada is already doing. If just one or two major U.S. companies become certified in the early years of the next century—as is probable—large sectors of the U.S. forest products industry will open up to third-party scrutiny of its forest management and supplies of certified wood will swell. Once larger volumes of certified product are flowing into the market, sustainability may well become a characteristic of products that premium customers simply expect.

NOTES

This chapter is adapted from "Marketing Products from Sustainably Managed Forests: An Emerging Opportunity," prepared by Diana Propper de Callejon, Tony Lent, Michael Skelly, and Rachel Crossley, Environmental Advantage, for the Sustainable Forestry Working Group, 1997.

1. Furstner, Wolfgang, Executive Director, General Interests Paper in the VDZ. "The German Magazine Publishers' Ecological Vision and Strategy," paper presented at the International Periodicals Symposium, June 1996.

2. Tuchman, E. Thomas, et al. "The Northwest Forest Plan—A Report to the President and Congress," U.S. Department of Agriculture, Office of Forestry and Economic Assistance, December 1996, 147.

3. Vlosky, Richard P., and Lucie K. Ozanne. "Willingness to Pay for Environmentally Certified Wood Products: The Consumer Perspective," June 1996. Unpublished paper.

4. Vlosky, Richard P., and Lucie K. Ozanne, "Forest Products Certification: The Business Customer perspective," August 1996. Unpublished paper.

5. This material is adapted from "J. Sainsbury Plc. and The Home Depot: Retailers' Impact on Sustainability," prepared by James McAlexander and Eric Hansen for the Sustainable Forestry Working Group, 1997.

REFERENCES

Canadian Forestry Certification Coalition (1997). Bulletin, vol. 3, no. 1, January.

Canadian Pulp and Paper Association. Reference Tables 1996, July.

Crossley, Rachel (1995). *A Review of Global Forest Management Certification Initiatives: Political and Institution Aspects.* October, p. 16.

Environmental Advantage database, 1997.

FINNPAP (1994). Internal study of 300 German consumers.

FINNPAP (1994). Interviews with 30 German individuals involved in purchasing paper for German printing and publishing industry.

Food and Agriculture Organization of the United Nations, Rome (1995). *FAO Forest Products Yearbook, 1993.*

Furstner, Wolfgang (1996). "The German Magazine Publishers Ecological Vision and Strategy," paper presented at the International Periodicals Symposium, June.

Johnson, Brad, Procter & Gamble (1996). Personal communication, August.

Juslin, Hikki, coordinator of Finnish Certification Initiative (1997). Personal communication, January.

Koski, Teppo, senior consultant for environmental strategy, Jaakko Pöyry (1997). Personal communication, January.

Krantz, John, Minnesota Department of Natural Resources (1997). Personal communication, February.

Laishley, Don, director of forest strategy, Weldwood of Canada, Ltd. (1997). Quoted in *Sustainable Forestry Bulletin* January.

Lapointe, Jerry, Canadian Pulp and Paper Association (1997). Personal communication, February.

Maezawa, Eishi, World Wildlife Fund Japan (1996). Personal communication, December.

Magazine Publishers of America and American Society of Magazine Editors (1996). *The Magazine Industry and the Environment—Findings and Recommendations of the MPA/ASME Task Force,* September.

Mansley, Mark (1996). "The Demand for Certified Wood Products." Delphi International, unpublished paper.

McCloy, Brian, Canadian Pulp and Paper Association (1997). Personal communication, January.

McCoy, John, Blandon Paper Co. (1997). Personal communication, February.

McNulty, John, and John Cashwell (1995). "The Landowners Perspective on Certification." *Journal of Forestry* April.

Peck, Aw Beng, "Asian Timber" (1997). Personal communication, February.

Pennsylvania Department of Conservation and Natural Resources (1997). Press release, January.

Pöyry, Jaakko (1996). *Solid Wood Competitiveness Report,* Tarrytown, N.Y.

Taylor, Donald, Champion International (1997). Personal communication, January.

Vlosky, Richard P., and Lucie K. Ozanne (1996). *Willingness to Pay for Environmentally Certified Wood Products: The Consumer Perspective* June.

Vlosky, Richard P. (1996). *Forest Products Certification: The Business Customer Perspective* August.

Western Wood Products Association (1993). "Research Shows Lumber Retail and Wholesale Customers More Concerned about Environmental Impacts of Purchasing Decisions," April.
Weyerhaeuser internal study (1994). Used with permission.

Chapter 4

New Mandates for Technology

The Sorbilite composite molding system seems to possess all the characteristics of a technology "hit." In the 1990s sales of composite panel products, such as MDF and OSB, soared because they can be produced from sawdust, chips, and other lower grade fiber sources to create less expensive alternatives to plywood and solidwood products. But those production technologies are generally adapted to large-scale producers, even though most wood processors in North America are small- to medium-sized companies. The Sorbilite system, however, is designed to provide cost effective, efficient compression molding for smaller-scale use. The technology merges several manufacturing steps to produce a variety of high-quality composite products—everything from furniture parts and building parts to toys and caskets—wasting less wood fiber and using less energy than other comparable production technology.

Traditional compression technology requires that fiber have 8 percent or less moisture content, but the Sorbilite process uses wood fiber with up to 50 percent moisture. It can process a combination of fibrous materials—even carpet fibers, peanut shells, or a plastic and wood combination. The products it creates are as strong and dense as wood but cost much less to produce. To make a wooden back for a secretary's chair out of plywood costs about 90 cents: The Sorbilite process molds that chair back for about 30 cents. With a starting price of $350,000, buyers can earn back their equipment investment in less than 30 months. The compact system needs no special site preparation and is easily installed close to the source of fiber waste, so it is suitable for rural or isolated regions.

Despite its merits Sorbilite technology has hit a few roadblocks in North America—almost 70 percent of Sorbilite Inc.'s existing customers are in South America and Asia. In the United States, large-scale commodity composite panel producers with investments in conventional technology are resistant to change. Wholesalers and manufacturers are also reluctant to be the first to try out new composite products on the public. Nor do large wood waste producers, such as sawmills, necessarily want to manufacture value-added products. And often the employees with the vision to see the potential in the technology "are not the decision makers in large companies that can keep an operation alive," explained Deanne Beckwith, former Director of Marketing with Sorbilite Corp.

Innovative technologies like the Sorbilite process that slash costs, turn waste into usable product, make leaps in efficiency, and are suitable for small operations and for developing countries are as important for the commercial success of SFM as demand in the marketplace or the right forest management practices. The growing experience with SFM throughout the world has produced one incontrovertible conclusion: SFM costs more to implement in the forest than conventional forest management practices. This one factor continues to fuel the debate over whether SFM practices are economically feasible for the forest products industry.

In the long run, successful SFM will depend on new technologies that can increase the value of wood products produced using equal or less resource and enhance the bottom-line results of SFM. Fortunately, a number of those promising technologies are in the pipeline. Unfortunately, the difficulties the Sorbilite system encountered in its bid for acceptance in the U.S. market are typical of those faced by emerging technologies. For SFM to succeed, it may be essential to enlist financial institutions, nonprofit funders, and environmental groups to help forge links between technology developers and the marketplace, if these new technologies are to gain visibility in the marketplace.

Technology with a Difference

SFM technology is not a "business as usual" approach for technology development in the industry. SFM-related technologies may share some similarities in objectives and intended results with traditional technologies, but they are fundamentally different. Historically, new traditional technologies offer the industry equipment that helps forestry and wood operations process raw material with increasing speed and with higher volume, using traditional resource standards. Related technologies have optimized machinery and manpower efficiencies to achieve those goals.

SFM technologies might be defined as those that help to create a better balance between sustaining natural resources and economic development (Figure 4.1). While increasing the rate of raw resource output or production is a traditional industry objective, SFM technologies are designed to enable operations to process a variety of raw resource grades or quality. A lack of consistent size and quality of logs can be a major constraint in wood processing operations. Those operations that can buy and profitably process a range of log quality are more likely to have a direct and consistent access to supplies of raw resource.

Similarly, increasing the volume of raw resource production is a traditional industry objective. SFM technologies are oriented toward increasing the value of wood processed through effective waste recovery, value-added production, and the development of custom grades of wood. Enhancing the value of products through secondary processing has direct economic bene-

Traditional Objective	Intended Results	SFM Objective	Intended Results
Increase _rate_ of raw resource output	Faster processing of raw resource done over time	Increase _readiness_ of operation to process traditional and custom grade wood output	Increased access to and reliability of wood supply due to ability to process both high and low grade material
Increase _volume_ of raw resource output	Higher volume of raw resource processed over time	Increase _value_ of output using existing wood volume input	Increased dollar value per unit of resource processed due to waste recovery, value-added, and custom grade development
Optimize machinery efficiencies for producing traditional wood grades	• Decrease downtime • Decrease wood waste • Increase _volume_ of resource output • Decrease number of employees and labor cost	Optimize machinery efficiencies to produce both traditional and custom grade material	• Decrease downtime • Decrease wood waste • Increase _value_ of resource output • May actually increase number of employees
Optimize manpower efficiencies for producing traditional wood grades	• Increase employee safety • Increase worker training skills • Decrease material production time	Optimize manpower efficiencies for producing traditional and custom wood grades	• Increase employee safety • Increase worker training skills • Provide more stabilized job security

Figure 4.1.

The Difference between Traditional and SFM Objectives and Results.
Source: Mater Engineering

fits by creating new jobs. Every one million board feet of wood that is converted from logs to commodity lumber creates three full-time, family wage jobs. If that same wood is further processed into components for furniture, for instance, another 20 people have jobs, according to the Oregon Competitiveness Council. And many value-added products can be made from underused species, smaller pieces of wood alone or joined together, and wood with "character" that is often considered waste or defect material in traditional wood processing.

Barriers to Promising SFM Technologies

Even though innovative technologies, such as Sorbilite, show great promise, many will have trouble gaining visibility and acceptance in the wood products industry. A multitude of constraints, many rooted in the industry and the process of technology development, exist for new SFM technologies. Gaining visibility for emerging technologies among small- and medium-sized wood product producers located in rural or isolated geographic regions is an ongoing challenge. Yet, these producers have significant impact in the forest products industry. In the United States, small producers with fewer than 20 employees constitute almost 80 percent of all wood products operations, including pulp and paper. Often located in rural areas with limited access to industry trade association activities or publications, this audience can easily miss information on new technology offerings. Small landowners and mill operators also often lack the capital to invest in new technology or the management know-how to exploit it successfully.

Volume versus Value

New techniques and technologies in the forest products industry are most often developed by academic institutions, small private organizations, and individuals. In academic institutions, the research tends to be concentrated on technologies that are more likely to benefit the major wood product producers in the solidwood and pulp and paper industries because these producers fund the research. While the focus of research differs at each institution, much of it is directed toward more traditional industry objectives, such as increasing the volume instead of the value of wood produced or toward primary wood product production over value-added wood production. Academic institutions also typically have long lead times to complete research, which makes it difficult to produce market-timely results.

Technical innovations that come from small organizations or individuals—like many SFM-related innovations—are at a disadvantage. Small organizations are more likely to have limited financial capability to market the technology and lack the marketing acumen to sell the technology to industry. Frequently, small companies want to retain ownership of the technology and sell the product produced from it. But they often can sell more technology or product than they can produce or service. Given the limits on management, capital, and manufacturing at small companies, an industry rule of thumb holds that it is the second owner of a technology that makes it a commercial success, even though the original designer introduced it to the market.

Who Goes First

The mind-set of the traditional wood products industry also works against emerging SFM technologies. The wood products industry and the construction industry that it serves are conservative when it comes to new products, partly out of fear of liability. In a litigious society, makers of products that are expected to last for decades are slower to adopt new technology than companies that make consumables, or other products, which are not expected to last. The "not invented here" syndrome also thrives in the industry. This mind-set is expressed is several ways in relationship to the purchase of new technology. Any performance improvement that new technology delivers must be marketed to a daily production regime. For example, a new dry kiln technology may cut material lost to overdrying by 35 percent. But for the production facility to invest in the new technology, the performance must reach 50 percent waste reduction so that the mill can add a second shift. In general, the industry also requires a return on investments in new technology within 18 to 24 months, with rates of return in the 22 to 25 percent range—a rate many technologies do not meet.

Wood products producers are also wary of a new technology unless a major traditional equipment maker offers it. But before traditional equipment manufacturers will invest in new technology, they need to see documented interest in the technology from manufacturers. The technology designer is forced to serve as a middleman between the supplier and producer—a process that requires expertise, time, and money, which are usually in short supply.

The lack of computer literacy in the industry also hinders the adoption of SFM technologies. Much of the new processing technology is computer-based. The average employee in a wood products operation is unlikely to have the computer skills to adapt willingly to the new technology. Mill managers, who often lack computer skills themselves, complain of too much downtime when working with new technology. That one factor alone is a significant deterrent to the purchase of SFM technology. The problem highlights a need for training to make employees proficient with computers, as well as the importance of persuading employees to support new technologies to make them a success in the workplace.

A Shortage of R&D Financing

Successful marketers of emerging technologies typically need to focus first on sales to large producers, so they can quickly capitalize research and development (R&D) investments from both public and private funders. SFM technology designers may not be able to do that. A technology better adapted to value-added process improvements may find a more ready audience with small- to medium-sized product producers, but it would require a few major producers to invest in the technology to meet the payback periods required to secure the initial R&D financing. Traditional equipment manufacturers, however, are often unwilling to participate unless a major producer is involved. The dearth of research on market opportunities for underused and lesser-known species of wood in product development and the use of lower grades of lumber in higher-end product manufacturing and "character wood" for custom grades also hamper emerging technologies by fostering a business-as-usual approach to production.

Different Needs and Constraints with SFM Technology

In SFM each link in the chain of forest products development, whether a landowner, logger, primary or secondary manufacturer, faces a different set of technology needs and constraints (Figure 4.2). Landowners, for instance, need access to information systems and services on SFM practices. But the computerized systems that can provide that information are often difficult to access and too expensive for the small U.S. landowners who own the majority of hardwood forests. Wood producers, on the other hand, need to slash wood waste during production but may be unfamiliar with affordable waste reduction options.

A rich variety of emerging technologies, techniques, and programs are available to address the SFM issues in each part of the industry. In 1997 Mater Engineering Ltd. of Corvallis, Oregon, identified 10 emerging SFM technologies and programs that addressed specific bottlenecks to SFM (Figure 4.3). All of the technologies had completed product/service testing, were on the market, needed higher visibility to gain market acceptance, or lacked significant market presence in North America. Most target SFM opportunities to convert wood waste into profits or to make more wood product with equal or less resources, which can be adapted by both large and small wood product manufacturers worldwide. All represent the kind of innovations needed to help make SFM a commercial success. SFM and its technological requirements also create another need: to train loggers in SFM logging methods (see Box 4.1). Since the logger is integral to the success of SFM, this program represents a successful attempt to expand SFM knowledge to a key part of the industry.

	Key Needs	Key Constraints
Landowners	• Access to information systems and services providing sustainable forest management practices	• Land management systems and services difficult to access and often too expensive for the smaller landowner
	• Identifying and accessing knowledgeable logging contractors	• Systems not adaptable for use in isolated locations
		• Must rely on word-of-mouth recommendations for logging contractors
Loggers	• Decreasing the cost of doing business (increase profits)	• Increase in worker's compensation costs
	• Complying with environmental regulations for logging in the forests	• Lack of information on new logging technologies to match environmental concerns
Wood Producers (Primary and Secondary)	• Decreasing wood waste during production	• Lack of information on affordable value-added production options
	• Increasing the value of product per unit produced	• Lack of information on affordable waste reduction options
	• Stabilizing raw resource supply	• Lack of access to capital for systems improvements
	• Accessing affordable chain-of-custody solutions for moving certified wood	• Lack of information on chain-of-custody tracking options save for barcoding and separate production runs

Figure 4.2.

Targeted Needs and Constraints.
Source: Mater Engineering

94

	Key Bottlenecks	Targeted SFM Solutions
Landowners	• GIS capabilities not adapted to personal computer technology used by consumers	• New GIS software especially designed for personal computer use
	• Neighbor referrals for logger contacts unreliable	• Logger's certification program
Loggers	• Log harvest machines of the size, weight, and function which create heavy impact in forest ecosystem	• Low-impact harvesting technology
	• Increased worker's comp. costs	• Logger's certification program
Primary Producers (Lumber)	• Slabs, trim ends, and lumber shorts recovery and use in product development	• Scrap recovery system • Trim block drying rack system
	• Conversion of lower volume sawdust and chips to value-added product (especially for smaller producers)	• Sorbilite systems
	• Waste factor due to defect cutout in production	• New scanning technologies
Secondary Wood Product Manufacturers	• Utilization of short pieces in production development	• Fingerjointing technology: – Greenweld – Soybean-based adhesives
	• Lack of wood resources with preferred characteristics for product development	• Wood-hardening technology
	• Waste factor due to defect cutout in production	• New scanning technology

Figure 4.3.

Bottlenecks and Solutions.
Source: Mater Engineering

Box 4.1. The Certified Logging Professionals Program

Maine's 19.7 million acres of forestland covers 89 percent of the state, supports a wood fiber production industry that employs close to 30,000 people, and produces about 30 percent of the state's total manufacturing output. Maine is the nation's largest supplier of paper products, housing 16 paper mills and 15 pulp mills that produce about 12 tons of paper daily. The logging industry initiated the Certified Loggers' Program (CLP) in 1991 to address four key concerns. The number of high school and vocational training programs in the state that were the natural place to train loggers had declined. The industry had an aging workforce, and fewer younger workers were choosing logging as a way to make a living. At the same time, public outcry over clear-cutting raised the threat that the state might step in to regulate the environmental practices of loggers.

The CLP was designed to give logging a professional status that would encourage the development of skill, knowledge, and pride among loggers; attract young people to the profession; and promote sustainable forestry. In the late 1990s, the program consisted of an initial hands-on training program that included a curriculum on sustainable forestry principles, forest management, safe and efficient harvesting, fish and wildlife preservation, felling techniques and mechanical harvesting, and first aid. After attending the training program the logger must document six months of paid experience using the techniques and practices taught in the program, and must pass an in-field inspection to evaluate how well he or she applied those techniques, before receiving certification. Loggers who pass the course are issued certification cards and numbers by the Certified Logging Professional Program, which is sponsored by The Maine Tree Foundation.

By 1997 the CLP program had met its objectives. The program was so successful at reducing injury rates that a special insurance code classification for certified loggers was created to separate certified from noncertified loggers. The injury rate of Maine loggers fell from 20 injuries per 100 full-time employees (FTEs) compared to the state average of 14.5 per 100 FTEs in 1990 to 8 per 100 FTEs compared to the state's average of 10.5 per 100 FTEs in 1994. In response to declining injury rates, workers' compensation rates for CLP-certified loggers also dropped. By May 1997 the workers' compensation rate for CLP-certified loggers was 53 percent lower than for noncertified loggers. Certified loggers also missed less work: The number of days of work lost for certified loggers was 4.8 days per 100 FTEs in 1994. The SFM training in the CLP program is a primary reason that large forest products companies in the state prefer certified loggers. In 1997 International Paper announced a goal of having all its logging contractors earn CLP certification by early 1998. Between 1991 and 1997, almost 2000 loggers were certified under the CLP program.

Edward Berry of Robert Berry and Sons, a logging partnership in Norridgewock, Maine, credited the program for separating loggers who work safely, are effective, and are environmentally responsible from those who are not. "A lot of loggers don't deserve the name," Berry commented. "They've created messes that everyone is embarrassed about." The CLP program, he said, is raising the performance standard for all loggers. Eventually certification will become necessary because "you're not going to sell any wood if you're not certified."

Information for Landowners from Easy-to-Use GIS Software

Sustainable forest management is information intensive. It can require mapping a forest area tree by tree, identifying habitat for endangered species, and planning buffer zones or riparian set-asides. Geographic Information Systems (GIS) and Global Positioning Systems (GPS) technologies have been invaluable in helping the forester develop and gain access to such information. A variety of GPS collection units gather and enter information from the field about land under assessment. The control points collected by the GPS units become the data that can be used to create images that produce inventories of timber and identify wetlands, recreation trails, habitats for wildlife, vegetation, and forest landscape schemes, which foresters then use to plan harvests, estimate timber volume, and so on.

The GIS systems that create the maps, however, can be complex and expensive. Usually, the landowner brings the information he has collected in the field to a company that performs the GIS services. This information can be in the form of topographic maps, satellite images, quadrant maps, or even city maps. The GIS service company then generates specific maps of the area being assessed or inventoried. The service is expensive. Until recently, the technology was not readily available or affordable to the small landowner or individual consulting resource manager.

ForestView, a standardized software package that runs on a personal computer with the Windows 95™ operating system developed and marketed by Atterbury Consultants, Inc., of Beaverton, Oregon, is an affordable tool that can create maps and perform land assessment and inventory for the small landowner independent of expensive outside services. In 1996, the year it went on the market, the software package sold for $3,195, which at the time was at least 50 percent less than the cost of the outside services that provided the same land-mapping results.

Armed with the software, some ancillary equipment such as a scanner and several related software programs, a landowner can collect and enter data from his land right in the field. The system can also integrate a variety of other data, including scanned images, such as aerial photos, quadrant maps, and data from GIS and Computer Aided Design (CAD) sources to create maps. Other programs, including FLIPS 96 (Forest Level Inventory Planning Systems) linked to ForestView enable the landowner to compute volumes of sustainable harvest, calculate timber growth, and generate projections of future timber inventory on a given piece of land.

The ability to collect, plot, process, and manipulate such information can make small landowners better managers. Mark Schroeder, land manager of 25,000 acres of timberland on the Oregon Coast for Miami Corp., based in Chicago, discovered that ForestView gave the company "a much better knowledge of our tree farm," which enabled it to implement better management plans for long-term productivity. Schroeder buys satellite imagery of the tree farm and puts that into the ForestView program along with aerial photos of the 25,000 acres, which gives him the entire tree farm on the monitor. He can zoom in to look at an individual tree, and he knows precisely where the streams are, and even the location of the spotted owl that lives on the property. On his computer Schroeder lays out harvest plans, plots roads, calculates how much timber to cut, and inventories timber, which saves "immeasurable" time. And cost—Schroeder estimated that the system paid for itself within six months.

A Kinder, Gentler Harvesting Machine: Ponsse Systems

Equipment that does less damage in the forest—log harvesting machines that minimize damage to the forest floor, for instance, and systems that can perform selective cutting or thinning—are part of the new technology needed to implement SFM. Low-impact log harvesting is not a new technology concept. Designs for equipment that does less damage to the surrounding forest during harvesting continue to improve. However, in North America, especially for smaller-scale log harvesting operations, awareness and use of this low-impact harvest technology remains low.

Typically, trees that are "thinned" under SFM are small in diameter and considered lower grade because they have a variable diameter and defects in the log. In standard practice this material is typically treated as pulpwood. Because the entire pulpwood log will be chipped for pulp, the logger has no incentive to harvest any quality wood that might be obtained from that log. A classic example would be a small-diameter log with a heavy taper at the top. The lower portion of the log, although of smaller diameter, might have wood that could be used in shorter-length logs called boltwood that can be manufactured into value-added products, such as furniture and flooring.

The Ponsse log harvesting system developed by Ponsse Ltd. in Vierema, Finland, was designed to address those issues. In 1983 the company, under pressure from the Finnish public and government over intensive logging practices used in Finland, designed a series of fast, portable, efficient log harvesting and forwarding systems that could cut logs to length on site during first and second forest thinning. These machines do much less damage to plant life and forest ecosystems as the machinery travels through the forest than does conventional equipment.

In particular, the Ponsse HS 10 log harvesting unit is an excellent performer in first and second forest thinning operations. The system harvests the whole log and initiates a highly efficient cut-to-length process while still at the logging site. Loggers scan an individual log at harvest to determine the amount of sawlog or boltwood that can be extracted from the pulpwood volume of the tree. Simply put, the system cuts the log on stump, de-limbs the log while scanning, then cuts the log according to the sawlog grade or boltwood grade anticipated from the log scan. The entire process is accomplished within a few minutes from the time the tree is first cut from the stump. For first and second thinning of trees between 2 inches and 25 inches in diameter, about 50 percent of the wood is converted from pulpwood to more valuable log grade by using the system.

Brown Trucking and Logging of Rhinelander, Wisconsin, a small harvesting operation that bought a Ponsse system, reported a 20 percent increase in the value of logs it harvested. The value had increased, according to the company, because more boltwood could be converted from the traditional pulpwood harvest, and the maneuverability of the machine allowed loggers greater access to forest areas over more difficult terrain. Although the Ponsse HS 10 system has a $435,000 price tag, the machine earned a payback on its initial investment within two years.

Manufacturers Technologies that Use Waste and Boost Efficiency

Sawmills and other wood manufacturing operations produce a variety of wood wastes—sawdust, chips, short ends of lumber, and pieces of wood with defects. Until recently, most of that material was considered almost worthless. But as the price of logs rises processors have intensified their search for ways to make more product with less wood and to use species once considered weeds. A variety of new technologies developed for primary and secondary wood processors address these issues.

Solid Wood Waste Conversion:
Scrap Recovery Systems (Auburn)

Efficiency and value-added production may be a priority in the industry, but many primary and secondary processing operations still treat wood trim ends, lumber shorts, lumber jackets, edgings, mis-machined material, and low-grade or defect cuts as scrap. Existing technology to improve the recovery of scrap is limited: Typically the operation cannot process lumber slabs and short wood pieces in a single-pass, or it cannot process pieces of wood as short as 6 inches. To process short pieces with existing technology typically requires more than one pass on the machine, which is inefficient.

Auburn Machinery Inc. of Lewiston, Maine, manufactures a system that combines scrap recovery with the ability to cost-effectively process pieces of wood as small as 6 inches. The Yield-Pro machines convert both softwood and hardwood in a single-pass process that works difficult-to-process material such as lumber slabs, odd-shaped pieces, wood with defects, and trim ends and shorts. The machines convert this material into component parts—blanks for fingerjoints, glue blocks, and molding—or pallet stock. In 1996 Mater Engineering Ltd. contacted sawmills that were prospective buyers of the technology to see how much waste they thought they might recover if they used the Auburn scrap recovery system. The manufacturers estimated a total percentage added recovery of 8 percent of their volume; they calculated that if they bought the equipment with an initial investment ranging from $60,000 to $100,000, they would get a return on the investment within two years.

Wood Drying Technology:
Trim Block Drying Rack System

Sawmills typically process green or wet lumber. Widespread adoption of the Auburn technology could be facilitated by technology that would effectively and affordably dry wet lumber trim ends and shorts. For this reason Auburn offered its technology in a product line that included the Trim Block Drying Rack system developed by Carter-Sprague Inc. of Beaverton, Oregon. The system, a one-of-a-kind emerging technology, provides a solution to a long-standing industry problem: how to convert wet wood trim ends and shorts into dry material so it can be processed into value-added products.

Until recently, the wood products industry had no economic incentives to pay close attention to this type of waste recovery. The lack of an affordable drying process also deterred efforts to recover slab, trim ends, and lumber shorts. The new Trim Block Drying Rack system, however, combined with scrap recovery processing equipment such as the Auburn system, and fingerjointing technology, enables the industry to use its green trim ends and shorts for product instead of treating it as waste or converting it to low-value chips or boiler fuel.

The Trim Block Drying Rack, which can be adapted for both hardwood and softwood mills, was tested on sugar pine, ponderosa pine, Douglas fir, and white pine commodity lumber trim ends, as well as hardwoods. The results showed that it reduced moisture content from 19 percent to 7 percent. The system also reduced labor costs over a nonrack method for drying trim ends and shorts, produced dried material with fewer documented defects, and was more energy efficient than conventional drying technology.

The technology has money-making potential for hardwood and softwood operations. In an analysis, Mater Engineering Inc. found that a softwood manufacturer that produced 6.4 million board feet of rough green trim ends a year could recover a net value of over $1 million by using the drying system and producing fingerjoint material instead of converting the waste to chips. A similar calculation for a hardwood producer that produced about 500,000 board feet of rough green trim ends a year added up to more than a $500,000 net gain annually.

One softwood manufacturer that processed about 60 million board feet of lumber a year predicted that by using the Trim Block Drying Rack system in 1996 it would gain an estimated 3.6 million board feet annually in wood volume. The company also found that it could add a second shift with just a 20 percent increase in total manpower. With the ability to dry short pieces, the operation could not only convert its own waste into product but could also buy green and partially dried trim ends and shorts from other nearby operations for processing. The manufacturer expected to recoup its investment in the equipment in one year.

Short Piece and Scrap Utilization: "Wet" Wood Fingerjointing Technology

The industry has had the ability to join pieces of wood together in a fingerjoint fashion with an adhesive since the late 1980s. The technology increases efficiency by recovering wood that would otherwise end up in the waste bin or in low-value uses and offers compelling financial benefits for both hardwood and softwood operations. Ponderosa pine is worth just $160/mbf when used for pallet stock. But use the same material for fingerjointed softwood molding and it fetches $1,250/mbf. The gains are comparable with hardwoods: Hardwood made into pallets sells for about $200/mbf. As fingerjointed hardwood molding blanks it can command as much as $1,350/mbf.

Fingerjointed material is popular for a variety of products: moldings that will be painted, door and window parts, vertical-use-only studs, and full structural rated products, such as floor joists. By 1997 fingerjointed material had captured a commanding share of some targeted markets—including molding and millwork products—and producers reported receiving up to 30 percent premiums for fingerjointed products. Product buyers prefer fingerjointed products over solidwood products, primarily because fingerjointed material maintains its dimensional integrity when stored. Solid lumber with its long continuous grain tends to cup and warp when "yarded" over time. Export customers in Japan have lost as much as 32 percent of lumber in yards due to cupping and warping. Fingerjointed material has an average yard loss of only 3 to 4 percent.

Fingerjointing is not an emerging technology, but adhesive applications that allow wet or even frozen wood to be fingerjointed are new. It is standard practice to dry all wood destined for fingerjointing first, then cut out any defects before the wood is glued, which is a costly process. The technical specifications for fingerjoint systems that can process wet material are demanding. They must able to join wet softwood trim ends and shorts with a moisture content of 30 to 100 percent to produce defect-free lumber that can then be dried for value-added product use. The systems also have to allow fingerjointing of partially dried softwood lumber trim ends and shorts with an 18-to-20 percent moisture content to produce defect-free lumber for further drying to produce value-added products.

New adhesive technology that can process wet wood for fingerjointing has both environmental and economic benefits: Between 30 and 50 percent of kiln-dried lumber used to produce fingerjointed product has defects that must be cut out and discarded. Removing defects before drying could save up to 50 percent of the drying costs incurred by traditional milling operations. And, of course, the short lengths of wet lumber can be upgraded into marketable product. A wet process also cures faster than standard dry wood fingerjoint adhesive technology—five minutes versus 24 hours for conventional adhesives.

The Greenweld Process

Two emerging adhesive technologies, the Greenweld Process and the Kreibich System, accomplish these objectives. Greenweld Technologies, Ltd., a wholly owned subsidiary of the New Zealand Research Institute Ltd. of Rotorua, New Zealand, has pioneered fingerjointing of wet and frozen wood. The technology has been used in New Zealand and Australia since 1993. In 1997 the company opened its first U.S. operation in Oregon through a partnership with WTD Industries Inc., a large U.S. lumber producer, to market the technology in North America.

The Greenweld process has demonstrated its advantages in tests on a variety of softwoods and hardwoods. Material fingerjointed by the Greenweld process can perform to required building code and lumber standards in multiple countries, including the United States. The technology can process wood of some species with up to 200-percent moisture content. Typically it takes 7 to 8 days before quality-control product testing can be completed for traditional dry wood fingerjoint technology. The Greenweld process cuts the waiting time to 24 hours. Problem areas can be detected during, not after, production so the final product can be ready to ship after 24 hours. The process can produce 80 blocks per minute and can process pieces from 5 inches to 30 inches in length, 3 inches to 6 inches in width, and up to 4 inches thick. To get these advantages, manufacturers do not even have to buy new equipment: The Greenweld system uses the same machinery as the dry wood fingerjointing process.

In 1997 a group of companies that had invested in the Greenweld process reported to Mater Engineering Ltd. that they had reaped financial benefits. At the time, sales prices for green (wet) fingerjointed studs matched or exceeded green (wet) solidwood studs. The net profit per mbf of green fingerjointed product ranged, at the time of analysis, from $125/mbf to $130/mbf. Green fingerjointed lumber also cost about $35/mbf less than dry fingerjointed lumber, which is a sizable benefit to building contractors who are not only attracted by the lower cost but also by the material's ability to take nails more easily without splitting. The companies each invested about $3.5 million to buy the technology, but they realized a payback within the industry's acceptable limit of 18 to 24 months.

The Greenweld process, however, does have environmental drawbacks. The process uses phenol-resorcinol formaldehyde (PRF) resin, a toxic. The resin accelerator used in the adhesive product produces fumes in the work environment that must be removed with ventilation systems or mechanical cooling of the accelerator. And wastewater generated from the fingerjoint process must be treated to produce "land fillable sludge." New Zealand Research also requires a royalty payment of $7/mbf of production to use the Greenweld technology.

The Soybean-Based Adhesive (Kreibich System)

The Kreibich System, developed by Roland Kreibich of Kreibich and Associates with funding from the United Soybean Board, is a two-part glue system based on the reaction of soy hydrolyzate and phenolic compounds.

The combination delivers some advantages over existing systems for gluing wet fingerjoint material: The soybean adhesive costs less than other adhesives used to fingerjoint wet wood, and the system cures rapidly without the need of external energy. In tests the two-part glue system worked on a variety of North American softwood species at moisture contents up to 150 percent. The simplicity of the system—it requires no high-technology equipment or electricity, the glue can be applied by hand, and the pieces of wood can be held together with pressure applied by simple lever action—makes it a promising one for developing countries.

Two-part adhesive systems, however, are uncommon in the United States, although they are the norm in European fingerjointing operations. The Kreibich System requires some adaptation. The adhesive application process cannot be used with standard fingerjointing technology: It requires manufacturers to switch to a two-part process that requires the addition of a second spray head. The process, which uses toxic phenol-resorcinol formaldehyde (PRF) resin for just half the glue, represents an improvement over many current systems. But resins containing formaldehyde, which offer the most rapid curing, still pose an environmental challenge.

Willamina Lumber Co. of Willamina, Oregon, first used the system commercially in late 1997. But the Kreibich system still needed additional research and testing before it could be widely marketed. The technology has so far been certified just for vertical-use studs and substructural components, but certification for full structural is expected soon. Because of the unusual nature of the process, industry standard tests had to be reviewed and adapted to accommodate the use of green fingerjointed dimensional lumber as structural components. The adhesive system also needed testing in other engineered-wood products manufacturing such as laminated veneers and I-beams before it could be marketed for those applications.

Waste Reduction: The Multi-Talented Robo-Eye

Computerized scanning systems are commonplace in the forest products industry. Historically mill managers have used them to figure out how to cut the highest value of wood from a log based on characteristics such as the size and the grade of log. These characteristics are often determined by the amount of defect—knots, splits, or fluctuating grain configurations—in the log. By using the appropriate scanning systems, wood processing operations can increase the value of logs as much as 20 to 50 percent.

Getting greater value from each log is increasingly important to the industry. For hardwood operations, the price of logs represents most of lumber production costs—between 60 and 80 percent. For both hardwood and softwood operations, raw material is also simply harder to find due to tighter harvesting restrictions. And as old-growth forests are cut, the wood supply in many regions is coming from smaller logs 12 to 18 inches in diameter. Value-added production operations everywhere are emphasizing the use of less-known and underused species in product development as the well-known commercial species become harder to find. But buyers are wary of using species that have unfamiliar quality and performance characteristics. Scanning systems that improve processing efficiencies and consistency can help raise buyers' confidence in the quality and consistency of products made from these nontraditional species.

Until recently most scanning technology was dedicated to scanning logs. Saboteurs who pounded spikes into trees where logging is controversial, such as the Pacific Northwest, were partly responsible for the development of better X-ray scanning systems in the early 1990s. These more sophisticated systems detect foreign objects such as ceramic spikes, nails, and barbed wire that have been inserted into the trees, and which pose serious processing hazards to loggers and mill workers. Demand in the late 1990s was rising for better scanning systems in lumber production and secondary manufacturing. The existing leading-edge scanning systems for lumber and value-added wood product production used color imaging sensors, laser-based ranging sensors, or X-ray scanners. Each performed a different role in increasing the value the lumber processed:

- Color Scanning: A system that uses color line scan technology readily identifies knots, holes, splits, checks, stains, and decay during processing. It is also effective in evaluating color in the wood, an important function for lumber sorting and creating custom grade material.
- Laser-Based Ranging Systems: This technology detects characteristics of shape in lumber—wane, thickness, splits, checks, holes, and various types of lumber warping.
- X-Ray Scanning Systems: This X-ray technology, similar to that used in airport systems that scan luggage, can accurately recognize the full extent of knots and decay in lumber and find internal features such as knots and honeycomb that cannot be detected by the other systems.

The challenge for scanning systems has been to develop one reasonably priced system that incorporates all three functions. A team of Virginia Polytechnic and State University researchers in electrical engineering, wood science, and forest products, and the U.S. Forest Service Southern Research Station developed just that technology. A multiple sensor machine (MSM) developed at Virginia Tech—christened the "Robo-Eye" by the CNN television network—is setting the stage for the next generation of integrated scanning systems in lumber production. Equipped with all three scanning technologies, the general-purpose machine can automatically check lumber for surface features (knot holes, splits, decay, color, grain orientation), for board geometry (warp, crook, wane, thickness variations, voids), and for internal features (honeycomb, voids, decay). The first commercial MSM systems, which cost between $250,000 and $500,000, were installed in 1998.

Transforming the Characteristics of Wood: Indurite Technology

Traditionally when the industry confronts limits on the harvesting of a particular species, it adopts one of three options to relieve the immediate impact of the restriction. Producers identify another species with similar characteristics that can be substituted in product development. In many cases, this may mean searching for wood supplies in other regions of the world that have less stringent harvest requirements. Manufacturers might convert the material used in product development from all wood to a partial-wood based or nonwood based material—engineered products represent one solution. Or manufacturers might switch from using a hardwood to a softwood. When North American hardwood prices escalated during the 1980s, the furniture industry switched from using hardwood in frames to good quality softwoods.

In the future, the industry will increasingly be forced to exploit species of trees that were once underused or considered "weed" species. Using these nontraditional species is essential for successful SFM. Conventional forestry in the tropics is based on a handful of commercially valuable species that in many areas have been logged out. Tropical forests, however, contain a variety of more plentiful species with commercial potential. To sustain forests over the long term, landowners need to harvest multiple species, instead of just one or two, so they can maintain large enough harvests to be economically viable without depleting any particular species.

Technologies that can alter the characteristics of wood will be invaluable in making nontraditional species acceptable in the marketplace. Researchers at the New Zealand Forest Research Institute in Rotorua have developed one of those technologies called the Indurite process. Rimu, a hardwood with a beautiful reddish-brown color and grain used for furniture, flooring, and paneling, once grew plentifully in New Zealand's native forests. In 1994 after decades of overharvesting, Rimu production plunged from 120,000 cubic meters of wood harvested annually to 20,000 cubic meters. The Indurite process hardens radiata pine, an economical, abundant, easy to process softwood species grown on plantations in New Zealand, so that it performs and looks like Rimu. The process makes it possible for faster-growing softwoods that are typically not used in many value-added products to replace more valuable and increasingly threatened hardwoods.

In the late 1990s, Evergreen Indus-Trees Limited of Auckland, New Zealand, offered the trademarked Indurite process to wood product manufacturers worldwide. An associated company was constructing a processing plant in Tauranga, New Zealand, to process radiata pine for remanufacturing into flooring for the Japanese market. A similar plant was also under construction to convert Fijian-grown Caribbean pine into glue-edged laminated panels in New Zealand, as well.

Essentially, the Indurite process increases the density of the treated lumber by impregnating a largely cellulose-based formulation into the wood—in effect, pouring wood into wood. The process has been successfully applied to all other pines, aspen, birch, poplar, coconut, and eucalyptus. After combining six components in a batching tank, lumber that is pre-dried to 30-percent moisture content is then impregnated in a vacuum pressure cell for about one hour, which increases the weight of the lumber up to 50 percent. The lumber is then dried in a standard medium-temperature kiln in line with the normal drying schedules for appearance grade lumber.

The impregnated radiata pine becomes harder than such desirable hardwoods as oak, teak, and African mahogany, according to tests. But it retains a natural wood color and easily takes dyes and stains. Important characteristics for manufacturing products—stability, how well the material takes glue, and ease of machining—are all enhanced in the treated wood. Indurite-treated softwood has other advantages, as well. During impregnation the lumber can be stained, which means that the wood can be machined after the treatment—an important benefit for wood with enhanced machining characteristics. And manufacturers can produce veneers.

On a per unit basis, the cost of introducing the Indurite treatment to softwoods and lower-quality hardwoods runs about $250 per thousand board feet of lumber processed. This means that, coupled with the normal purchase price of dried lumber, the economic benefits to the value-added manufacturer in using Indurite-treated lumber can be substantial. In 1997 Mater Engineering compared the cost of a variety of Indurite-treated pines from multiple locations across the United States to the cost of traditional higher-quality hardwoods that are preferred in product development: On average, a 40 percent cost-savings could be realized by using the Indurite-treated wood. The lower cost, and the ease of staining Indurite-treated lumber, make it an effective, economical substitute for many native hardwoods in flooring, furniture, doors, joinery, and a variety of other interior finishing and decorative applications.

Appropriate SFM Technology in the Tropics

A huge gap in technology transfer between developed and underdeveloped countries complicates efforts to introduce promising SFM technology in the tropics. Latin America, according to forestry experts, lags as much as 15 to 20 years behind North America in technology. The reasons are complicated—fragile political structures and forest policies incompatible with SFM deter investments. Tariff and nontariff barriers designed to protect local industries create disincentives for using the latest imported technology. In Brazil, for instance, the tariff for imported technology usually runs between 16 and 20 percent, but it can be as high as 60 percent, according to Robert Petterson, vice president of the Latin American Region Caterpillar, Inc. Large multinational companies have the capital and know-how to invest in the technology they need. Smaller companies and community forestry efforts, which are essential for SFM in the tropics, however, often lack the knowledge, capital, and skills to invest in appropriate SFM technology, much less implement it successfully. The educational levels of workers also tend to be low, which hinders the use of advanced technology.

In several key areas technology and know-how are critical for SFM in the tropics. Greater knowledge of tropical ecosystems is needed to learn how these rich ecosystems work and how to reproduce and regenerate an economically and ecologically viable forest, according to Amantino Ramos de Freitas, president of CPTI Technology and Development, a research cooperative in Sao Paulo, Brazil. Technology to minimize adverse impacts of forest operations—road building, river crossings, and harvesting techniques—and reduced-impact logging systems, such as the Ponsse system or one developed by Caterpillar Inc., are also needed, but not widely implemented.

The failure to use low-impact logging, however, is rooted less in technology than it is in lack of trained personnel and lack of supervision and planning, according to Geoffrey M. Blate, Low Impact Logging Program Coordinator with the Tropical Forest Foundation in Alexandria, Virginia. The logger, according to many Latin American forestry experts, is the weakest link in tropical SFM. During the early 1990s IMAZON, a private research institute in the eastern Amazon, studied unplanned logging operations using a bulldozer and planned logging operations using a bulldozer and skidder operations where all harvest trees were marked and mapped in advance. Access to the trees was also planned in advance, and loggers used low-impact logging techniques such as directional felling. The planned operations did 30 percent less damage to the overall forest than the unplanned operations and lost no downed logs. The unplanned operation left behind as much as 7 cubic meters of wood per hectare because operators could not locate the downed logs.

In the tropics, small-scale SFM operations—cooperative ventures among small landowners, community forestry, for instance—will play an important role in determining the health of tropical forests and the spread of SFM. But these ventures need far more than the latest in forestry technology and techniques, logging machinery, and efficient secondary manufacturing technology to make SFM viable. SFM limits the amount of wood owners can harvest per hectare from any given species to assure a stable or growing inventory. Under those circumstances, to make SFM economically viable, small landowners must "work with abundant species that right now don't have any market recognition," according to Robert Simeone, president of Silvania Forestry. These small producers need access to technology to study the properties of abundant, secondary species in their areas and the marketing skills to develop value-added products that will earn them acceptable margins and promote those products in local, regional, national, and international markets. It can take 12 to 16 months, according to Simeone, to identify market opportunities, locate buyers, and develop products, which needs to be done ahead of investing in manufacturing operations. Yet few landowners have the knowledge of secondary species or anything beyond local markets.

BOLFOR

Efforts in Bolivia to develop an industry based on SFM illustrate the magnitude of the technical, market, and social challenges for tropical SFM. Until recently, the Bolivian forest industry in the Santa Cruz region sold primarily sawn mahogany. But in the early 1990s, mahogany production plunged as overcutting made the wood scarce. To survive, the industry was forced to shift from selling commodity mahogany lumber to marketing lesser known, less valuable species to which they need to add value through manufacturing to make a profit. In 1996 the government passed a new forestry law that, for the first time, required producers to present and operate under long-term management plans and to pay an annual fee per hectare for the concessions, which set up conditions more favorable for SFM.

But SFM ventures face difficult challenges in Bolivia. Some 400 different species of trees grow in Bolivia's forests. In any given area about 20 species will be abundant. In most cases, though, the mechanical characteristics of just a few are known. Bolivia has 52 million hectares of forests, but just 180 trained foresters. Most Bolivian producers produce under 5 million board feet of lumber a year; they lack efficient technology for primary processing as well as the capital, technology, and technical skills to develop efficient secondary processing. According to some experts, the lack of access to appropriate technologies causes as much as 30 to 50 percent of hardwoods initially processed to end up as wood waste.

In 1993/1994, USAID and the Bolivian government set up the jointly funded and implemented $20 million Sustainable Forest Management Project, known as BOLFOR, to reduce the degradation of forests and protect biodiversity by helping Bolivian producers make the transition to SFM and value-added processing. BOLFOR provides assistance to transfer forest management technologies and to develop markets for alternative species by encouraging joint ventures, the marketing of "green" wood products, and matching producers with buyers, primarily in Europe. By 1998 BOLFOR's courses in forest management technologies had reached 2500 Bolivians. BOLFOR had designed a cost management and budgeting package to help businesses manage forestry, manufacturing, and distribution operations more efficiently. A training course for decision makers in 1995 that addressed forest management, certification, and values of the forest was so influential with the nation's senators and other public officials that BOLFOR became a major player in developing new forestry laws, regulations, and technical standards.

BOLFOR's efforts have also been a catalyst for FSC certification. By 1998, eight of BOLFOR's 14 forestry clients had been evaluated for certification. By early 1999, the group expected that nearly 1 million hectares of forest would have FSC certification. As part of BOLFOR's efforts to promote certification, Simeone, a consultant, worked with six Bolivian producers who together harvest 250,000 hectares of certified forest to market products made from alternative species. In 1997 the group sold about $650,000 worth of products; in 1998 they expected to sell over $3 million. Programs similar to this one that offer technical and marketing assistance for SFM, according to forestry experts, are needed throughout Latin America and in other developing countries.

Solutions That Invite Change

The BOLFOR project demonstrates one approach to spreading appropriate SFM-related technology and know-how in the tropics. If emerging technologies that make SFM more economically viable are to get a fair testing in the marketplace, barriers to market entry and technology transfer to developing nations need to fall. Gaining visibility and establishing credibility for emerging technologies are critical parts of that process—so is support from industry and the environmental community, which is currently difficult to come by.

Gaining credibility for emerging technologies may well depend on having access to independent testing of the technology by an accredited organization or institution. This testing can facilitate market entry by easing producers' concerns of legal liability and providing outside validation for the performance of technologies and products. But it is also costly. Few programs exist to help finance that hurdle for technology designers, especially for technologies directed toward smaller-scale uses that are so needed in rural regions. The total estimated annual public and private investment in forestry and forest products R&D in the United States in the mid-1990s was about $1.3 billion. For forest products research, public funding contributes approximately $40 million, while private funds supply about $825 million.

Industry and universities have a solid track record of cooperative research projects, but most university links are with large corporations whose research interests are focused on either biological problems, such as genetics and vegetative management, or physical problems, such as soils and water quality. In 1987, 51 U.S. universities had research cooperatives of this nature with wood-based industry. The cooperatives were investing approximately $5.2 million in R&D, over 50 percent provided by industry.

Promoting Technology as Part of an SFM Strategy

In the United States, federal programs do exist to help introduce new technologies to market, but access to these funds often has additional requirements. For example, the Small Business Innovative Research (SBIR) program has a three-step process that usually spans several years before funding is provided to introduce a new technology to the market. Expanding the SBIR program to include a Small Business Emerging Technology (SBET) arm might better help new technology get to market within the existing SBIR program.

Similarly, universities that offer sustainable forestry programs could identify opportunities to collaborate in obtaining public and private R&D dollars to provide independent testing for emerging SFM technologies. As part of university technology testing, SFM technology rental programs could be established that would allow industry to test a technology on-site to verify its performance before investing in it. The rental program would also give the university the opportunity to test the technology in a laboratory setting, yet monitor and record its performance in an industrial environment.

Environmental organizations and the funders that support them also need to change their attitudes and strategies to support emerging SFM technologies. Environmental organizations, which have been instrumental in promoting SFM, have concentrated their efforts on the forest or on changing government policies. If sustaining communities in balance with forest ecosystems is an objective of SFM, in their current activities environmental organizations are missing an important link. Few of these organizations or those that fund them have made effective wood product production, or defining market opportunities for products made from less known species and for certified products, a part of their sustainable forest ecosystem strategy.

As a matter of policy and funding, encouraging the global environmental community to become actively engaged in evaluating "up-stream" production solutions that can help stabilize community impacts when implementing forest conservation efforts should rank as a top priority with those who provide financial support to environmental groups. An annual list of "Top Ten" emerging technologies each year might be one way environmental groups could raise the profile of promising SFM technologies. Such a list might encourage unusual political and financial partnerships between environmental groups and segments of a forest products industry, who usually find themselves sitting at opposite ends of the negotiating table on these issues.

Gaining visibility for emerging technologies usually requires following two marketing paths simultaneously. The first path leads to introducing the technology to the industry; the second, to getting the product produced by the emerging technology accepted in the marketplace. It does little good to introduce a new wood hardening technology like the Indurite process if the newly created hardened wood is not recognized as acceptable under traditional building codes. Gaining visibility and access to the market in these areas is difficult for all the reasons discussed earlier, and little assistance is available to the technology producer.

Forging Financial Partnerships

A few programs demonstrate the kind of assistance that could make a difference in marketing new SFM technology. During the early 1990s, Key Bank in Seattle, Washington, earmarked a $50 million dollar commitment-to-loan for value-added wood products technology that was marketed for export. The program, a first of its kind for the banking institution, would be valuable if it was directed toward SFM technology. The program, however, was aimed at conventional value-added technology, which underscored the reluctance of financial institutions to work with new, emerging, or nonstandard technologies that are backed by the assets of smaller-scale producers or individuals.

A comparative review of the U.S. environmental products and services industry illustrates the point. In 1991, the industry consisted of 207 public companies each averaging $198 million in annual revenue, compared to 58,700 privately held companies each averaging $1.3 million in annual revenue. Smaller companies have a harder time getting financed—especially for exports. The lack of up-front capital may be exacerbated by the lack of focus in the United States in general, and in the environmental services segment in particular, on export opportunities. The 1992 data for the environmental products and services industry show that only 5 percent of the total U.S. environmental services industry production is exported, compared to 24 percent for Japan and 31 percent for Germany.

Under typical lending guidelines, nonstandard technologies are often classified as too high-risk for funding consideration, especially when backed by small-company assets. Moreover, the technology generally lacks testing in the market, which further deters banks from supporting technology developed by small companies. Nevertheless, a partnership between a traditional banking institution such as the Key Bank, which has taken steps to focus on the export of value-added wood product technology, and the World Bank, for instance, could be an effective way to make emerging SFM technologies more visible and accessible in world markets.

Private nonprofit foundations also have a role to play as a catalyst for financial partnerships that could encourage the development and adoption of emerging technologies. They have the financial and organizational wherewithal to create opportunities to bring the various parties together to exchange information and highlight the importance of considering the entire SFM chain of production—from the forest to the end-user—when targeting investments in emerging technologies. Creating links—in the production chain, between companies, between the industry and technology developers—is essential to move innovative, emerging SFM technologies into the marketplace. It is, for example, far easier to market a Trim Block Drying Rack system that efficiently dries short pieces of lumber when it is linked with a short-piece lumber recovery and sizing technology such as the Auburn system than it is to market either system alone. Marketed separately, these emerging technologies offer the product manufacturer no more than a technology option. Combined they offer the manufacturer a technology solution. The distinction matters—it can determine whether or not these new technologies are successfully marketed.

NOTE

This chapter is adapted from "Emerging Technologies for Sustainable Forest Management," prepared by Catherine L. Mater, Mater Engineering Ltd., for the Sustainable Forest Working Group, 1997.

Chapter 5

Lessons from a Pioneer

Long before most of the industry, the owners of the Collins Pine Co. recognized the value of sustainable forest management. The company traces its origins to 1855 during the "cut and run" era of logging when Truman D. Collins bought forestry and milling operations in Pennsylvania. In 1940, the founder's grandson, Truman W. Collins, adopted sustained yield forest management on company lands near Chester, California. The system he embraced, based on U.S. Forest Service models under research at the time, emphasized selective cutting, a practice that creates stands of uneven-aged trees similar to those found in some natural forests. The Forest Service later switched to techniques that promote even-aged stands of trees, but Collins Pine retained and improved its uneven-aged management, which remains the foundation of the company's western operations.

Throughout the company's 142-year history, the Collins family has maintained its ownership and involvement. In 1997 Maribeth Collins, widow of Truman W. Collins, was the chair of the board. The company's values and philosophy reflect those of family members, past and present. Stewardship anchors the Collins corporate philosophy, which the company defines as a commitment "to the long-term management of our forest resources and to the responsible utilization of these and other resources to produce the finest-quality finished products."

When environmental certification of forests started, Collins Pine was one of the first companies in the world to have some of its timberlands certified by an independent organization. The company was also among the first to attempt to market certified wood products. In 1993, Scientific Certification Systems (SCS) of Oakland, California, certified Collins Pine's Almanor Forest in Chester as a "State-of-the-Art Well-Managed Forest"; in 1994, SCS awarded the Kane Hardwoods forest in Pennsylvania its "Well-Managed Forest" designation; and in 1998, the company's sawmill and timberlands in Lakeview, Oregon, were certified.

Headquartered in Portland, Oregon, the Collins Pine Company manages timber holdings in conjunction with manufacturing operations that produce a variety of lumber products in California, Pennsylvania, and Oregon. Collins Resources International, Ltd. (CRI) markets the company's products internationally, traditionally in western Europe, but in the mid-1990s the company targeted the Pacific Rim for expansion. The privately held company also operates three retail stores under the name of Builders' Supply in California. In 1996, Collins Pine expanded into the production of plywood, hardboard, and particleboard with the acquisition of Weyerhaeuser Company's Klamath Falls operation, which nearly doubled the size of the company—revenues in 1997 for consolidated operations totaled $205 million.

As a pioneer of sustainable forest management and certified wood products, Collins Pine has confronted a variety of challenges. Several early products were disappointments, and the company has had difficulty ringing up premiums for its certified products. Collins initial marketing problems, however, are indicative of those that may confront other wood products companies as they try to produce and sell products in emerging certified markets. By the same token Collins has also reaped advantages from its early foray into SFM. During the 1980s and early 1990s dozens of small and mid-sized mills closed in the Pacific Northwest as large trees disappeared and federal lands were taken out of production. But Collins forests continued to provide its mills with quality timber, and its certification status helped opened up new markets and improve the company's business practices.

Management by "Principal and Interest"

At Collins Pine, timberlands are managed with multiple objectives. Forest management is designed to maintain and enhance diversity in the forest (among species and sizes of trees), improve forest health, and increase the production of high-quality timber to feed the company's production facilities. The broader goals of maintaining the forests' functions as watersheds and habitats for wildlife are also part of forest management planning.

Management's overriding objective, however, is to keep forest management options open for future generations.

Forest management revolves around the company's system of "principal and interest." Management considers the company's timberlands a resource base, the "principal," while growth is considered "interest." Managers are free to draw from the interest, but the principal must remain stable. Collins Pine managers confirm that they consciously act in ways that will retain options for future managers, promote forest diversity, allow the forests to regenerate naturally whenever possible, and protect wildlife habitat and watershed functions. Forest managers tailor their management practices to each site, using a variety of silvicultural techniques dictated by tree species, age, and other characteristics. The company uses both even-aged and uneven-aged practices to mimic the natural processes that create diversified tree stands and promote natural regeneration of trees.

Strategies Related to Sustainable Forestry

In addition, the company operates under six strategic priorities that are related to sustainable forest management and its corporate values:

1. Quality: Collins Pine management and employees recognize product quality as the company's paramount competitive advantage. "When consumers are walking down the alleys in The Home Depot looking for lumber, they are not looking for a sticker that says certification," commented Lawrence Potts, general manager at Chester. "They are looking for a board of quality." The company's forest management strategy facilitates quality because it produces larger, higher-quality logs; the manufacturing operations follow through by maintaining high levels of technical sophistication. Collins Pine pays close attention to customer concerns, which contributes to the quality of its products. The Chester operation routinely brings customers to the mill and asks them to evaluate how effectively Collins supplies quality products. Collins Pine also surveys customers quarterly to determine their satisfaction with the products.

2. Price: Collins has adopted a long-term strategy to develop markets for certified products. It tries to establish relationships with customers at market prices and has an informal agreement with one customer to share profits when a premium is realized. Although the company would like to realize price premiums for certified products, in 1997 it had not yet required a price premium as a prerequisite for offering certified products.

3. Distribution: Traditionally, Collins Pine sold to commodity markets. In the early 1990s, the company shifted its efforts from com-

modity markets toward higher-margin markets, such as furniture and specialty shelving, that would be more likely to pay for high-quality products. Since then Collins Pine has streamlined its distribution channels by selling more products directly instead of through brokers or wholesalers. The shorter channels allow Collins to deal more effectively with niche markets, help offset the added costs of selling in smaller volumes, enhance communication with customers, and facilitate quality improvements.

4. Company Image: Collins Pine works diligently to maintain a respectable corporate image and will tell its sustainable forest management story to anyone who is interested. Certification has generated numerous positive articles in newspapers, magazines, and forest industry and environmental publications. The company was also recognized for its efforts by the President's Council on Sustainable Development, which awarded Collins Pine the President's Sustainable Development Award in 1996. This type of promotion, company executives pointed out, could not be purchased at any price. By 1996 customers had also begun to recognize the company's brand name, "CollinsWood.[R] The First Name in Certified Wood Products," even though they may not have recognized the Collins Pine name, according to R. Wade Mosby, vice president of marketing. A well-recognized brand name may prove valuable in the future to build demand for Collins Pine certified wood products.

5. Competition: The company encourages competition in certified products, particularly from larger companies, to overcome the limited availability of products and the poorly developed distribution channels that have inhibited the nascent market. One company executive estimated that to make a market for certified products work efficiently about 10 percent of the wood consumed should be certified. In 1997 just one-half of one percent came from certified production. Low consumer demand contributes to Collins Pine's difficulty in establishing significant market share for its certified products. Most consumers remain unaware of sustainable forestry issues and do not understand what certified products are. Collins management expected that consumers' awareness would rise if retailers stocked more certified products.

6. Strategic Alliances: In 1996 Collins Pine was considering alliances with other companies to market certified products. One suggestion involved teaming up with suppliers of certified products for home construction. Management envisioned creating a package of certified products that could be marketed to the professional builder or final consumer building a home. Such associations, according to

management, would help to educate the final consumer—where demand needs to be generated—and start building demand for Collins brand name certified products.

Silviculture in Oregon and California

The success of Collins Pine's "principal and interest" forest management depends on accurate inventory and growth estimates. Different methods are used in each of its three major forests to gather the information. The Almanor Forest near Chester, California, has a long history of forest inventory. Permanent growth plots were established in the 1940s and now number over 550. Timber in these plots is managed identically to surrounding timber and the plots are remeasured every 10 years, which gives an accurate estimate of timber growth that is then used to determine harvest levels. The Kane and Lakeview locations, however, do not have the same amount of stand information on which to base harvest decisions, so foresters there have relied on their personal knowledge of the condition in stands to set harvest levels. Critics have faulted these methods for their potential lack of accuracy.

The 94,000 acres at Chester, California, and the 80,000 acres at Lakeview, Oregon, are located in relatively moderate terrain dominated by ponderosa pine/Jeffrey pine and white fir. The Lakeview operation produces about 35 percent white fir and 55 percent ponderosa pine/Jeffrey pine, with the remainder a combination of lodgepole pine and incense cedar. Trees grow slowly in this region and as a result produce high-quality wood. In both places, the company uses predominantly uneven-aged management and natural regeneration. Units are logged in 12-year to 20-year intervals. Trees are selectively cut, particularly those that have begun to decline in vigor. In recent years, the foresters have concentrated on removing white fir from stands, an action recommended by Scientific Certification Systems to improve sustainability. Collins Pine's practices of selective logging and suppressing natural fires on its western lands had created stands that were overstocked with white fir, a situation that increased the risk of fire and discouraged regeneration of the more desirable ponderosa pine/Jeffrey pine trees.

Foresters mark sections of forests to be logged, indicating which trees are to be cut and which are to be left standing. In general, those trees with the poorest health and/or form are removed, with the exception of those left for wildlife habitat. Foresters sometimes leave particularly old trees standing, out of reverence for their age and stature. These management activities create healthy stands that contain trees of a variety of sizes and species. Foresters rarely use herbicides because the partially shaded, managed stands tend to keep the density of undergrowth moderate. Chemicals are used primarily in disturbed forest areas, such as those that need planting or that have been damaged by wildfire.

Silviculture in Pennsylvania

The 122,000 acres of Kane woodlands in Pennsylvania's Allegheny Plateau differ markedly from Collins Pine's West Coast land holdings. The forests are dominated by black cherry and other hardwoods that have a limited tolerance for shade. They are managed in even-aged stands with relatively long rotation lengths of about 100 years.

To encourage the trees to regenerate naturally, foresters use a shelterwood method of management. They mark the trees in a given area that are to be left standing and those that will be used as seed trees, then cut all the others. This drastically reduces the density of the stand, allowing sunlight to reach the forest floor. Herbicides are used, when needed, to reduce competing undergrowth. The shelterwood trees are left standing until the seedlings have regenerated to the desired level. The overstory trees, unless they are designated as wildlife habitat or left for aesthetic reasons, are then removed and full sunlight is allowed to reach the forest floor. The process promotes rapid growth among the hardwoods. If the overstory is not removed at the proper time, competing vegetation will outgrow the young trees, requiring many more years for the area to adequately regenerate. Even then, the density of trees may be too low or the trees poorly formed.

In Pennsylvania, the company logs small (approximately 5- to 10-acre) plots that are distributed over a broad area to maintain diversity in tree ages, forest structure, and wildlife habitat in the larger landscape. Harvest levels in the Kane forests are conservative; foresters estimate that they remove only one-third of the annual growth. Until recently they based their growth estimates not on plots but on their own deep knowledge of the land base. The staff, however, installed a geographic information system to keep records after certification, as Scientific Certification Systems recommended, which will serve as the basis for future management decisions.

Remedies for Past Problems

Much of the land purchased by Collins Pine suffered from overcutting, poor regeneration, or lack of attention in the past. Some of those lands may require many years of growth and remedial actions to create the diverse tree stands that company managers want. On the western lands, remedial actions usually involve cutting down unhealthy trees, reducing the number of white fir, and thinning to promote the desired age and size distribution of trees. In Pennsylvania, past practices have left a number of overstocked stands dominated by small-diameter trees. As these stands are thinned and otherwise managed, the diversity of the stands, average tree size, and overall forest health will increase. Fire and grazing are not used as forest management tools to any great extent in the Chester or Kane locations. However, in 1997 the Chester forestry staff was interested in experimenting with these tools to control vegetation after wildfire and to encourage the regeneration of ponderosa pine/Jeffrey pine. Although company policy does not allow grazing near streams, foresters may try grazing cattle in other areas to help keep down the vegetation that prevents the growth of desired tree species.

Healthier Forests, Good Relations with Neighbors, and Satisfied Employees

Management credits sustainable forest management for benefits beyond timber, including healthier forests, better relations with the public, and lower employee turnover. The effect of Collins Pine's land management practices is visible in the forest. Typically, foresters log sites at 12-year to 20-year intervals. They leave standing and downed deadwood in significant quantities. On lands under active management, harvests are light. Good road maintenance, relatively moderate terrain and climate, and careful timing of logging minimize damage to soil and water quality. In its audit, Scientific Certification Systems documented that the company's management practices cooperate with nature and that the company's foresters have a "commendable level of recognition for all forest resources, including wildlife, water quality, natural biodiversity, and visual aesthetics."

The company also takes part in a variety of formal and informal activities that support the communities in which it operates. The company gives the public liberal access to its forests and directly and indirectly supports local land-use consensus-building groups made up of all types of individuals, including preservationists. Collins Pine encourages research and educational projects on its forestland, supports schools and hospitals, and, through its long-term commitment to stable employment, is recognized as a contributor to community economic stability. At both the Lakeview and Chester operations, employees are active in consensus groups organized to help communities resolve the competing demands on land so prevalent in the Pacific Northwest. The employees, ambassadors of Collins Pine to these groups, help educate and communicate with neighbors, who do not always favor logging, and are able to stay "in tune" with the needs and concerns of the communities.

Through these actions, forest certification, and the willingness of employees to spend time in the woods explaining their practices to the public, Collins Pine has earned sufficient credibility to work cooperatively with environmentalists on forest management issues. At the Chester operation, for example, this credibility enabled Collins Pine to become an active member of the Quincy Library Group, a local consensus group that has stalwart environmentalists as members.

On occasion, Collins Pine managers have made sacrifices to maintain the company's hard-won credibility. In 1995, for instance, Collins Pine was ready to bid on a salvage sale of burned timber on nearby federal land. Since the most feasible access to the salvage timber was through Collins Pine land, the company had an obvious competitive advantage to win the bid. Members of the Quincy Library Group, however, opposed the logging. Collins Pine withdrew from the bidding even though the sale made good business sense. In this case, the company preferred maintaining its relationship with the consensus group to proceeding with a deal of relative short-term importance.

Management also credits the company's SFM and its corporate philosophy for helping to keep employee turnover low. At one mill people had an average of 21 years seniority. The company's position as an early leader in SFM, according to Mosby, made it easier for the company to attract new hires during the 1990s, which were turbulent years in the region's wood products industry. "People knew we were a sustainable company," he said.

The Constraints of Sustainable Forest Management

Private ownership undoubtedly plays a key role in Collins Pine's ability to employ the conservative forest management style that helped it gain certification. Under the company's forest management objectives, the land cannot be pushed to its maximum production level. Harvest levels may not exceed growth and often are below this level. Rotation lengths are significantly longer than those used by competitors, and management costs are higher on a per-unit basis. To achieve those objectives, Collins Pine's owners consistently place less emphasis on maximizing short-term profit than do most publicly held companies.

This long-term outlook has its advantages. Mill managers are able to anticipate timber production from company lands well in advance and plan accordingly to supplement their own supply with outside purchases. However, SFM can increase production costs or diminish profits in a number of ways. Harvest plans tied to the status of the forests may hinder the company's ability to respond to fluctuations in market demand and/or price. Allowing trees to grow longer, leaving larger and more trees standing, and protecting nontimber resources often require lower harvest levels and the use of more expensive harvesting methods. Finally, the company's need for comprehensive timber stand information and significant control over harvesting operations makes forest management labor intensive.

It is difficult to fully quantify the cost of sustainable forest management for Collins Pine. A company adopting a Collins Pine style of forest management would likely recognize distinct increases in costs and perhaps decreases in profitability. Collins Pine, however, has operated under these constraints for years; any additional costs are an accepted price of its corporate philosophy. In the mid-1990s, Collins Pine invested from $16 to $36 per thousand board feet of logs to cover forestry costs. These include the costs of marking trees for sale, overseeing harvest contractors, measuring forest growth, and maintaining forest roads. Company managers acknowledge that their costs are higher in some areas than those experienced by many other industrial forest owners. But they counter that the long-term stability of their wood supply afforded by conservative forest management practices compensates for any sacrifice in short-term profits.

Collins Pine's annual allowable cut gives the company an average of 316 board feet of logs per acre per year on its Almanor Forest. Its Lakeview timberlands supporting the Fremont Sawmill in eastern Oregon produce an average of 125 board feet per acre. Because the Almanor Forest receives considerably more rain and has been under active management for much longer than the Fremont operation, this difference in productivity is not surprising. Other private industrial forest lands in eastern Oregon averaged 268 board feet of logs per acre in 1995. Statistics specific to industrial forest lands were not available for northern California, but the average for all private timberlands in the area, including Collins Pine's Almanor forest, was about 230 board feet per acre per year in 1994. This indicates that Collins Pine's Lakeview forests are producing well below industry averages, while its more-established Almanor Forest produces at a level that actually exceeds the area's average.

Eastern hardwood forests typically produce at a much slower rate than western softwood forests. Collins Pine's Kane hardwood forests are no exception. Annual allowable cuts there average 57 board feet of logs per acre per year, but recent measurements of the forests' growth indicate that cut could be considerably higher—perhaps even doubled. The average harvest for all of Pennsylvania's timberland was 102 board feet per acre in 1989. At that time, state estimates indicated that growth exceeded harvesting by 2.6 times.

These harvest estimates support Collins Pine's contention that it harvests at rates lower than the industry average. However, since many of Collins Pine's trees are allowed to reach greater age before being harvested, the overall quality of the trees and their resulting value should be higher. Logs originating from Collins Pine's Almanor Forest do, in fact, tend to be of higher grades and larger sizes than those the company buys from outside sources. However, Collins Pine may not be purchasing a uniformly representative sample of logs produced on lands other than their own. If an area company were aggressively seeking quality logs on the open market, the material obtained by Collins Pine might be skewed toward smaller, lower-quality logs. What is clear is that Collins' forest management practices provide it with higher quality raw material than what it is able to purchase on the open market.

The costs of Collins Pine's certification activities can be more easily quantified than its forest management costs. Each certified location had a pre-audit and a full certification audit. Each is charged a yearly fee to maintain its certification and will be re-audited five years from the initial certification date. Initial fees for the two certified locations totaled $60,000 to $80,000, and yearly fees for each location will cost as much as $7,200. The company, of course, expected to pay additional costs as its Lakeview and Klamath Falls locations went through the process. Collins Pine estimated that capital improvements made as a result of certification may cost as much as $250,000 per year from 1997 through 1999. These include setting up new systems to measure and document timber volume and growth in its Pennsylvania forests.

Forest management costs rose after certification—they roughly doubled in the Almanor Forest—as forest managers responded to suggestions made by the certification team. In addition, the increased materials handling costs associated with tracking certified wood from the forest through manufacturing may reach $150,000 per year, which represents less than one percent of Collins Pine's total sales. Company executives consider this cost modest, and they point out that many of these costs paid for improvements that were needed anyway and that those investments will return dividends through increased efficiency.

The Elusive Market for Certified Products

Collins forest management may be a success, but the company has been relatively unsuccessful in marketing its wood products as certified. Although management has invested considerable time and energy, including 35 percent of the vice president of marketing's time between 1993 and 1996, no significant markets for certified product have materialized. Failure to develop these markets, while frustrating for salespeople, is not necessarily surprising given Collins Pine's early entry into the market.

Company-owned timberlands at Collins Pine's three manufacturing operations supply about 50 percent of each location's raw materials. Each site then buys logs on the open market to supplement its own log production and maintain manufacturing levels, some of which come from sustainably managed forest lands. The combination of its own certified production and that purchased from the outside makes it likely that more than 50 percent of the total raw material volume used by Collins Pine's three solidwood manufacturing locations comes from sustainable forestry operations. Companies certified by a Forest Stewardship Council (FSC)-accredited certifier may incorporate the FSC logo into their marketing materials. Scientific Certification Systems, which certified Collins land, is accredited by the FSC; so by virtue of its SCS certification, Collins Pine has access to the marketing logos of both SCS and FSC. Collins, however, uses the FSC version because it is better known.

Both of Collins Pine's certified locations (Kane and Chester) segregate material from their forest land that can be sold as "certified" and track it through manufacturing and shipping, so that it will not get mixed with products coming from noncertified sources. But neither location markets more than 5 percent of its total production as certified, even though at least 50 percent qualifies. The limited market demand for certified wood accounts for the discrepancy between certified production and sales.

Collins Pine has identified specific geographic and demographic market segments that are receptive to certified products. Receptive consumers tend to be highly educated and have significant levels of disposable income. These geographic markets include Austin, Texas; Santa Fe, New Mexico; the San Francisco Bay area in California; and Vail and Aspen, Colorado, as well as the United Kingdom. In the United States, the company has found that areas with harsher climates often harbor more "green" consumers. The failure of the company's certified products to meet expectations in Portland, Oregon, is an indication, according to Collins Pine managers, that consumers are often more inclined to "talk green" than to "act green."

Collins Pine's evaluations of consumer demand come from the company's experience in dealing with their markets rather than from primary research. Salespeople often field calls from people interested in buying certified wood products, but those calls come mostly for consumer products, for which Collins Pine can provide only the raw material. Salespeople at corporate headquarters also get similar calls. This may happen partly because the company's 800 number is printed on the sticker that accompanies its certified product and because it has received extensive press coverage for becoming certified. In any case, Collins Pine has become a source of information for consumers trying to find certified products.

Market Barriers to Certified Products

In its efforts to market certified products, the company has identified five general barriers to development of the market:

1. Limited Market Demand: The actual demand for certified or otherwise sustainably produced wood products was limited

and segmented as of year-end 1996. As a pioneer, Collins Pine
has struggled to identify and serve these small niches efficiently.

2. Unfavorable Consumer Perceptions: Collins Pine sales and marketing personnel discovered that their customers often harbor the mis-
conception that certified wood must be inferior to wood produced
through "standard" industry practices. These individuals think that
companies sacrifice quality to reduce environmental impacts. This
belief was evident even when marketing to another environmentally oriented firm, The Home Depot. In Collins Pine's case, however,
the opposite is actually true. Trees are allowed to grow longer than
on comparable industry forests. These older trees tend to have a
higher proportion of clear, defect-free wood. Collins Pine personnel
have had to educate potential certified product customers by
demonstrating the relationship between their forest management
practices and the quality of the products they produce from that
wood.

3. Limited Distribution Channel Development: Existing wood products distribution channels are reluctant to carry certified wood
products because certification adds complexity and cost to the distribution process. These products must be tracked from the forest
floor to retailers' shelves, which requires sophisticated systems
unless certified product remains segregated during storage and
transport.

4. Difficulties in Meeting Specific Market Demands: In markets with
a significant demand for certified products, such as in the United
Kingdom, potential buyers have precise demands. They typically
require the highest-grade lumber of a specific species and thickness.
More often than not, the volume requested in the specific grade,
species, and thickness exceed what Collins Pine can produce or CRI
can get through other sources.

5. Limited Product Availability: Certified wood products were available only in extremely limited volumes, which has a number of
implications. Most wood products producers have neither sought
nor obtained certification, which makes distributors hesitant to
carry the certified products available. It is difficult for distributors
to find enough product volume to justify allocating floor space,
storage, and other distribution resources to certified products. In
turn, the dearth of readily available sources of certified materials
makes product specifiers, such as architects and engineers, reluctant
to use these products in their designs.

A Checkered Success in Marketing

By year-end 1996, Collins had mounted five significant certified product initiatives. Two of them, pine shelving and white fir lumber for furniture, were defunct. The company's experience with these five product initiatives prior to 1997 reflected the growing pains of the company's certified marketing efforts as well as the difficulties of selling in an embryonic market:

1. Pine Shelving: Collins Pine developed and sold pine shelving to The Home Depot of Atlanta, Georgia, which was stocked in six stores in the San Francisco Bay area. By selling directly to the retailer, Collins realized 15 percent more profit on the product than it would have through normal distribution channels. Concurrently, The Home Depot was able to lower its retail price and maintain profit margins. The shelving was sold as CollinsWood^R appearance grade, a proprietary grade designed to meet customer preferences and optimize the value of the raw material. The pine shelving sold well and store managers liked it, but in late 1996 The Home Depot dropped the product for reasons that remain unclear. Collins Pine managers attribute the action to the difficulties of warehousing the shelving. The Home Depot warehouse in Stockton, California, had to store the product separately to meet the chain-of-custody requirements, since Collins could not supply enough shelving to meet the demand for more than a few of The Home Depot's many stores.

2. White Fir Furniture Stock: White fir lumber sold to Lexington Furniture, part of the furniture maker Masco, for a line of designer furniture was a great success from Collins Pine's perspective. The company realized 40 percent more for the wood than if it had been sold as construction lumber. The "Keep America Beautiful" furniture line was featured on cable television's "The Furniture Show," which included footage of the Collins Pine mill in Chester along with interviews with the chief forester and general manager. The line also was covered in the December 1994 issue of *Furniture Design & Manufacturing*. But the line did not fare well with consumers for a number of reasons. The furniture was bulky and overpowered rooms in an average single-family home. More than 100 different pieces were available, but individual pieces were priced fairly high, and no suite prices were offered. Customers, who were more accustomed to hardwood furniture, did not take readily to white fir. The pieces were often damaged during shipping (if

dropped, white fir tends to split) and as a result the packaging had to be redesigned, which caused frustration at Lexington. During its first year, the line sold over $5 million. That level of sales might have been considered a success with a smaller company, but the cash flow was insufficient for Lexington and the line was discontinued. The fate of the pine shelving and fir furniture initiatives demonstrate that certification is only one of the many product attributes evaluated by consumers. Certification cannot serve as a crutch for an ill-conceived or poorly marketed product, nor will it as yet spark enough consumer interest to pull difficult-to-handle products through distribution channels.

3. Veneer Logs: The Freeman Corporation purchases high-quality, veneer-grade logs from the Kane and Chester operations. Freeman, which operates a veneer slicing operation in Kentucky, markets the veneer as certified. Freeman has agreed to share with Collins Pine any profits above a certain level that it realizes on sales of certified veneer. As of 1996 profit sharing had not reached significant levels, but Collins Pine benefits from a stable buying arrangement and alliance with a company that helps promote certification.

4. White Fir Construction Lumber: Collins Pine did find a ready market for construction grades of white fir lumber in Austin, Texas. Sales are directly related to the Austin Green Builder Program, which encourages the use of "sustainable" building materials. The white fir competes with southern pine in this market and in general costs less in the larger dimensions (2 x 8 and 2 x 10). But by year-end 1996, Collins Pine had not yet realized a consistent premium for the certified wood. Some months produced a premium as high as 2 percent, while in others the product was sold at a slight loss. The level of the premium was related to the fluctuating price of southern pine. While certification gave Collins Pine entry into the market, it was unclear whether any premiums can be attributed to certification or simply to the availability of larger-dimension lumber.

5. Hardwood Flooring: The Kane division sold about one truckload of low-grade cherry lumber each month to a company that produces flooring. Demand for the cherry outstrips what Collins Pine can supply. Traditionally, this low-grade material was sold as pallet stock. The rustic-looking lumber, however, appeals to certain segments of the building market. When sold into these niches, Collins Pine can sell the wood for almost twice what it gets as pallet stock.

Where Is the "Green" Premium?

Collins Pine uses certification as one component in the marketing of its total product offerings. It has had little success certifying an existing product line and recognizing a market premium, which makes it difficult to attribute any premium directly to certification. The company's experience did indicate, however, that certification was able to open up new markets for Collins Pine. In several instances, as with the cherry flooring, the profits from the company's products in these new markets exceeded those that the raw material would otherwise have generated if sold into its traditional markets. Collins' experience indicates that certification can positively influence market success if it is properly exploited, although—at least for this company— it is difficult to attribute a price premium to the certification itself.

Business Performance

Collins Pine's financial performance declined between 1994 and 1996, as lumber prices dipped in those years. In 1994 Collins produced a net profit of $8.83 million on consolidated sales of $50.95 million. By 1996 the company had a $2.34 million loss on revenues of $56.63 million, according to Business Information Reports from Dun & Bradstreet Information Services. During that period the composite price of random lengths of framing lumber fell from $402 to $329.

Several measures indicate that Collins Pine operates efficiently and competitively when compared with similar operations. A study conducted by The Beck Group in Portland, Oregon, in the mid-1990s compared the Chester operation to 16 other western softwood mills on a variety of performance indicators. The total conversion costs at Chester were lower than average, and the production volume per man hour was higher. Sales averages for ponderosa pine were much higher than average. Sales averages for Douglas fir, fir/larch, white fir, and hemlock-fir were the highest of any mill in the study.

Unintended Consequences of SFM

Collins Pine's early adoption of sustainable forest management gave the company certain advantages and benefits that became increasingly important in the 1990s—namely, reliable wood supplies and goodwill from stakeholders. SFM reduces the uncertainty of supply that Collins' mill managers typically face, to the extent that company-owned land supplies its own raw materials. It could be argued that Collins Pine's commitment to sustainable forestry allowed it to stay in business when others failed. During the 1990s, dozens of mills in the Northwest went out of business because they depended on harvests from public lands. As public forestlands were taken out of production for various reasons, those operations lost their supplies. Through sustainable forestry, Collins Pine is ensured that at least part of its supply will remain stable over the long run. The process also encourages long-range planning and long-term investment. Management's commitment to sustainable forestry and sustained communities appears also to engender significant goodwill from community members, employees (who may stay with the company for longer than the industry average), and people and organizations that do not typically support the industry. This goodwill should be more valuable in the future as restrictions on logging become tighter.

The company's long-term commitment to SFM also helped it work through the certification process with relative ease. Collins was not required to alter the management of its forests in any significant way to achieve certification. The certifying organization did, however, make a number of recommendations for improvements. Collins Pine responded with increased supervision of logging crews, more documentation of its forest management plans, increased road maintenance, and significant investments in forest inventory measurement and tracking systems.

Certification also yielded benefits for the company's operations. Some Collins Pine personnel claim that by increasing their understanding of the forestry and manufacturing operations, certification has made them better managers. The inventory control requirements to ensure the chain of custody for certified products, for instance, call for precise tracking of volumes, species, and points of origin of wood, which made the company's inventory systems more efficient and reliable. The process of marketing certified products helped the company shift from a commodity market orientation toward higher-value specialty products. If generally accepted marketing principles hold, that should enhance Collins Pine's financial performance in the future.

Lessons Learned

The Collins Pine experience in marketing certified wood products from 1993 through 1996 provides an indication of what companies might face if they enter the market early. Collins' experience indicated that market demand for certified products was limited and characterized by demographic and geographic segmentation. While future levels of demand cannot be predicted, it is clear that a variety of factors—such as unfavorable consumer perceptions, limited distribution channel development, and limited product availability—can dampen that demand. Certification, however, can provide a number of nonmarket, image-enhancing benefits that are difficult to account for monetarily, including public goodwill, credibility with environmental organizations, and interest from the news media. It can also offer competitive advantage as one characteristic of the overall product, and it can open up new markets and opportunities. But certification will not compensate for a low-quality or poorly marketed product, as the fate of the line of white fir furniture indicates.

Finally, the experience of Collins Pine indicates that the ownership structure of a company can have a significant influence on sustainability. Under the Collins Pine land management regime, foresters have tremendous authority to dictate to the mills what harvest levels will be, rather than the other way around. This is not the industry norm. It takes individuals committed both to the company's land and to its owners to earn a profit without impinging on long-term land management goals. By the same token, SFM as practiced by Collins Pine requires corporate owners who are willing to forego short-term gains for long-term sustainability. In this case, the family's ownership and values make such trade-offs possible. Publicly owned companies are typically driven by short-term profit. As a privately held company, however, Collins Pine can more readily forego short-term returns in favor of long-term objectives.

NOTE

This chapter is adapted from "Collins Pine: Lessons from a Pioneer," prepared by Eric Hansen and John Punches for the Sustainable Forestry Working Group, 1997.

Chapter 6

Harvesting Ecology and Economics

For forestry and milling operations, sustainable forestry raises fundamental issues. Can it make the forest healthier? What are the economic and management consequences for a mill that depends on wood from a sustainably managed forest? Obviously, the ecological and economic impacts of SFM depend on a variety of variables—everything from the type of forests involved and the structure of the regional economy to the business acumen of the executives who run an enterprise and their economic goals.

In the temperate forests of the United States and Canada where conventional logging has been controversial for more than 15 years, two enterprises are finding ways to make SFM economically viable and practical for their operations, even though it poses economic and operational challenges. The Menominee Tribal Enterprises (MTE), which manages the forestry and manufacturing operations of the Menominee Indian Tribe in Wisconsin, has an SFM system that has made the forest healthier while providing jobs for tribal members, even though it has required significant changes in operations at the tribe's sawmill. In British Columbia, the Ministry of Forestry (MOF) has found that even though SFM costs more to implement in the forest, it can recoup those added costs by getting higher prices for logs through an innovative log sort yard.

131

The Menominee Forest: Trees That Will Last Forever

Lawrence Waukau, president of Menominee Tribal Enterprises, sees little distinction between the well-being of the tribe and its 235,000-acre reservation of hardwood and softwood forests. "As the forest goes, so go the Menominee people" is how Waukau has described the relationship. Since 1854, the forest and the industry surrounding that resource have been the backbone of the Menominee Nation's economy. Since the forest is the tribe's primary economic legacy, maintaining its health and diversity while providing jobs and income over time is MTE's overriding priority.

MTE has developed a land management regimen that fits the Menominee's economic and social goals. Rigorous management and harvesting plans and an unusual system of incentives and penalties ensure that SFM is carried out in the forest. After 25 years of SFM, the Menominee have a robust forest where the trees have steadily increased in size—along with the volume of standing timber—in the face of steady harvesting. In 1997 the Menominee was the only Native American tribe whose forests were independently certified as sustainably managed, and the only forestlands operations that held dual environmental certification from both the Forest Stewardship Council-approved SmartWood and Scientific Certification Systems (SCS).

At the same time, MTE's experience also demonstrates that SFM can have potentially serious short-term repercussions for a mill's operations. MTE's forest management produces wide variations in the quality and volume of each species of timber harvested from the forest each year. For the mill to operate profitably in the wake of this changing mix—and value—of timber supplies, MTE must continually find ways to wring more value out of each unit of wood. The essential elements of that effort include a continued mill modernization, new value-added product development, processing efficiencies, and expanded marketing.

MTE's forest management choices and decision-making structure are clearly not appropriate for all forest products concerns, particularly large private forestry operations in North America and Europe. MTE's approach to sustainable forest management, however, may be relevant to other tribal owned or managed forestry businesses in North and South America that operate with similar objectives and to smaller privately held operations. The MTE management system may also have much to suggest for public forest management, particularly in the United States and Canada. Like MTE, the agencies that manage these federal lands must balance multiple uses, which often involve maintaining jobs in local communities and meeting public demands for environmental protection and improved forest health.

Management That Restores Ecological Diversity

In 1975 the Menominee became one of the few tribes to gain the right from the Bureau of Indian Affairs (BIA) to manage its own lands. That land, an area of greater ecological diversity than many other forest areas in the state, contains 11 of a total of 15 plant associations found in the entire state of Wisconsin and has more than 9000 distinct timber stands. Soils range from dry, nutrient-poor sites that grow poorer-quality tree species (scrub oak, jack pine) to moist nutrient-rich sites that support the growth of valuable species such as sugar maple and basswood.

The Menominee manage to maintain forest diversity, while maximizing the quantity and quality of sawtimber grown under sustained yield management principles. There is a saying in the Menominee oral history that captures the essence of the tribe's land management philosophy: "Take only the mature trees, the sick trees, and the trees that have fallen—and the trees will last forever." Tribal leaders recognize the economic need to harvest timber, but they will do so only at a speed or intensity that allows the forest to regenerate. Essentially, that means MTE decides to cut a tree when it has reached or is close to its growth potential, what foresters call "vigor," rather than a tree that has simply reached a saleable size—often the chief criteria in more conventional logging systems.

MTE has used The Forest Habitat Classification System as the basis for its management decisions for over 20 years. The system evaluates a piece of land to determine what tree species and groups of tree species will grow best in that particular environment, regardless of the types of trees that currently cover the ground. This perspective is decidedly different from traditional forest management decisions, which are based primarily on the current appearance or condition of a forest stand. The Menominee system recognizes that disturbances on a site, either by nature or by humans, may have destroyed native species and allowed other lower-value scrub species to take over prime forestland. The system identifies the most appropriate cover for a site by matching tree species to a particular habitat type, based on the potential for sawtimber growth, biological and ecological suitability to the site, and competitiveness with other tree species commonly associated with it. The system takes into account the ecological reality that many species, such as white pine, can grow on multiple sites, even though most tree species achieve their top form and quality in only one or two habitats.

In this way MTE has improved most of its 235,000 acres of forest. It is still, however, working on about 66,000 acres which, due to past forestry practices, are growing below their potential. Some of the sites on that land, which today are covered by aspen, white birch, red maple, and scrub oak, are ecologically better suited for more valuable pine and hardwoods. These more valuable species once grew there, but clear-cutting and uncontrolled slash fires altered the habitat. Waukau estimated that it would take 10 years to restore the appropriate tree groupings to their rightful habitats through judicious replanting and reseeding and an additional 150 years before the entire acreage is fully restored. In the meantime, MTE will selectively thin and remove overstock, or trees of low quality and vigor, on about 6000 acres of the 66,000 parcel a year.

Restoring the land to more nearly optimum forest habitats produces a number of economic benefits. The low-value material growing on the site is usually converted into pulpwood off the reservation, which provides no reservation jobs. Once the acreage is converted, however, the forest will consist of larger-diameter, larger-volume, more valuable trees. These higher-value pines and hardwoods will be processed in the tribe's Neopit, Wisconsin, sawmill. There, these more valuable trees can be directed toward higher profit-per-unit secondary wood product markets and markets that favor certified wood, which either require certification or might pay a premium for a certified wood source.

Impressive Aesthetics, Quality, and Productivity

Under MTE management, the forest has demonstrably improved over the past 30 years in aesthetics, quality, and productivity. In 1994, the evaluation of the Menominee Forest conducted for FSC-approved certification by Scientific Certification Systems and SmartWood concluded that "aesthetically, the Menominee Forest has no equal among managed forests in the Lake States region although its total productivity measured in the value of the products removed greatly surpasses the adjacent Nicolet National Forest, which has more than twice the acreage of commercial forestland."[1]

The inventory of standing timber has steadily increased under steady harvesting. Since 1884 that volume has risen 40 percent from 1.2 billion board feet to 1.68 billion board feet, even though over 2.25 billion board feet has been logged since 1854. Forest inventories conducted in 1963, 1970, 1979, and 1989 found that the Menominee Forest contained a greater volume of timber in each successive inventory, even though over 600 million board feet of timber was harvested during those years. During the same period, despite a decrease in the annual allowable cut (AAC) from 29 million board feet (mmbf) in 1983 to 27 mmbf in 1995, the annual harvest remained fairly constant (Figure 6.1). And in 1984 a report by the Wisconsin Department of Natural Resources stated that the Menominee Forest had a faster growth rate than neighboring forests—244 net board foot growth per acre per year versus 235 board feet for national forest systems in the area.[2]

Over the past 30 years, MTE forestry has also significantly improved the quality of the timber stands and the value/volume of wood growing in the

	Sawlogs	Pulp Conversion	Total
1962-69	20.2	4.9	25.1
1970-79	16.1	7.2	23.3
1980-89	13.4	8.4	21.8
1990-95	15.7	6.7	22.4

Figure 6.1.

Average Annual Harvest from the Forest (mmbf).
Source: MTE

forest. Between 1963 and 1988, some 17,373 acres covered by lower-quality, lower-value species such as aspen were converted to stands of native higher-value species, including over 10,000 acres of northern hardwoods. During the same period, the quality of almost all species growing in the forest improved, as measured in the volume of sawlogs versus less valuable pulpwood: Between 1963 and 1988, the total volume of sawlogs in the forest rose by over 200 million board feet.

Today, the Menominee Forest has a greater variety and volume of larger-diameter trees on a per-acre basis than surrounding forests, including nearby federal lands, according to Menominee Tribal Enterprises Annual Forest Inventory Data in 1995/1996. The quality and grade of sawlogs are often a direct correlation to the diameter of the tree. Within limits, the larger the diameter of a tree at breast height, the better the grade of lumber that will be processed from it with correlating less defect. By the criteria of diameter at breast height, the amount of acreage containing large northern hardwood sawtimber increased 30 percent on Menominee land between 1963 and 1988 (Figure 6.2).

These larger-diameter trees produce more higher-grade wood for processing. Grade 1 sawlog, for example, is often processed as higher-grade lumber for value-added products such as furniture. Lower-grade material from Grade 3 sawlogs, on the other hand, is likely to be manufactured into less expensive pallet stock. Between 1963 and 1988, MTE increased its Grade 1 sawlog volume from 25 percent of the total growing stock to over 30 percent, while the amount of Grade 3 sawlog volume fell comparably. On the basis of the rise in the number of larger-diameter trees, in 1997 MTE was considering the addition of three new classifications that exceeded standard Grade 1 status for logs with at least a 14-inch diameter and less than 13 percent defect.

Sustainable forestry also delivers more consistent flows of wood volume to the sawmill, even though the value of the flow can vary from year to year.

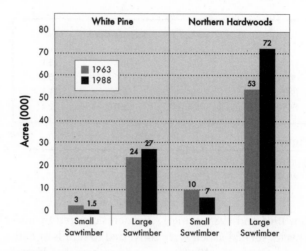

Figure 6.2.

Acres of Standing Timber by Size and Class.
Source: MTE

According to MTE data, the mill had a more even flow of sawlogs between 1970 and 1990, after MTE adopted the Habitat Classification System, than it did before. Under the Menominee forestry program, the mill knows well in advance what volume and grade, or quality, it will be processing for the year. These harvests do not take just high-grade material, as is often the case in traditional logging, but cut a balance of lower-grade and higher-grade materials.

Attractive Logging Incentives/Onerous Penalties

To ensure that loggers adhere to its harvest plans, MTE has enacted an unusual system of penalties and incentives for contract loggers. A logger who cuts 100 percent of a harvest area to contract specification, for instance, will be rewarded with a bonus of $2.50/mbf of logs. This is a critical inducement: With sustainable forestry, lower-quality material must be cut to provide growing space for higher-quality trees. But typically, loggers prefer to cut and haul only the larger-diameter, higher-grade material because they get a better price for it at the mill. Or they would rather log and haul straight pulpwood only because they can do that quickly without concern for any damage to the logs. Under conventional contracting methods, it is less profitable for the logger to do both. The logging contracts also provide a $5/mbf incentive for loggers to use new thinning techniques that extract

Action	Penalty Imposed
Unskidded or unmanufactured logs; logs manufactured into pulp	Double the logging rate of Agreement
Cut or girdled unmarked sawlog size tree	$250.00 per tree
Excessive damage to sawlog size tree	$125.00 per tree

Figure 6.3.

Examples of MTE Logging Performance Penalties.
Source: MTE

about 30 more "stems," or waste trees between 1-inch and 5-inch diameter, which provides better growth potential for seedlings.

Loggers who fail to measure up to contract requirements are hit with financial penalties. The contractor who does not log the entire area, barring natural forces or conditions beyond human control, for example, will forfeit all or a percentage of the performance guarantee on the basis of the percentage of the contract that is unfulfilled. Other financial penalties address the skidding and marking of logs and excessive damage to logs during harvesting (Figure 6.3).

The MTE system of penalties and incentives for loggers has helped to keep damage to the forest during logging low. A report to the Menominee Tribal legislature in 1984 conducted by the Wisconsin Department of Natural Resources found that logging in the Menominee Forest damages 1.9 trees/acre of logging, compared to 13 trees/acre on national forest systems in the region.[3]

The Logger's "Lottery"

The MTE also tailors its process for issuing contracts to improve loggers' performances. Traditional logging contracts operate through sealed bids or are assigned by the forest manager. MTE operates a logger's "lottery" to award its contracts. Officials release information on all the sites to be harvested to qualified loggers, then set a date for an open bid process. The open bid process determines only which site a contractor may want. If more than one logger desires a particular site, MTE officials put Ping-Pong balls, one with each logger's name, in a hat and draw a winning ball. Negotiations on price then begin with the winner and usually end in an agreement. If not, the process begins again. Loggers who gain a contract also have a set period of time to return selected contracts. When this happens, MTE also initiates new bids.

Menominee Indian contractors get preference to encourage tribal members to participate. In 1997 MTE had 29 loggers who were prequalified to bid on contracts—19 were tribal members. Even with the more difficult performance standards attached to logging practices in the Menominee Forest, however, the number of tribal and outside contractors who want to work with MTE grow is growing.

The bidding process offers a number of advantages. Contractors can discuss among themselves ahead of time who should take what contract on the basis of the capability of the contractor to meet performance requirements and avoid penalties. Because contractors select their own sites, they are more likely to perform according to contract stipulations rather than blame any failure to perform on being assigned a "bad site." The system also helps to ensure that full harvests are completed each year, which is critical for SFM and the sawmill operations.

The process has worked reasonably well. In recent years though, some contractors have returned contracts at the last minute in hopes of negotiating a higher price to harvest more difficult and/or lower-timber-quality sites. In 1995/1996, returned contracts left over 1 million board feet of timber standing uncut; this created a serious shortfall for the mill and set back MTE's forestry program. In 1997 MTE changed contract-return times and limitations to correct the problem.

Certification for Credibility

MTE's early adoption of SFM positioned the company to take advantage of third-party certification for its wood when the opportunity came along. In 1991 The Knoll Group, a manufacturer of high-end office furniture, proposed that the sawmill become independently certified, so that Knoll could develop a commercial line of furniture line made from MTE "sustainably managed maple." The Knoll Group paid SCS to assess MTE's forest management. SCS certified both the forestry operations and the sawmill. Then in 1994, the SCS confirmed its certification and SmartWood, another FSC-approved certifier, gave its stamp of approval.

Waukau, who was president of MTE at the time, pursued certification because "it goes hand in hand with marketing (sustainably produced wood)." Certification, according to Waukau, was valuable outside validation of MTE's forest management, which would give MTE credibility with customers and a stronghold in the new market for certified wood. "It's our assurance to customers that though we are not the biggest, we provide quality and we are in (sustainability) for the long haul." Sales of certified wood have paid for the company's certification, and "without doubt" MTE has gained market share through certification, according to Waukau.

Challenges for Manufacturing

The tribe's sawmill complex, the largest in the region, depends entirely on wood from Menominee land. In 1997 the operation employed 160 people in Neopit, Wisconsin—all tribal members—and a team of forest management professionals at its Forestry Center located in Keshena, Wisconsin. Hardwood and softwood commodity lumber products accounted for 75 percent of the mill's production. Sales of graded sawlogs amounted to 16 percent of the output, while high-grade sawlogs for quality veneers totaled about 7 percent. The sawmill pays no stumpage fees to the forestry operations for the logs, which is a typical operating cost for private forest products operations that do not own their own forestlands. The mill pays for all forestry costs through product sales. Since the sawmill receives no federal subsidies, its success depends on a steady flow of timber.

Sustainable forestry may produce more valuable trees, but it can also wreak short-term havoc on the sawmill. Converting forestlands to higher-grade native species and foregoing high-grade logging practices that take only the most valuable trees, both essential for SFM, can produce substantial variations in the mix of species and in the total volumes per species that reach the mill each year. That variability, even though predictable, can affect revenues and profits significantly year to year. Between fiscal years 1995 and 1996, for instance, the harvest of high-value hard maple dropped 17 percent, while the harvest of much lower-value aspen soared 47 percent. The market price for hard maple was about $1,100/mbf higher than for aspen—a difference in value that had repercussions for sales and marketing. Such variations are common (Figures 6.4 and 6.5).

The differences in the mix of species, indicated by Figures 6.4 and 6.5, can also be meaningful. Typically, prices for white pine lumber are higher than prices for hemlock lumber. Similarly, maple and basswood are preferred hardwoods for product manufacturing, so those species command higher prices than other hardwoods, such as aspen. And in general, hardwoods are worth more per unit than softwoods. Therefore, in a year such as 1996/1997, when supplies of white pine fell, softwood lumber sales for the fiscal year could be expected to drop in overall dollar volume, but hardwood lumber sales (depending on the grade being offered) should increase rather significantly. These year-to-year variations make it essential for the milling operations to maximize efforts to market and add value to the wood available for harvest in any given year to remain profitable.

	1995-1996	1996-1997 (Projected)
Total Production:	10,798,482 (bf)	9,266,940 (bf)
White Pine	26.6%	7.4%
Hemlock	4.0%	14.6%
Hard Maple	14.1%	36.9%
Basswood	6.7%	16.3%
Aspen	6.3%	4.2%

Figure 6.4.

MTE Species Mix as a Percentage of Production.
Source: Mater Engineering based on MTE data

		% Difference from Previous Year			
	1991-92	1992-93	1993-94	1994-95	1995-96
Hard Maple	Base line yr.	-41%	+5%	+51%	-17%
Aspen	Base line yr.	+49%	+7%	+<1%	+47%

Figure 6.5.

MTE Species Volume Harvest Variations.
Source: Mater Engineering based on MTE data

The Gulf Between Supply and Demand

Despite its lead in third-party certification status and the benefits that Waukau is convinced that certification has brought to the company, MTE has had difficulty selling certified wood. Even though 100 percent of MTE's products come from certified wood, in 1996 it sold only 4 percent of its hard maple veneer quality sawlogs as certified. Those certified veneers logs, however, did command a 10 percent free-and-clear (above cost plus standard markup) premium, according to invoices. And in 1996/1997, although certified lumber sales made up about 5 percent of MTE's total annual lumber sales according to company records, it appears that only hardwoods commanded the premium, which amounted to about $50/mbf or about 4 to 5 percent for certified hardwood.

Orders for certified wood are not the problem. In 1996/1997, for example, sales managers of veneer and lumber reported an increase in the number of inquiries for certified wood from all over the world. But MTE had trouble filling those orders for different reasons. With veneer logs, MTE simply had more orders than supply. Certified veneer logs are so important financially—per unit a veneer sawlog is worth 2:1 over lumber—that they are usually negotiated a year in advance, with preference given to certified veneer sawlog buyers. But MTE's practice of pulling set volumes of high-to-low grade material from its forest each year on the basis of what is best for the forest prevents it from making substantial sales of certified wood that could fetch 10 percent premiums. In 1996/1997, MTE received a sales inquiry from one U.S. veneer operation that wanted to purchase 4 million board feet of certified veneer logs annually. That one inquiry alone would have exceeded MTE's 1996/1997 total production of 814,550 board feet of veneer logs by almost 400 percent.

Orders for certified lumber presented a different problem. Many of the inquiries were too small for MTE to supply cost-effectively—under 1000 board feet. Without evaluating more creative—and economic—product transportation options, such as freight load matches, U.S. manufacturers typically prefer to ship full standard truckload orders, which are 11,000 feet for hardwoods and 20,000 feet for softwoods, to keep prices competitive.

MTE's inability to supply consistent volumes of certified wood per grade on a year-to-year basis deters customers that want to use certification as a marketing strategy for their products. Since MTE is only one of a handful of certified timber suppliers in the northeastern United States, a lack of other suppliers in the area that could help ensure consistent supplies of certified wood is an obstacle to MTE's own marketing. Long-term, however, the fact that the volumes and grades of material that will be available are known in advance under SFM could work to MTE's advantage.

The Bottom Line for Certified SFM

Despite the constraints imposed by sustainable forestry on the mill, MTE has operated at a profit since 1991, except in 1996 (Figure 6.6). That year weather conditions, unfulfilled logging contracts, and lower than expected prices for hard maple caused sales to drop to $11.2 million, a loss of $400,000. For 1997 Waukau projected a $2.5 million net profit on gross sales of $4.5 million. Historically, MTE has performed well below the traditional industry standard of 20 to 25 percent net profit as a percentage of total sales for comparable sawmill operations: Between 1991/1992 and 1995/1996, net profit ranged from a low of 8 percent to a high of 14.5 percent.

There are a number of reasons for these lower than average profits. MTE's long-term approach to sustainable forest management precludes it from capitalizing on market peaks and dips and can require it to forfeit short-term gain for long-term sustainability. In 1995/1997, Waukau pointed out that sugar maple and yellow birch were "hot," but MTE did not increase

	1991-92	1992-93	1993-94	1994-95	1995-96
Total Sales ($)	$9,388,258	$10,840,269	$11,528,901	$12,610,480	$11,214,027
% increase/(decrease) in total sales from previous year		+15%	+6%	+9%	(–12%)
Net Profit/(Loss)	$718,942	$776,849	$1,679,780	$1,274,083	($402,507)
Net profit as % of total sales	8%	7%	14.5%	10%	0
Volume bf produced	10,909,368	12,081,244	10,065,446	10, 460,992	10,798,482
Sales price/bf of production	$.86/bf	$.90/bf	$1.15/bf	$1.21/bf	$1.04/bf

Figure 6.6.

MTE Financials.
Source: MTE

production because logging more of those species at that time was inconsistent with its SFM plan.

The Menonimee's objective of providing sustainable, full-time, competitive-wage jobs for tribal members also adds to costs. In 1996, for instance, even though projected volumes to the mill were down 1 million board feet, MTE continued at 100 percent employment. The sawmill, which produces between 10 million and 12 million board feet of lumber annually, employs about 160 people. Privately held hardwood milling operations of that capacity generally would employ between 80 and 90 full-time people. Since labor costs typically constitute as much as 40 percent of total mill operating costs, this one deviation in operations can make a substantial difference in a mill's financial viability.

Other factors related to MTE's forestry and employment commitments also contribute to lower profits. The financial incentives for loggers increase logging costs. The variation in grade of logs that MTE must harvest from year to year can affect profits. Although the volume of higher-grade timber has increased under SFM, between fiscal 1994/1995 and 1995/1996 the volume of Grade 1 (good quality) hard maple sawlogs actually dropped 15 percent as a percentage of overall log volume delivered to the mill. During that period, Grade 3 (poorer quality) logs increased 15 percent. The drop in availability of Grade 1 logs, which are used in high-value products, had clear repercussions on revenues. While other factors, such as efficiencies in the mill and the accuracy of log grading, affect the overall mill production, such short-term variations in harvestable grades essential for long-term sustainable forest management can be a constraint for the mill operations.

New Ways of Doing Business

Long term, MTE management recognizes that the ability to continue with its "forest first" management hinges on generating more value from each board foot of wood during manufacturing. By Waukau's assessment, "the more value-added we get from operations, the more likely that we won't have to push sustainability over the edge." For MTE that means finding profit-oriented business solutions such as increasing mill efficiencies, adding value to commodity products, developing custom grade material, and better marketing of products made from certified wood.

In 1995 as part of an effort to improve its business performance, the mill commissioned a survey of 20 percent of its customers. Mill managers wanted to expand market opportunities for certified wood and identify ways to incorporate added-value in the mill's manufacturing to increase the value per unit of timber processed. They also wanted to explore whether it made sense to create custom grades of lumber from wood that would otherwise be classified as defective according to standard grading rules, because MTE was scheduled to harvest older-growth hard maple. The wood has a unique brown color configuration running like ribbons through its sapwood. This "brown-stain" wood is traditionally graded as defect material on the basis of its appearance only.

The results of the survey showed that customers were interested in all three options: marketing certified wood, receiving value-added products, and working with the brown-stained maple. A full 55 percent indicated immediate interest in value-added products, especially fingerjointing and/or edge-gluing. They wanted to buy more kiln-dried lumber products and brown-stain maple for flooring, door parts, and panel products. Fifty-five percent of them also said they wanted to work with MTE to develop visibility for certified wood products—even though most customers did not realize that they were buying certified wood from MTE. Customers rated high-value products such as face veneers, flooring, furniture, and specialty products such as cutting boards and tool handles as those products offering the best potential for certified wood.

The survey results and its own experience supported management's optimistic outlook for market opportunities for certified wood. In 1997 the company planned to hire a marketing specialist who would spend 70 percent of his or her time on certification efforts. MTE customers were also reporting high-profile orders for certified materials. In 1997 Connor AGA, an Amasa, Michigan-based manufacturer of sports flooring certified by the Smart-Wood program for chain-of-custody certification and a customer for MTE's high-grade hard maple, secured a contract from Walt Disney World, Inc., for hard maple wood flooring for Disney's new facilities in Florida. Company officials said that the certification status of its product was a prime asset in winning the contract, which will use over 165,000 board feet of MTE SmartWood-certified hard maple.

Exploiting market opportunities for certified wood, however, is likely to require MTE to shift a percentage of its product offering to a new customer base in national and international markets, which is a risky proposition for any operation. Most of MTE's customers are regional. Only in the past several years has it sold to customers in other parts of North America, Asia, and Europe. Since the demand for certified wood product is greatest in European markets, MTE will need to develop its marketing capabilities for that part of the world.

Greater Efficiencies in the Mill

In 1995, partly on the basis of the survey's feedback from customers, MTE mounted a mill modernization program that included an evaluation of its manufacturing productivity and a focus on implementing value-added manufacturing at the Neopit milling operations. In the first phase of that effort, the company planned to invest some $4 million between 1995 and end of fiscal year 1999 to improve efficiencies in the existing sawmill. In 1996/1997, the company overhauled lumber inventory control systems to increase production efficiency and provide better information for consistent product delivery to customers. MTE also conducted energy audits and related development improvements in energy system designs and development to cut overall production costs and provide additional revenue for its certified wood products operation.

In the second phase, which will take place between 1999 and 2001, MTE planned to spend about $4.5 million to maximize the processing of harvest timber by adding a log sorting operation, a whole-log chipping operation for pulp manufacture, and a sawmill that will process logs less than 10 inches in diameter. The last stage of the modernization plan will consist of launching a value-added manufacturing center between 1999 and 2001 that is expected to cost $5 million. As part of that effort, MTE will add fingerjointing and edge-glue technology. The entire modernization program is expected to create about 58 new jobs in a 5-year period and will increase MTE's ability to produce more product with less wood and convert traditional wood waste to profits.

Evaluating Pricing Policies

As of 1997, MTE also had additional potential opportunities to increase sales and improve production. In the past, the company has appeared to sell some products at prices below those published, and it has sold primarily lower-value green lumber instead of higher priced, kiln-dried wood. And it had not yet developed the custom grades that its customers said they wanted or added substantial value-added production on site.

Evaluating pricing policies could enhance MTE's profits. Many factors cause product prices to vary on a per-unit basis. Long-term relationships with customers, the distance between a mill and the customer, and fluctuations in current market values are just a few significant factors that can affect pricing decisions. In some categories of products, however, MTE seems to be consistently underselling published prices. A comparison of MTE's pricing data and that published in the Hardwood Review during fiscal years 1994/1995 and 1995/1996 for high-grade lumber showed that MTE prices for higher-grade hard maple were some 6.4 to 11 percent lower than the published prices for that type of wood. The difference is meaningful since hard maple has represented a significant portion of MTE's annual hardwood lumber sales—about 40 percent of its total sales in 1996/1997.

Opportunities in Kiln-Dried Lumber

Adding value to raw resources is critical for MTE, as it is for most mills. The mill sells most of its lumber as green, or nondried, rather than a higher-value dried lumber product. MTE charges clients for drying as a service at an average rate of between $50/mbf and $100/mbf for lumber dried, which is then added on to a green lumber price. The mill installed two new dry kilns in 1996, which increased drying capacity and quality, and planned to add six more computer operated packaged kilns by 2001. In 1996/1997 MTE sold about 33 percent of its lumber as dried. With the new kilns, the company doubled the amount of dried lumber sold in 1997/1998.

Selling more of its lumber dried represents a significant opportunity to increase gross revenue to the MTE mill. A comparison of prices published for green hardwood lumber versus those for kiln-dried lumber for hard maple, red oak, and basswood indicates that if in 1996/1997 MTE had sold all its lumber dried, instead of green, it would have realized an additional gross profit of approximately $300/mbf (Figure 6.7). That is a 40 percent average increase in gross revenue per thousand board feet of hardwood lumber sales. The projected average $100/mbf that MTE added for lumber drying as a service in 1996/1997 appears to represent only a fraction of the actual income that the dried lumber might generate if it were offered as a product.

	Green (FAS; #1)	Kiln-Dried (Sel & Btr; #1)	% Difference
Hard Maple:			
Sel & Btr/FAS	$1,126	$1,454	+29%
#1C	744	1,079	+45%
Basswood:			
Sel & Btr/FAS	756	1,0571	+40%
#1C	380	607	+59%
Red Oak:			
Sel & Btr/FAS	1,340	1,693	+26%
#1C	869	1,161	+34%

Figure 6.7.

Nondried (Green) vs. Dried Lumber Comparison ($/mbf; 1″ thick lumber).
Source: Mater Engineering

More Value from Custom Grades

As of 1997, the mill had not yet explored the development of custom grades that could convert traditional lower-grade, defective material into higher-value character wood. These custom lumber grades can present significant revenue opportunities for a sustainably managed forest products operation that must rely on varying grades of material from year to year. The hard maple "character" wood that MTE customers expressed an interest in during the 1995 survey is often used in pallet production and usually sells for between $125/mbf and $200/mbf. Converting the wood into a custom grade for flooring markets, for example, could increase prices to over $500/mbf.

Raising the Profile of MTE and Certified Wood

In the past several years, MTE has stepped up initiatives to promote the tribe's sustainable forest management. In 1995 MTE was one of a handful of companies to win the U.S. President's Award for Sustainable Forestry Development in the award's inaugural year. Waukau pursued the honor because he thought it would put MTE "on top of the list in sustainable forestry." That same year MTE held its first annual marketing workshop for customers to provide information about the products manufactured by its customers, the growing international markets, and product distribution systems for certified wood products. Other efforts have included videos for customers and the public on MTE forest management, public demonstration sites for sustainable forestry and certified wood products from MTE, and an MTE site on the World Wide Web.

But MTE officials have pursued public recognition for other reasons, as well. The visibility of the Menominee's sustainable forestry program has become a source of pride and has opened up new revenue-generating opportunities for the tribe. In 1993 MTE and the College of the Menominee Nation jointly established the Sustained Development Institute (SDI) to promote the Menominee Forest and its management and to educate the public, particularly the Menominee children, about the resource. In 1996, MTE conducted its first sustainable forestry demonstration workshop for public and private forestland managers, which drew over 120 participants. The College of the Menominee Nation offers college courses based on Menominee Forest management cases that address the cost/benefit issues involved in sustainable forestry for Menominee tribal students, Wisconsin's Department of Natural Resources forestry staff, and U.S. Forest Service personnel.

Sustaining Forests and Providing Jobs

In the context of the Menominee's goals—sustaining the forest and creating stable jobs—MTE's SFM program represents a success. Through decades of active logging, the Menominee have generated hundreds of long-term, full-time jobs for tribal members while improving the quantity and quality of remaining standing timber. But MTE experience also illustrates the significant consequences that sustainable forestry can have for a conventional sawmill operation. Even when planned, the year-to-year variations in volume, quality, and species harvested under SFM can have short-term financial impacts and can require a business to sacrifice short-term market opportunities for long-term sustainability goals. These constraints suggest that MTE's future success depends on combining SFM with sound business practices to improve efficiencies, add value, and maximize profits. For MTE that may mean:

- Reevaluating product pricing strategies
- Increasing product diversity through new product offerings (such as dry-kilned lumber)
- Developing custom grades that fetch higher prices for character wood
- Offering value-added products that convert wood waste into wood profits
- Implementing a log sort yard to take advantage of multiple log buyers' needs

Despite the harvesting constraints and raw material fluctuations for the mill imposed by sustainable forestry, SFM is economically viable for MTE. By virtue of their status as an Indian tribe, the Menominee receive an annual $1.3 million government subsidy and pay no taxes, which obviously contributes to the financial success of the wood products operations and are advantages that are unavailable to most forest products companies. The Menominees, however, have made specific choices in the way they run their business, which include operating at double the typical level of employees needed for similar-sized wood processing operations. Since wages for labor in a typical mill operation the size of MTE account for about 30 percent of total operating costs, while taxes account for about 8 percent, the increased employment levels are a more significant influence on MTE's financial performance than is its tax status.

Given the history of certified wood sales at MTE, however, it is unclear whether any premiums for certified wood and wood products are, or will be, a growing or even consistent trend for the company or others marketing certified wood products. But by 1997, for certain wood products, the demand for certified wood had already exceeded supply. Increasing the amount of available certified wood is a prerequisite to the development of robust markets for certified wood products.

The Vernon Project: Log Sorting for Profit

During the 1980s and early 1990s, British Columbia was the stage for a classic tug-of-war between economics and environment played out over the management of federal lands. The public was fed up with some of the more negative aspects of large-scale logging—bare hillsides, mudslides, and deteriorating water quality. Many communities adopted a "not in my backyard" posture toward clear-cuts and lobbied the government for alternatives. Environmentalists clashed repeatedly with industry over clear-cutting Crown lands. Forest communities, frustrated over dwindling job prospects as more logs harvested at home were shipped outside for processing, wanted to develop local wood manufacturing operations to create jobs. Yet nearly 85 percent of the province's timber is sold to large corporations, while just 13 percent goes to small businesses.[4]

Against that backdrop, in 1993 the Small Business Forest Enterprise Program (SBFEP) of the Ministry of Forestry (MOF) financed the Vernon Project to help resolve the tension between the province's traditional value of timber production and new public values that support environmental protection, recreation, and communities. The project, carried out in the Vernon District of the Kamloops Forest Region in south-central British Columbia on 38,000 hectares of predominantly softwood forest, set out to answer three questions: Were alternative logging practices technically and financially viable; could innovative methods of selling logs through a log sort yard give small manufacturers better access to wood; and could the log sort yard boost the financial returns on sales of timber harvested under alternative logging practices?

Less than five years later, officials pronounced the experiment a success. According to Thomas Milne, the MOF official in charge of the log sort yard, the project "proved that alternate harvesting systems are more expensive, but the cost can be justified by sorting and selling on an open log market." While a log sort yard does not solve all of the problems, "it is a very big part of the solution to both environmental concerns and log supply for small operators."

MOF foresters evaluated five alternative logging practices at Vernon: clear-cuts with reserves; group, strip, and single-tree selection systems; seed tree systems; small clear-cuts; and uniform and group shelterwood systems. None of these methods involved clear-cutting large areas, and all were selected to minimize the visual damage of logging. Foresters chose the harvesting method they wanted for each site on the 38,000 hectares, depending on which of a variety of environmental and forestry objectives they wanted to achieve, including salvaging diseased timber, protecting soils, water, or wildlife habitat, and regenerating commercial tree species.

Between 1993 and 1997, MOF cut about 216,247 cubic meters of wood, an average of 54,000 cubic meters a year. Greenpeace Canada, an effective adversary of clear-cutting in Canada, conducted an environmental assessment of the alternative logging practices used in the project. This was Canada's first independent, third-party certification of the environmental performance of certain alternative silviculture practices. But the group, through the Silva Forest Foundation, evaluated only two sites that used group and single-tree selection logging in 1995/1996. These sites, which Greenpeace certified as "Ecologically Responsible Wood," yielded 4000 cubic meters of certified wood volume, or less than 2.5 percent of the total wood harvested during the evaluation period.

New Ways to Sell Trees and Market Logs

The success of the Vernon Project hinged on new procedures in two arenas—selling timber and marketing logs—because the traditional practices favored large-scale wood products producers. Typically, the MOF sells timber on the stump. The MOF gets paid an appraised value of the standing timber or stumpage plus any additional amount (bonus bid) the purchaser elects to offer to secure the award of the timber contract from the MOF. The successful purchaser then assumes all of the costs for logging and transportation of the logs to a designated mill. The large size of these contracts puts small manufacturers, which cannot afford the up-front costs, at a disadvantage in bidding for timber. And it all but forces logging contractors to work with the major wood product producers to secure jobs.

Conventional log sales systems also favor large-volume commodity buyers. It is standard practice to separate veneer-quality logs from logs of lower grade that may be manufactured into commodity construction lumber. It is not standard practice, however, to recognize the value of lower-grade logs with defect or "character wood." But these woods are often prized for making higher-end products, such as furniture or log homes, that many small manufacturers produce. Log sales yards located near wood product manufacturers and log brokers that use this type of wood can increase their profits by adding log sorts for these character logs. These special sorts not only sell for more per unit over lower-grade logs to interested manufacturers but they also provide small, local buyers a source of wood they would not otherwise readily have.

For the Vernon Project the MOF retains ownership of the timber until it is sold at a log sort set up near Vernon, B.C. Logs harvested under the alternative methods are brought into the MOF yard where they are scaled, sorted, and sold in lots based on species, potential products, and grades to the highest bidder. Between 1993/1994 and 1994/1995 the average volume through the yard was approximately 54,000 cubic meters per year. In 1995/1996 volume reached 57,500 cubic meters per year, about 5 percent of the district harvest volume. Market demand and buyers' requests determine the number and type of log sorts at the yard. These requests come in by multiple product application, species, length, top and butt diameter, and quality.

A Store Front for Local Manufacturers

Nearby buyers are essential for a successful log sort yard. Sorts based on customer requests are the primary method of marketing since all logs must be sold to the highest bidders. Fortunately for the Vernon yard, a variety of primary and secondary manufacturers operate within the area. Requests from these small operations, which include log homes, furniture, flooring, and other specialty products manufacturers that form the competitive base for log sales, quickly escalated the number of sorts at the yard. Within two months after the yard opened, the number of sorts rose from 11 that were planned to 23. Within a year that number doubled to 42. Nearby manufacturers requested a plethora of special sorts—balsam peelers, firewood, dry white pine, character logs for log homes, pine shake logs, and even "acoustic" logs that are used to make musical instruments—which yard manager Milne obligingly filled.

Although large commodity manufacturing operations still bought 73 percent of the logs that went through the Vernon yard in 1997, by doing the special sorts for small manufacturers, the Vernon yard has become a "store front" where they can buy as little as one log of ready-to-work wood in grades tailored to their needs. By doing the special sorts, over the four years of operation the yard had steadily increased the participation of small value-added manufacturers. In 1997, for instance, small companies purchased 91 of 138 sorts. Buyers came from as far away as 250 miles.

For many small operators, the Vernon yard is the only source of nearby wood. Two guitar manufacturers can find the small number of special "acoustic" Englemann spruce they need to make their instruments only at Vernon. A number of the area's estimated 30 to 40 log home builders, who had trouble convincing other log buyers to sort long, straight timber for them, have become regular customers. Many of these small operators add enormous value to the material: A saddle-maker, for instance, buys just two dry Douglas fir trees a year from which he produces between $50,000 and $60,000 worth of saddle trees annually.

The Vernon yard has simplified timber supply issues for Oyama Forest Products, Inc. "It is virtually impossible" for small operators to buy wood from major forest products companies in the area, according to Edward Tarasewich, a vice president of Oyama. Yet stumpage and royalty fees are too high for most of these operations to compete for logging contracts, and logging costs are higher for small-scale operations. Tarasewich estimated that it cost Oyama $650 per thousand board feet to log timber, while large operators can log for $500 per thousand board feet. For 30 years Oyama, a specialty maker with 146 grades and items in production and the need for a variety of species, had scrounged for small-business sales and salvage timber in the area for its wood supply. The log sort yard gives the company the opportunity to "pick out the type of wood we require for a specific product, and it's available right now," said Tarasewich. In 1998 Oyama planned to buy up to 5000 board feet from the yard.

Tangible Benefits, but Higher Costs

The use of alternative harvesting methods led to tangible environmental and social benefits. The MOF evaluated the environmental and economic impacts of the alternative silviculture systems for the first year of harvest in 1993/1994, which produced 53,030 cubic meters of timber. That evaluation found that the alternative methods met all of the MOF's immediate environmental objectives—preserving views for surrounding communities, enhancing deer ranges, salvaging diseased trees, creating habitat snags, protecting soils, encouraging the regeneration of desired tree species, and preserving higher-quality tree species.

The project also cooled antagonisms and built new relationships among forestry officials, environmentalists, community members, and loggers. When the harvest block for strip shelterwood logging was selected near Cherryville, for instance, the Vernon MOF asked the community to help design the harvest area. Residents liked the result: Their forest views were preserved. The increased sensitivity of MOF foresters to the environment and the surrounding communities created a new level of trust and credibility for the agency in the Vernon district, according to environmentalists and other outsiders. Greenpeace supported the project, which increased the credibility of the project with local communities. In some places, according to participants in the project, public opposition would have scotched logging entirely without the use of the project's alternative harvesting systems and the consideration that MOF officials showed for the concerns of the neighboring communities.

The benefits of sustainable forestry, however, came with a price tag. (See Figure 6.8.) On the 11 harvest tracts where the MOF calculated equivalent

| Track Species | Harvest Cost/m³ | | % Difference |
	Alternative	Clear-Cut	
Fir Cedar	Group Selection $24.15	$20.38	+18%
Fir Larch	Group Selection $29.95	$24.36	+23%
Spruce Lodgepole Pine Balsam	New Forest $32.64	$24.44	+34%
Lodgepole Pine Fir Larch	Strip Shelterwood $39.95	$28.06	+42%
Fir Lodgepole Pine	Selection $34.55	$26.60	+30%

Figure 6.8.

Harvest Costs: Clear-Cutting vs. Alternative
Practices.
Source: Vernon Forest District

costs for conducting large-scale clear-cutting, alternative harvesting practices cost an average of 20 percent more to implement than conventional clear-cutting methods. For clear-cut-free alternative logging practices that may have met Greenpeace certification standards, the average cost over clear-cutting methods was about 30 percent.

Ironically, however, by adopting alternative harvesting methods the MOF was able to offset a portion of those increased costs. The use of alternative methods allowed the MOF to harvest more land than it would have otherwise been able to do by clear-cutting. The Forest Practices Code of British Columbia forbids clear-cutting on steep slopes or sensitive areas. By using alternative logging practices, the MOF gained access to an estimated 10 to 20 percent more land base, according to participants and observers, without citizen protest, according to Milne.[5]

The alternative logging practices also created job opportunities. The MOF opened up the logging contracts for the project sites to smaller independent operators who normally would have been excluded from bidding because of the requirement to buy the timber up front. It required a 10 percent security on the bid, instead of bonding, which was advantageous to small contractors. The alternative logging systems also supplied more jobs because the logging and site protection restrictions of the project lowered productivity.

A Profitable Venture

Despite any extra costs, the alternative methods turned out to be more profitable than clear-cutting. Both an internal report produced by the MOF in September 1994 and an independent evaluation conducted by Price Waterhouse in May 1995 concluded that the project was at least as profitable as conventional logging. On the basis of costs and revenues analyzed by the MOF and Price Waterhouse, both determined that the project earned a $2 million net profit after paying all stumpage fees. The two reports, however, differed in their calculation of the level of estimated net profit that would have been realized from conventional harvesting methods: Price Waterhouse projected a $2.1 million profit; the MOF estimated a $1.1 million profit. The reports used different estimates of what bidders would have bid as a "bonus bid"—the amount bid over the minimum-bid price set for stumpage—for the timber had it been offered under the conventional method of bidding.

The log sort yard clearly made the difference between profit and loss at Vernon. In almost all cases, the special sorts that catered to customers' product interests fetched a higher price than if the logs had been left in the original sort, typically as sawlogs or chip logs. In some cases, as Figure 6.9 indicates, those price differentials were considerable, ranging from 11 to 42 percent more.

New Sort	Old Sort Price	New Sort Price	Differential (%)
Balsam Fir Peeler	$74.03	$105.15	(+42%)
Spruce O/S (Dry)	$73.92	$93.09	(+26%)
Spruce Peeler (Dry)	$89.13	$111.19	(+25%)
Pine Peeler (Dry)	$79.17	$88.14	(+11%)
Spruce Bldg. Log (Dry)	$79.17	$111.75	(+41%)

Figure 6.9.

Price Differentials Due to Added Yard Sorts.
Source: Vernon Forest District
Notes:
1. Old sort is where the logs were placed when the log yard first started up.
2. Old sort price is the price of the old sort at the time the first logs were sold in the new sort.

Sales of lower-grade logs, which included a large portion of dry logs and logs that died on the stump, provided some of the largest increases. Milne, the log sort yard manager, who was knowledgeable of the markets and the value of dry sorts, knew that many of these logs could be used as peelers, for shake manufacturing, and for log houses. Dry spruce logs were also used to make chopsticks and for window and door stock. A typical dry sort sold at $110 per cubic meter. After deducting an average cost of $55 per cubic meter for logging, sort yard, and stumpage costs, these special sorts provided a return of $54.75 per cubic meter more than if the logs were sold for traditional lower-grade uses.

A Project with Influence

The Vernon project, in its fifth year in 1997, had become a catalyst for other related developments. Hundreds of visitors from governments, companies, and forestry groups from Canada, the United States, and other regions interested in setting up log sort yards have come to see the Vernon operations. Not only has the project helped resolve tensions between environmentalists, forestry officials, and local communities—but the successful use of alternative harvest practices for lower-volume harvests and the ability to offset the increased costs of SFM through the log sort yard has become a benchmark for future activity in this arena.

In 1995/1996, the MOF operated a log sort yard in Prince George. In 1996 Weyerhaeuser set up a log yard next to the Lumby facility to assist in merchandising logs. MOF officials estimated that similar log yards could be justified in other British Columbia locations, including the East and West Kootenays as well as the Cariboo Region. In addition, the MOF was considering other programs to increase small manufacturers' access to wood. One scheme would allocate logging bids to small manufacturers partly on the basis of how much value they would add to the wood during processing. Finally, the log prices supplied through the open bidding at the Vernon yard, which are published weekly in *The Forest Industry Trader,* have become the barometer of the log market for the interior of British Columbia, providing the only available up-to-date information on log prices and current market conditions.

Despite the kudos for Vernon, in 1997 the MOF had no plans to set up a bevy of imitators in British Columbia. Some environmentalists and small operators asserted that large corporations opposed the expansion of the small business program because they were reluctant to adopt SFM logging techniques, or they were concerned that any expansion might increase the costs of logs. Whatever the politics, Canada's timber industry is traditionally based on volume, while the Vernon project is based on small- and medium-sized operations. Any significant expansion of the small business programs would pose restructuring challenges both for the industry and for the MOF. Even with no plans for expansion, the MOF continued to separate certified wood in the sort yard. The yard's 1997/1998 performance, predicted Milne, would match that of the previous year, with about 70 percent of the more than 55,000 cubic meters harvested that year cut by alternative systems.

A Tool for Community Stability

In British Columbia and other regions where conflicts simmer over logging on public lands, the Vernon Project suggests that a log sort yard combined with SFM may be a useful tool to help land management officials balance environmental protection with economic sustainability in forest communities. While the use of alternative logging methods in this project did increase logging costs, those very same methods enabled forestry officials to expand the land base by logging in some sensitive areas that would otherwise have been off-limits to conventional clear-cutting. This proved to be a significant equalizing factor in the project's economic success. However, the results of the Vernon Project also underscore the importance of finding ways to offset the added costs of implementing SFM in the forest farther up the production process. In the case of Vernon, a log sort yard proved to be the vital link.

Certification played a minor role at Vernon, which suggests that the third-party certification of forestlands is not essential to successful SFM. The certified wood, a one-time harvest, was separated and sold separately. But only 1 percent was purchased based on its certification status—by Rouck Brothers Sawmill Ltd. for flooring—and the company did not pay a premium. The failure of the yard to sell its certified wood on the basis of certification indicates that a consistent supply of certified product is essential to meet market demand, which was not the case in this project.

Despite some concerns of government officials and business people at the outset that the Vernon experience might prove to be an anomaly that would have little relevance elsewhere, much of what was learned at Vernon proved transferable. Activities in the province of British Columbia—new agreements between environmentalists and industry, creation of new log sort and sales yards, and government reviews to reevaluate stumpage payment prices and practices for harvesting on Crown-owned lands—bear this out.

NOTES

The section on the Menominee tribe was adapted from "Menominee Tribal Enterprises (MTE): Sustainable Forestry to Improve Forest Health and Create Jobs," prepared by Catherine L. Mater, Mater Engineering Ltd. for the Sustainable Forestry Working Group, 1997.

The section on the Vernon Project was adapted from "Vernon Forestry: Log Sorting for Profit," prepared by Catherine L. Mater and Scott Mater of Mater Engineering Ltd. for the Sustainable Forestry Working Group, 1997

1. "FSC Certification Assessment Report for Menominee Tribal Enterprises," *Scientific Certification Systems,* 1994

2. "The Menominee Forest: Its Condition and Management: A Report to the Menominee Tribal Legislature." Prepared by the Department of Natural Resources, State of Wisconsin; Kenneth R. Sloan, Menominee Liaison Forester, Keshena Ranger Stations, Wisconsin DNR, April 1994.

3. Ibid. "The Menominee Forest: Its Condition and Management."

4. "Bringing Ecoforestry to the B.C. Forest Service," *Global Biodiversity* vol. 7, no. 2 (Fall 1997), 17.

5. A variety of sources among government officials and environmental groups agree on the 10 percent to 20 percent estimate.

Chapter 7

Opportunities for Private Timber Owners in Sustainable Forestry

The private forest owners who control 58 percent of the commercial forests in the United States are indispensable for the U.S. wood supply. West of the Great Plains, the majority of forests are publicly owned, but east of the Mississippi River private forest ownership accounts for more than two-thirds of the region's timberland. Private woodlot owners supply 49 percent of the timber harvested in the United States, according to the U.S. Forest Service. Some large wood products manufacturers depend heavily on small forest owners. Weyerhaeuser Co., for instance, harvests 58 percent of its timber supply from NIPFs (nonindustrial private forest landowners) nationally. The 261 million acres of nonindustrial private forests are also a reservoir for valuable ecological services and natural capital, protecting watersheds, providing habitat for wildlife, and offering scenic beauty and recreation for communities.

The 10 million NIPF owners—a diverse group that includes individuals, partnerships, estates, trusts, clubs, tribes, corporations, and associations—confront a variety of challenges that can complicate the practice of sustainable forest management (SFM). Many are not well informed about the economic value of their resource or the importance of consulting professional foresters when making management decisions. Annual property taxes and capital gains taxes can be disincentives to sound, long-term forest management. Without proper estate planning, owners can be forced into making decisions that may prevent them from passing forestland from one generation to the next and may lead to its conversion to other uses. In the long run, the objectives of the owners in the context of their financial circumstances are the critical factors that determine whether NIPF properties will be managed sustainably.

Seven NIPF ownerships are evaluated here. They are located in widely different timber-growing regions—the Northeast, the Pacific Northwest, and the Southeast—and illustrate the rich variety of NIPF owners' backgrounds, objectives, and financial circumstances. They demonstrate how a diverse group of private landowners can successfully address issues of forest sustainability. However, the process of certifying sustainable forestry is still too expensive and cumbersome for most of them. Three new approaches to adapt certification to NIPF ownership are examined at the end of this chapter—that of a certified resource manager, a chain-of-custody certified manufacturer, and a single forest owner seeking certification. Any approach to sustainable forestry of NIPFs must accommodate the diverse management and objectives of the owners that these cases capture.

An Account Book with Varied Entries

Given their great diversity, it makes little sense to assess the likelihood for sustainable management of NIPF lands solely in terms of monetary returns or sociopsychological benefits. The NIPF account book needs to balance some financial costs against financial gains. But it also needs to acknowledge that many owners would invest some wealth and personal energy in their lands even if they realized no financial gain. NIPF owners often own and manage land for the same reasons that people take care of their homes and gardens, invest in community projects, or attend church. The inverse is just as true—most NIPF owners will trade these nonfinancial benefits for a financial return when the need or opportunity becomes compelling. Any tally of the costs and benefits of NIPF lands must also recognize the benefits of these lands to society at large. Owners create recreational opportunities, educational sites, stable soils, and aesthetic views for the surrounding community.

A Portrait of NIPF Lands and Owners

The size, type, and quality of NIPF lands vary widely, and the owners are equally diverse. The NIPF category includes those properties not held by government or forest products manufacturing firms. As Figure 7.1 indicates, 90 percent of the NIPF owners hold less than 100 acres. These small

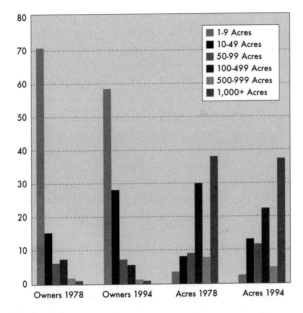

Figure 7.1.

Distribution of Private Forest Ownerships, by Size and Class of Ownership in the United States in 1978 and 1994.
Source: Birch, 1996

parcels account for 30 percent of all NIPF acreage. Just 3 percent of private owners hold about 29 percent of the private forest acreage in parcels greater than 1000 acres. This includes forest products companies and some large NIPFs.

The number of NIPF owners increased 27 percent from 1978 to 1994, according to a study by Thomas Birch of the U.S. Forest Service, Northeast Forest Experiment Station in Warren, Pennsylvania. More than 40 percent of NIPF owners acquired their property since 1978. However, during the same period the number of large tracts—those over 1000 acres—dropped, which indicates that private forestlands are becoming increasingly fragmented. This poses a challenge for sustainable forest management on NIPF lands. Since small forest tracts produce less timber, owners may be forced to log more timber in the short term to meet immediate financial needs.

In general, studies by Stephen Jones, director of the Alabama Cooperative Extension System at Auburn University, show that NIPF owners who have acquired land since 1978 are younger, better educated, and have a higher income than the average owner of 1978. Birch also found that the proportion of retired owners has increased. About 20 percent of owners are retired, which raises questions about the continuity of management philosophy and the need for estate planning. Smaller parcels cannot be managed as efficiently as larger ones. And the disposition of lands to several heirs or outright sale to pay estate taxes are major contributors to the fragmentation of forest lands and the parceling of larger tracts into smaller ones.

NIPF owners hold forestland for a variety of reasons. About 40 percent cite recreation or hunting as the primary reason for owning forest land, according to Birch (1996) and Jones et al. (1995). Ownership may be incidental to other uses if, for example, forest land is part of the farm. In suburban areas, forests are often conveyed with homes as part of subdivisions. Many, however, are forest owners by design. Nine percent of NIPF owners (10 percent of the NIPF acreage) purchased their land as an investment (Figure 7.2). The reason for ownership plays a determining role in an owner's forest management decisions. But landowners do not always behave consistently with their attitudes. Even though they may cite other objectives as more important, about 50 percent of owners, representing about 75 percent of the acreage, have harvested timber at some time during their land-owning tenure, according to 1994 U.S. Forest Service estimates.

Knowledgeable landowners and those who consult natural resource professionals tend to make decisions that are more consistent with principles of sustainable forestry. A 1993 study by Anthony F. Egan, assistant professor of Forestry at West Virginia University, and Jones found that landowners'

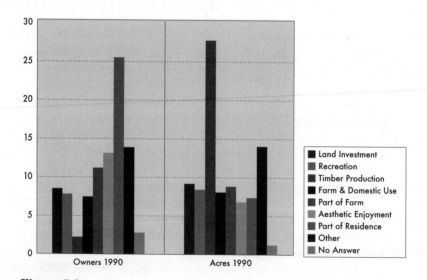

Figure 7.2.

NIPF Owners Reasons for Owning Forest Land.
Source: Birch, 1996

forestry decisions vary directly with their knowledge of forests and forestry: Informed landowners are more likely to make decisions that result in sustainable practices.

The Sustainable NIPF Owner

Although studies such as these characterize the "typical" NIPF landowner, each real landowner brings a unique set of demographics, motivations, understanding, and objectives to managing an individual property. The seven properties discussed here each encompass a unique set of circumstances. Collectively, they express a wide range of ownership characteristics and justify some general conclusions about the economic and ecological viability of SFM on these properties. The properties represent a wide spectrum of NIPF ownership (Figure 7.3). At one end the Brent tract, owned by a retired brother and sister, exemplifies low-intensity management on 171 acres for scenic values, wildlife habitat, and timber production within the context of maintaining continuous forest cover. The VanNatta family, in contrast, runs its 1728-acre tree farm to support four generations, while the 12,768

Property/Total Acreage	Ownership Type	Owners' Objectives
Brent Oregon, 171	family, brother, and sister	scenic values, wildlife habitat, timber
Freeman Pennsylvania, 639	family, husband, and wife with three sons	forest stewardship, community education, timber
Trappist Abbey Oregon, 1,350	community, 37 brothers	self-sufficiency, timber, aesthetic quality
Cary Florida, 2,634	individual	maintain ownership through land and timber sales
VanNatta Oregon, 1,728	family, four generations	sustaining family farm business
Lyons Pennsylvania, 2,000	family, two brothers	timber, investment, recreation
Frederick Alabama, 12,768	corporate trustee	asset growth, investment, estate planning contingencies

Figure 7.3.

Key Characteristics of the NIPF Properties.

Frederick property in Alabama is maintained as a trust for asset growth, investment, and to provide long-term income for the owner's family.

The Essential Role of Professional Advice

The experiences of these landowners indicate that professional advice is essential for sustainable forest management on small woodlots, even though, according to recent studies, fewer than 20 percent of NIPF owners tap professional advice. Most of the seven successful forest owners discussed here use formal professional advice to set and implement their forest management strategies. The source of advice varies but includes consulting foresters, forest stewardship advisors, and management committees. If owners like the VanNatta family do not seek formal advice, it is because they have the technical expertise themselves. For those without the expertise, professional advice is especially important in planning and managing for financial returns: A properly managed forest is more profitable, in the short and long terms, than a forest growing randomly and harvested by people concerned more with what is cut than what is left. Informed landowners make better decisions because they recognize alternatives that sacrifice future potential for short-term gain.

These seven NIPF ownerships also make it clear that sustainability will look different on small properties than it will on large tracts and that conditions on any area of land will change through time. If the areas is "large," the "average" condition over the entire area is likely to be similar through time; but for a small area, the "average" condition will be much more variable. For NIPF lands averaging a few dozen to a few hundred acres, many events can force changes to the land that will give the appearance of unsustainability. A storm, fire, or disease may make unplanned harvests necessary, as might the need to generate income. However, the long-term sustainability of the area can be preserved if the land is maintained in the same ownership and if the long-term motivations of the owners remain unchanged. The definition of sustainability must allow for the dynamic condition of small properties.

The Importance of Nonmarket Benefits

For small woodlot owners, benefits of forest ownership such as recreation, scenic beauty, the pleasure of working in the woods, the desire to maintain the land within the family can be just as important as any economic return. George Freeman, a retired Quaker State Oil Co. executive, and his wife, Joan, work in the woods almost every day. They both cite "sanity" as the most important output they get from managing the farm. And, according to

George Freeman, "Some of the best relationships come from working together in the woods." The Lyons family manages its 2,000 acres for income, but family members also hunt on the land.

NIPF owners typically experience long periods of low cash yield and are acutely conscious of opportunity costs and options foregone for other, more lucrative uses of their land. At any time they could cash in their land and timber for another investment. Yet the owners have refused opportunities to put the land to other uses that would generate greater income. The Lyons family, which originally purchased a 237-acre tract in 1933 for $500, turned down an unsolicited offer in 1996 of $500,000 for the parcel. The Freeman family refused an offer to locate a radio transmission tower on their property, and they refuse to strip-mine the land for coal.

By keeping their land in forest, NIPF owners provide ecological benefits as well as benefits for the surrounding communities. These NIPF owners have improved wildlife habitats and protected watershed and the aesthetics of the forest, which would otherwise have been lost. In the future, as development claims more forestland these benefits will become more important. The 1350-acre Trappist Abbey forest in Oregon, for instance, helps preserve the aesthetic values of Yamhill County by keeping its forests and fields in an attractive and productive state. But since the area has little nonfarm development, the aesthetic qualities probably add little monetary value to surrounding properties today. In the future, though, projected population growth throughout the Willamette Valley promises to increasingly fragment forest areas and convert them to other uses. As development spreads, the Abbey's productive forests, species diversity, and maturing tree stands will increasingly be an ecological benefit for the area. And as forest management intensifies in the region, rotations are likely to shorten and diversity diminish, which will also make the Abbey forest more unusual and ecologically valuable in time.

The seven landowners profiled here provide good evidence that despite widely varying circumstances, over the long-term private forest owners can manage their land for income, appreciation, and as a hedge against inflation. None of the properties profiled here have mortgages. All have been owned for at least 25 years, so the purchase price does not enter into the financial equation of annual costs versus income. In general, these landowners can be less concerned about the flow of income from their properties than someone who is making annual mortgage payments.

Four of the NIPF properties—the VanNatta tree farm, the Freeman farm, the Lyons property, and the Trappist Abbey forest—have each developed a sound management strategy for very different properties that has improved the health of the land while meeting the differing economic objectives of the owners.

The VanNatta Family Tree Farm:
A Living for Three Generations

The 1728-acre VanNatta tree farm lies in the center of what is sometimes termed temperate coniferous rainforest in the northern Oregon Coast Range region of Columbia County, about 40 miles northeast of Portland. It is one of the nation's most productive timber-producing regions. The VanNattas have used conservation-based timber management strategies for over 50 years. The family does not necessarily embrace "sustainable forestry," but family members do advocate progressive management techniques. In a landscape dominated by intensively managed Douglas-fir plantations, the VanNatta property is one of the few examples of uneven-age management of Douglas fir. The family's success demonstrates that sustainable forestry can provide an alternative forest management system to conventional clear-cutting.

The family has clear management objectives: Family members want forestry to produce enough income to support the family and allow them to perpetuate the tree farm and the family business indefinitely. In the process, the family wants to set an example of good sustainable family tree farm management. In 1998 four VanNatta generations lived on the property in three households, supported primarily by timber revenues.

A Combination of Progressive and Conventional Forest Management

About 1650 acres of the 1728-acre farm are suitable for commercial timber. The mild wet maritime climate of the region produces superb growing conditions for the Douglas fir, Western hemlock, and Western red cedar that dominate forests in the area and that give the VanNatta property an annual growth potential of over 1000 board feet per acre. George VanNatta, a retired lawyer, moved to the property in 1940 and started a cattle ranch. Before that the land was logged in the 1920s, then burned by wildfires, and was extensively grazed by sheep. In 1940 Douglas fir was beginning to seed in from a remnant stand on a distant ridge. Gradually, the forest began to dominate the landscape again, and in 1965 the VanNatta family began to shift the farm from cattle to forestry. Initially, the family replanted understocked areas and commercially thinned overstocked areas. Early thinning consisted primarily of cutting the largest trees, which were among the few saleable ones at the time. This type of thinning released the second tallest size class in the stand from competition, so they then grew rapidly into high-quality timber. Encouraged, the VanNattas continued to practice "high thinning," or removing the tallest, largest trees.

Forest management combines progressive and conventional management techniques. Most harvests in healthy young stands are conducted at 6-

to 12-year intervals and use selective high-thinning methods. At harvest, competing vine maple is mechanically cut and/or uprooted, which helps fir and hemlock to regenerate naturally. In stands thinned repeatedly and/or heavily, natural regeneration and isolated enrichment planting lead to uneven-aged stands. To combat the laminated root rot (*Phellinus wiereii*), a native disease common on rich coast range sites, clear-cuts are used to salvage infected and at-risk stands. These sites are then replanted with resistant species, such as cedar, white pine, and alder.

The family does not operate under a formal management plan or rigid harvest schedule. Harvest levels are maintained at 50 to 75 percent of annual growth, with a constant goal of increasing the vigor of tree stands and their growth rates. Annual harvest levels fluctuate based on current family financial needs, long-term management of the lands, market prices, and outside employment. Each year the harvest priorities are influenced by which market may be paying premium prices—domestic logs, export logs, pulp, or hemlock/alder, for instance. In the late 1980s and through much of the 1990s, rising log values aided sustainable practices by offering dependable markets for forest products. Markets for a wide variety of forest products, including log export facilities, are located at Longview, Washington, 25 miles north.

The VanNatta family members rarely seek outside professional expertise in logging, forestry, or legal matters because they believe they possess the expertise they need within the family. Two family members are lawyers, and two have completed Master Woodland Manager training from the Oregon State University Extension Service. K.C., the oldest of three sons, holds a degree in biology, lives on the farm, and manages the VanNatta Bros., the family timber management company that carries out the forestry activities on the land. The company has been certified under the Professional Logger program of the Associated Oregon Loggers (1995). Family members frequently give advice about forestry, logging, and legal issues to others.

Estate planning is informal but thorough. Five family members own The VanNatta Tree Farm in divided interests. They have structured land ownership and plan harvest operations to best meet the family's long- and short-term financial needs. In the late 1990s, for example, logging concentrated on George's (father) land, because he had greater financial needs and needed to keep his asset value down for estate tax purposes.

Growing Harvests and Rising Timber Values

The VanNatta family has sold timber every year since 1966, and nearly every year the harvest volumes and values have increased. A 1984 timber inventory estimated net timber volume of 27.5 mmbf with annual growth of 1.3 mmbf. Since 1984, 7.8 mmbf have been harvested, an average of 649 mbf/yr.—50 percent of growth. As of September 1996, the standing timber volume was nearly 36.2 mmbf. Timber value in 1940 was near zero. At a 1996 stumpage rate of $500/mbf, timber on the land was valued at about $18.1 million, with an annual growth value of $650,000–$850,000. Since 1966, sales have totaled 12,561 mbf. During this period, the sales price for Douglas fir leaped from $50/mbf in 1966 to $650/mbf in 1995. Sales in the period 1990–1995 averaged 730 mbf/year. Historically, 25 percent of the harvest volume consisted of higher-priced export quality wood. In the mid-1990s this percentage rose, as the VanNatta timber increased in size and quality and export specifications loosened, and bare land value has appreciated from $1/acre in 1940 to over $300/acre in 1996.

Higher Costs for Uneven-Age Management

The family takes full advantage of all timber- and business-related tax provisions and times and structures business activities to minimize the tax consequences. Since the property is paid for, property taxes are the primary fixed cost. They average $4.35 per acre for undeveloped forest land, or approximately $7,200 per year. The bulk of property taxes on timberland is paid as a severance tax of 3.8 percent on net stumpage proceeds at harvest. The severance tax, a one-time tax paid when timber is logged, benefits owners because it delays the bulk of tax payment until the time of harvest, when cash is available. It also helps to reduce loss in the case of a natural disaster. Timber volume and value growth rates exceed comparable alternative investments, so the family has little incentive to liquidate and reinvest. The annual volume growth is 3.5 to 4 percent, log grade appreciation 1 percent, and real stumpage value appreciation 1 to 2 percent.

In 1994, timber sales grossed $481,213, from 742 mbf of export and domestic sawlogs and 136 tons of pulp (12 mbf equivalent), as indicated in Figure 7.4. That year, payroll, logging operations, and depreciation costs were $191,137, an average of $235/mbf. Records of harvest acreage have not been kept, however, so accurate per acre cost estimates are unavailable. The Oregon Department of Revenue uses $190/mbf as a standard for regional cutting/yarding/hauling costs, which suggests that the VanNatta's uneven-age management adds at least 10 percent to the costs.

An Alternative to Clear-Cutting

The VanNatta farm is a prime example of a sustainable forestry enterprise that is maintained from normal family income. Conservative harvesting strategies have enabled the family to significantly increase timber volumes, and rising timber values have made the family's woodland assets far more valuable. Certainly, the history and physical condition of the VanNatta land contributed to the success of the family's management. The mixed-age

Payroll	$70,110	wages, taxes
Logging operations cost	$97,777	supplies, repairs, fuel, etc.
Depreciation	$23,250	
Stumpage paid to landowners	$71,220	all to VanNatta family
Payments to partners (VanNatta Bros.)	$72,000	most reinvested in equipment
Ordinary income (pretax profit)	$36,650	distributed to partners

Figure 7.4.
Costs for VanNatta Bros., 1994.

nature of the existing stand and the VanNatta's ability to use natural regeneration is based, to some extent, on the fire and grazing history of the property. Douglas fir regenerates well on mineral soil (as after fire) and when brush is controlled, through grazing, for instance.

It is unlikely that the VanNatta efforts would have been so successful without the careful control of both overstory stocking and competing vegetation. In addition, much of the VanNatta land is gently sloping, which makes it well suited to ground-based logging. But most of the industrial land in the Northwest is not, so wholesale application of the VanNatta's management methods would not necessarily be so successful everywhere in the region. Even so, the management techniques used by the VanNatta family do offer alternatives for conventional management in the Douglas fir region, which is synonymous with clear-cuts. Their experience shows that natural regeneration and selective thinning can work well on high-quality sites, if thinning intensity and competing vegetation are controlled.

The Freeman Farm: Stewardship in Action

The Freeman Farm is a fine example of stewardship forestry in hardwood forests. George Freeman and his wife, Joan, have been sole owners of the 639 acres in Clarion County, Pennsylvania, since 1971. The Freemans have a clear mission: They intend to sustain the land as forest and pass it on to their three sons by way of a well-crafted estate plan. The Freemans own the land because they derive great enjoyment from caring for the forest and developing a high-quality timber stand. The diversity of wildlife, a primary source of the Freemans' enjoyment, provides an added incentive to practice sustainable management. The Freemans' objectives of recreation and occasional and incidental income from timber are most common among NIPF owners.

But educating others about sustainable forestry is also an integral part of the Freemans' mission. Each year the couple hosts hundreds of visitors at workshops that cover topics ranging from timber taxation to forest stewardship. The goal, said George Freeman, is to help others "learn to preserve these resources for future generations." The farm has a 12-acre Stewardship Demonstration Area managed with Pennsylvania State University and the Pennsylvania Bureau of Forestry that shows how six different forest management techniques are applied on two-acre parcels. The farm is also part of the Pennsylvania Forest Stewardship Program sponsored by the U.S. Department of Agriculture to encourage SFM on NIPF lands and of the National Tree Farm system, a private program started in the 1940s to ensure future timber supplies. The couple has received several awards for its conservation efforts, and George Freeman belongs to a variety of forestry and environmental groups.

Sustainable Forestry in a Hardwood Forest

The owners realized early that the property had the potential to pay for itself. Timber production is a primary objective, but the Freemans have effectively managed timber while enhancing wildlife habitats and aesthetic values, protecting watersheds, and maintaining biodiversity. The management strategy is aimed at keeping the farm intact and forested. Timber harvests are sporadic. They are scheduled based on several factors—when commercial quality timber is ready to harvest, when improvements are needed to encourage the growth of commercial species, and when market conditions are favorable.

The land consists of highly productive stands of mixed hardwoods with a high proportion of red pine and Japanese larch. Some 497 acres are in commercial timberlands, which the Freemans have actively managed since 1970. The site quality is good over most of the property. The Freemans manage the forest to encourage the growth of high-value commercial species and to regenerate these species after harvest. They conducted their first thinning in 1973 to remove "mature and defective" stems and held five timber sales between 1981 and 1996. Clear-cutting was used in the past, but new selective cutting techniques such as crop tree harvests, which retain the better growing stock to mature and regenerate on the site, are now implemented to regenerate oaks and black cherry while maintaining the aesthetics of the property. In addition to native stands, the farm has several plantations of red pine and Japanese larch. These stands are pruned regularly to encourage the growth of more valuable clear wood. Competing vegetation, especially the grapevine growth, is controlled by hand sawing and the direct and judicious use of herbicides.

During harvest, the Freeman family protects water resources and trees that provide habitat for wildlife. Roads and trails are designed to minimize erosion and sedimentation in streams. Culverts are strategically placed to protect existing water courses. Harvesting is done with chain saws and skidders. Roads and landings are reseeded after harvest to minimize erosion and to benefit wildlife.

The Costs and Rewards of Stewardship Forestry

Between 1980 and 1995, expenses on the farm varied considerably. Variable costs include machinery maintenance, fuels, forester fees, vehicle and facilities maintenance, and insurance. The single largest fixed cost is payment on an equipment loan. For the period 1993–1995, annual payments averaged $3,203.61. Combined taxes on the property are the second largest fixed expense. For the period 1990–1995, payments averaged $2,703.07. Taxes on the property increased by $132.05 in this five-year period, which translates into a cost of $9.24 per acre per year (based on 1995 figures). The family operates the forest as a small business, which allows building and equipment to be depreciated.

Proceeds from the five timber sales between 1980 and 1995 totaled $111,857.84, or about $11.67 per acre per year of gross timber revenue. Roughly $1.31 per acre each year was paid to a consulting forester to administer the sales, which leaves $10.36 per acre per year to the Freemans. The estimates were calculated by averaging the totals over 15 years and do not account for expenses other than the consultant's fees. Timber sales have exceeded variable costs over the span of ownership, but they do not exceed fixed costs. This indicates that as a sustainable business venture, the farm and all its educational activities would not survive in its current form—but timber production alone would be self-sufficient.

The Deciding Factor for Forest Stewardship

Not all NIPF owners have the financial flexibility that the Freeman family does. The couple does not rely on timber as a sole source of income, and the couple is willing to accept a disparity between income and costs. Although timber income does not cover all of the costs of running the farm, it allows the owners to enjoy their retirement, practice sustainable forestry, and provide environmental education, which they consider worthwhile.

The Freeman family is committed to sustaining the forest. The couple has also invested a lot of its own time in forest management, which many

NIPF owners cannot do. The Freemans have also assured that their efforts can continue through careful estate planning. By viewing their land as a forest, as opposed to simply a financial asset, they have consistently made decisions that promote sound forestry and have chosen to invest their own capital in making the farm work. The farm is not threatened by urban sprawl or rising land values. If it were, the Freemans' commitment would be the deciding factor between sustainable forestry and something else.

The Lyons Family Tract: Income through Sound Forestry

Unlike the Freeman and VanNatta families who live and work on their land, the Lyons family members, which own 2000 acres of timberland in northwest Pennsylvania and southwest New York, are absentee owners who expect their land to primarily provide income and appreciation. As far as Paul Lyons, a businessman, is concerned, "you'll never find a better investment than a good piece of timber." Mr. Lyons acquired the tracts largely in the 1940s and 1950s as an investment. To ensure a smooth ownership transition and reduce the estate tax burden, he has transferred ownership to his four sons, all of whom have other unrelated careers and live in Florida. The sons have no firsthand knowledge of timber management, but they share their father's dedication to sound forestry. The sons are currently managing the property aggressively for maximum income from timber production using the best forestry principles, selling when the market is strong, and cutting with an eye to future timber harvests. The four sons intend to pass the land to their heirs and have prepared a formal estate plan, which involved setting up a trust to hold the land for Paul Lyons' six grandchildren. The Lyons family has been so successful at generating income from sound forestry that its members were featured in the *Forbes Magazine* Money Guide issue of June 17, 1996.

Management for Regular Income

The 2000 acres are owned in 11 tracts that are located in western Warren County and eastern Erie County, Pennsylvania, and in southwest Chautauqua County, New York. The properties contain largely northern hardwoods, mainly sugar and red maple, tulip poplar, black cherry, American beech, and ash. A stream running through the Blue Eye tract, listed as an "exceptional quality" stream by the Pennsylvania Fish and Boat Commission, has a significant red oak component. Site quality is very good over all of the properties. A high percentage of the sawtimber is of veneer quality, and nearly all of the acreage grows commercial-grade timber.

The family will harvest no areas (except salvage) unless it is silviculturally prudent and market conditions are favorable. They use SILVAH, a computerized stand inventory and analysis tool, that shows how various levels of harvesting will affect a stand. The program accounts for species composition, wildlife habitat quality, and the ability of the stand to regenerate to a new forest. Prior to cutting, a SILVAH stand analysis is taken, and the subsequent recommendation is used as a guideline in marking trees to be cut. While being aggressive, the family would rather err on the side of caution, cutting less and waiting for markets to improve.

Harvests are designed to minimize disturbance to the site and ensure appropriate reclamation. Locally rare species of trees have been left in certain areas. Wet areas and stream seeps are protected. Typically, rubber-tired skidders and chainsaws are used for harvesting. On some wet sites, they use horses to minimize the impacts. Temporary plank roads have been built with rough-cut wood to protect soil. Potential and certain den trees and snags are identified and retained for wildlife. The Lyons make sure that each stand is in a healthy and vigorous condition following treatment. When some of the stands near maturity, the owners plan to use proper regeneration techniques including herbicides to control undesirable competing vegetation and to protect the tree cover from an invasion of ferns and grasses.

A Strong Financial Return

Between 1983 and 1996, there were 10 silvicultural treatments on the property, nine of them commercial. The revenue has varied from a low of $12,000 on 35 mbf in 1984 to a high of $220,015 on 340 mbf. The only fixed cost of ownership is property taxes. In the period 1990–1995, property taxes averaged $5 per acre per year ($10,000/year total). Variable costs include site inspection visits, consultant fees, and miscellaneous expenses for management activities. Three of the four brothers travel to the property twice a year at a cost of approximately $1,200 per year total. During these trips, they conduct activities such as boundary line confirmation, painting, and grapevine removal. These expenses are estimated to reach $500 per year. Consultant fees for the same period totaled $79,812.00, or about $7.98 per acre per year for the period.

The property has appreciated dramatically. One 237-acre tract was purchased in 1933 for $500: In 1996 the family turned down an unsolicited offer of $500,000. The Blue Eye Tract was purchased at a tax sale for $5/acre. The Pennsylvania Game Commission (the adjoining owner on the east) will pay as much as $400/acre for cut-over land. Nor is it uncommon for individuals and clubs to pay more just to own a property for hunting in this part of Pennsylvania.

The Lyons' success demonstrates that small woodlot owners can earn a strong financial return from timber while practicing sound forest management. Many owners, however, might have accepted the offer for $500,000 for the land without regard for the future of the forest resource or generations that will depend on it. Although a strong financial return is the prime motivator for this landowner, the Lyons have an enviable intergenerational link and the satisfaction of seeing the positive results in the form of a strong financial return as well as recreational enjoyment.

The Trappist Abbey Forest: Effective Conflict Resolution

The Trappist Abbey of Our Lady of Guadalupe, a monastic community of men that owns 1350 acres of forestland about 30 miles southwest of Portland, Oregon, manages its forest for income, too, but the group has distinctly different objectives than other landowners. The forest is an essential part of the monastic community that provides the physical and spiritual setting for the community, acts as a buffer to neighbors, and allows for cloistered retreat. Forestry is also one of the four cottage industries that members run, which enable the Abbey to be self-supporting. To date, each of the other industries (book bindery, fruitcake bakery, and wine storage) has provided more income than forestry. In the future, though, the Abbey will depend on the forestry program to provide a rising share of its income, as members age and income from the other industries declines. In 1996, 37 members with an average age of 67 lived and worked at the Abbey.

The 30-year effort of the Abbey has produced successful sustainable forestry. But the experiences of the Abbey illustrate the difficulties involved in developing successful SFM from scratch on NIPF lands. A lack of knowledge, the lack of a thorough management plan, and, until the mid-1990s, dissension over the forestry program hampered the Abbey's efforts to achieve an integrated sustainable management program for the entire ownership.

Forestry on Abbey Lands

The Abbot of Trappist Abbey, Inc. (ATA) is a 501-C3 not-for-profit corporation. The enterprises of the Abbey are organized under the Trappist Monks of Guadalupe, Inc. (TMG), a 501-d corporation. Similar to a partnership, each monk, as a member of TMG, receives shares of income and pays individual federal and state taxes. TMG receives timber harvest income and pays expenses, including property tax. Conscientious stewardship is a tenet of the Abbey community in all of its endeavors. In the forestry program, maintaining the spiritual and aesthetic values of the forest is a top priority. But the program is also designed to increase understanding of forest management within the community and to pass the program on to future members. Annual timber harvests are planned to generate income and to minimize conflicts with other uses of the forest.

The Abbey's 760 acres of commercial timber consist of maturing fir stands, mixed oak-fir stands, 330 acres in 10- to 26-year old plantations, and 50 acres in plantations less than 10 years old. Rainfall on the site averages just 40 inches per year. The site quality on average is fair to moderately good, with a 50-year site index of 100 to 125, which means that on average trees grow 100 feet every 50 years. In 1996 timber volumes exceeded 7.8 million board feet, with annual growth of 500,000 board feet. At a 1996 stumpage rate of $550/mbf, timber values were near $4.3 million, with annual growth of $275,000.

Most of the property was logged just before the Abbey purchased it in 1953. In the mid-1960s, the Abbey started active forest management, and a forestry crew and chief were appointed. Initially, the brothers concentrated on replanting cut-over stands and surplus agricultural lands. From 1969 to 1981 all suitable timberland was planted, totaling 320 acres. The first commercial harvests began in the late 1960s. In the 1960s, several plantations failed on droughty sites because there was a relative lack of knowledge of how to reforest harsh sites, and few private woodlands in the area were under any type of stewardship management. The Abbey forestry committee learned by trial and error, refining its methods while gaining more experience. In later plantings other species, such as ponderosa and radiata pine, Leland cypress, and poplar, which were better suited to the adverse sites, were used in addition to Douglas fir.

Conflict Over Forest Use

Early management needs were easy to recognize and, with the exception of droughty sites, relatively easy to implement. Only in the late 1980s and early 1990s, however, were all parts of the property brought under management. As harvests intensified in the mid-1980s, small clear-cuts (5 to 6 acres) became more common, and resistance among the brothers to the visual and aesthetic impact of logging mounted. In 1989 when a site near the edge of a common picnic area was clear-cut, the brothers became openly divided over the forestry program. Some insisted that the cutting stop; others wanted the program reconsidered. From 1989 until 1994, small clear-cuts continued, but they were out of sight of the residence. In 1994 a new abbot, Father Peter, took over. He created the forestry committee, and the community hired an outside forest manager to draw up a long-term plan for the Abbey's woodland and to oversee future harvests.

In 1995 the Abbey contracted with consulting forester Scott Ferguson (ITS Management) to help develop a forest policy, prepare a detailed long-term management plan, and assist in future timber sales. The Yamhill

County extension forester also became a member of the Abbey forestry advisory committee. The next year the Abbey adopted specific forest policies for SFM and integrated them into a long-term management plan to minimize conflicts between the community's financial needs and its needs for the forest's other amenities. Specific policies addressed ways to maintain diversity in tree species and ages; preserve unique trees and areas near streams; implement suitable harvest methods, such as thinning; limit the size and use of clear-cuts; and maintain the productivity of the soil. Harvests in 1996 used a combination of individual tree selection, high thinning, and regeneration cuts using group selection. Scarification was used to mechanically expose mineral soil, which helps the forest regenerate naturally. The advisory committee oversees forestry activities to ensure that it conforms to specific policies and to resolve any conflicts.

The Forestry Balance Sheet

Annual sales on the property began in 1987. Between 1987 and 1997, harvests ranged from a low of 87 mbf to a high of 144 mbf. Gross income averaged $49,711 (gross). In recent years, per acre harvest proceeds (mill delivered values) were $2,306 (1994) and $1,419 (1995—includes significant hardwood). Between 1987 and 1997, operating expenses averaged $30,204 annually.

The 1995 harvest is representative of the revenues and costs of the operation. That year the Abbey harvested 102 mbf. Based on a harvest of 30 acres (26 acres thinned, 4 acres clear-cut), operating expenses totaled $29,721 or about $1,142 per acre harvested. However, the costs include timber stand improvement and maintenance expenses for the entire property. With low annual harvest levels, expenses per harvested acre are high—$292/mbf in 1995. Income from rent for the agricultural land produces an average annual income of $6,000.

The property taxes on the 885 forested acres totaled $1,800 in 1995, or $2.03 per acre. All forestland is enrolled under the Western Oregon Small Tract Optional Tax (WOSTOT), a program for NIPF owners that has a low annual property tax based on forest soil productivity for lands dedicated to timber production and no severance tax. Taxes are paid by TMG, which also pays a modest rent of $750 per year ($0.85 per acre) to ATA for use and operation of the forestlands.

The value of the Abbey land has grown considerably. In 1953 bare land was valued at $125 per acre; in 1997 it was at least $500 (as zoned for forestry, actual market value probably higher). Timber value on the tract in 1953 was zero. By 1997 stumpage values exceeded $5,500 per acre (10 mbf/acre × $550/mbf).

A Sustainable Future for the Abbey Forest

The Abbey's tenet of stewardship and its structure were conducive to the development of successful sustainable forestry. The members of Trappist Abbey live within a defined structure, share many basic beliefs and philosophies, and have made a long-term commitment to living and working together, all factors that helped the Abbey overcome its early lack of forestry expertise and later conflicts over forest use and to adopt SFM. But 37 individuals still must agree on a forest harvest plan each year. The brothers vote the plan either up or down at an annual meeting. The forestry committee's oversight and a concerted educational effort have made this process more effective in recent years. But having a professionally prepared sustainable forestry management plan was also instrumental in reassuring the Abbey community that its resource management was balanced and defensible, and it enables the brothers to make more informed judgments about the future of the forestry program. Conflicts are inevitable in forestry settings where several individuals or groups have different interests in a piece of forestland. The Abbey through its forestry committee was able to resolve those disagreements. Policies based on sustainable forestry were an instrumental part of that process and the solution.

Long term, the prospects are favorable for the Abbey forestry program. Both timber values and sustainable harvest levels will increase significantly in the future. The Abbot is excited about third-party certification for the Abbey's sustainable forestry practices as a way to validate the goals and achievement of the forestry program. In 1998 the Abbey foresters were working to certify their forest through their consulting forester, who had been recognized by SmartWood as a "certified resource manager." Ultimately, though, the success of the Abbey forestry program will be determined by its ability to provide a greater proportion of the community's financial needs as the brothers get older. Meeting those needs could eventually force changes in the forestry program.

The Highly Variable Intensity of NIPF Management

The VanNatta family farm, the Freeman tree farm, the Abbey forest, and the Lyons property indicate just how highly variable sustainable forestry can be among NIPF owners, depending on the type of forest, location, and the owners' circumstances and objectives. The properties profiled here include lands tended daily by a resident owner, such as the Freeman farm, and those that are visited and manipulated several times a year. The intensity of management relates to a variety of characteristics. For absentee owners of relatively large tracts, for whom financial profit is the major motivation, such as the Lyons, land management is often confined to assuring regeneration, performing standard silvicultural treatments, and harvesting. For owners of relatively small tracts, though, land stewardship is more hobby than business. For owners such as Matthew Brent and Virginia Picht, who hold the 171-acre Brent Tract in Oregon's Willamette Valley, the satisfaction they have gotten comes largely from their personal relationship with the land.

The Brent Tract: A Perpetual Forest of Douglas Fir

Brent and Picht, a brother and sister, grew up on the farm in the Willamette Valley and inherited their 171 acres from their parents. Picht, age 76, who lives in Corvallis, manages the property. Brent, age 78, lives on the farm. Picht's daughter, who lives in Germany, hopes to live on the farm one day. They value the forest primarily for its scenic beauty and its ecological and recreational services. Their priority is to maintain a perpetual, healthy forest cover and to keep the property in the family. Both feel a responsibility to be good stewards of the land; they take a long-term view of management, even though they realize that neither may live to see the final results of their actions.

Regular Thinning for High-Quality Growth

The Brent property lies in the western Oregon Coast Range where forests are typically dominated by Douglas fir, with Grand fir (*Abies grandis*), bigleaf maple (*Acer macrophyllum*), and Oregon white oak (*Quercus garryanna*) as common associates. Growing conditions are good for the 40- to 80-year old high-quality Douglas fir that cover most of the 111 acres of commercial timber land. The average site index for the property is 125 (50-year basis), which means the land has the potential for timber growth of nearly 1000 board feet per acre annually.

To maintain forest cover, the owners use selective thinning and natural regeneration practices. They avoid clear-cutting and other industrial-style techniques practiced on neighboring properties. Regular thinning has kept the stands relatively open by conventional standards, but it has also encouraged very high-quality growth. Thinning has also allowed the firs and hardwoods to regenerate naturally and develop significant vertical stand structure.

Prior to the 1960s, forest management of the Brent Tract was limited to isolated small sales for farm use or for income when the largest trees were cut. The property was commercially thinned in 1964/1965, then again in 1980, 1985, and 1987, each time under the supervision of consulting foresters. These thinnings used high-thinning practices with limited use of shelterwood and overstory-removal harvests to promote natural regeneration. High thinning, which removes a small number of the dominant trees on the basis of how saleable they are and how much space their removal would give to neighboring subdominant trees, tends to produce more consistent income and sustainable harvests. The property has been managed sustainably since 1965, but only recently have the owners understood effective uneven-age management techniques. Previously, low log prices and lack of knowledge and experience with alternative management regimes kept the owners focused on commercial thinning. Rising timber values, however, gave the owners a strong incentive to pioneer high-thinning practices.

Ms. Picht and Mr. Brent want periodic income from timber, but no family member relies on the property as a primary income source. The owners have no formal written management plan or timber inventory. As long as the forest stands are healthy and attractive, timber volumes and values increase, and harvests occur when income is needed, the owners are satisfied. Regular meetings with the forester keep the landowners apprised of actions needed and the progress taken toward their goals.

Sustainable forestry has created a healthy forest on the Brent tract. Generous riparian set-asides provide wildlife habitat, add to diversity within the landscape, and protect the water quality. Stand management practices enhance the diversity of the tree stands and the landscape. Hardwoods and minor conifers are encouraged, although fir is clearly the dominant species. The owners tolerate brush and shrub growth. The trees are grown to large sizes (larger than 24 inches in diameter) and old ages (more than 80 years old). Selective thinning has produced a forest that in some places resembles the uneven-age conditions characteristic of a natural forest. One stand on the Brent Tract was among four sites chosen for a 1995 study of selective thinning and uneven-age management for Douglas fir. The preliminary results confirmed that 30 years of selective thinning and natural regeneration are creating uneven-age stand structures.

The Benefits of Conservative Management

The management used at the Brent Tract since 1985 has generated more than $47,000 of net stumpage income. The total property value in 1996 was estimated to be more than $1.5 million (standing timber $1,085,000; bare land $171,000; location value $250,000). The total annual growth of 95 mbf was valued at $66,500 at 1996 stumpage values ($700/mbf). With the increases in stumpage prices, a conventional regime of heavy cutting and conversion of "underproductive" stands would not have performed nearly as well as the more "conservative" regime chosen.

Annual property taxes on the woodland portion, which were between $600 and $700 through the 1990s, are the main fixed cost. The property is enrolled in Oregon's "timberland deferral" classification, which allows a lower annual property tax, with a severance tax of 3.8 percent on the net stumpage proceeds at harvest. Timber sale administration represents the sole variable cost of management. Forester fees were initially 10 percent of the gross stumpage proceeds. Since log values have risen, however, those fees have dropped—they were 7 percent in 1996.

Low log prices through the 1980s probably deterred harvesting on the Brent land. The resulting conservative management has been beneficial. Thinning stands that might otherwise have been candidates for clear-cutting has allowed both regular harvests (and income) and a significant increase in the volume of standing timber. The growth in timber and the rising prices for stumpage have greatly increased the value of the property. At the same time, the property remains an attractive place to live and enjoy, while generous stream area set-asides promote wildlife diversity. Careful management of the timber and estate planning could enable Picht's daughter to live on the land one day. But the family does not yet have an estate plan. Nor do the owners fully understand the monetary value of their forest asset or the significant role that estate planning plays in the ability to maintain and perpetuate that asset.

Tax Laws—A Major Impediment to Sustainable Forestry

The seven landowners profiled here unanimously agree that tax laws—property taxes, capital gains taxes, and estate taxes—are probably the greatest impediment to sound long-term forest management on NIPF lands. "With the existing tax laws, there's not an awful lot of financial incentive to take out defective trees and leave the best," said K.C. VanNatta. Tax laws put such a burden on the NIPF owner that, said VanNatta, " you manage your land for taxes, not timber."

The tax laws encourage owners to cut timber aggressively in several ways. Although many NIPF owners are not making mortgage payments, they all must pay property taxes annually, even though they generate income from timber only occasionally. For most NIPF owners, annual property taxes are the largest fixed cost and the fundamental impediment to sustainable management in the short term. When cash flow is restricted for a landowner, standing timber may be the most liquid asset, so harvesting occurs at levels that are suboptimal in the long term. Current-use tax plans tailored to the income potential of sustainably managed forests are in place in some states and locales, but forest-friendly taxes are still the exception rather than the rule. Capital gains taxes are a similar disincentive for sound long-term forestry. The appreciation of timber over time represents a major capital gain for a forest landowner, but capital gains taxes hit timber owners harder than do capital gains taxes on investments in stocks, for instance. Typically, money is made from stock investments in just a few years, but the gain in timber values comes over decades. Over that long period, much of the gain is actually inflation, so, in real dollars, the gains are much lower.

The Importance of Proper Estate Planning

Proper estate planning is an absolute necessity for sustainability. Estate taxes have the greatest potential to wreak havoc for long-term sustainability on NIPF lands. In general, sustainability depends on continuity of ownership and management strategy. Often failures in sustainability can be attributed to failings in estate planning. An owner, for instance, who anticipates an estate tax in the near term has no incentive to keep timber standing: In fact, there is an incentive to keep the inventory low, so the tax is not severe. Commonly, when an owner dies, properties are logged, or sold to pay inheritance taxes, or divided into smaller parcels without regard to the long-term impact on the forest. Careful estate planning can prevent such problems, but the legal and land management activities may require decades to implement. And most landowners need competent legal assistance to assure smooth intergenerational transfers.

The Cary Property: The Impact of Estate Taxes

The Cary property represents a classic example of how inadequate estate planning can disrupt forest management and income production. Mrs. H.T. Cary, the wife of a career military officer, inherited the 2364 acres of pine forest near Pensacola, Florida, from her parents in 1947. When Mrs. Cary inherited the land, the timber volume was light and the stocking poor. For nearly 50 years Mrs. Cary managed the 13 separate parcels to produce regular income and to build a productive forest.

Cutting plans were based on forest condition, silvicultural needs, and the owner's changing cash flow requirements. Sometimes, timber was cut each year. At other times, several years of logging significant volumes of timber were followed by periods of little or no cutting.

The land did well for Mrs. Cary. During those 50 years stumpage prices and land values for the property increased enough so that Mrs. Cary has had a compound annual rate of price appreciation, unadjusted for inflation, of about 5 to 8 percent. In addition, the timber grew and income was produced. Excluding mineral income, nontimber income—from hunting, grazing, and so on—has amounted to about 5 percent of the income derived from the land. Over this period, overall costs increased from less than $1 per acre to $5 or more per acre.

When Mrs. Cary died in 1992, the average volume of timber was near optimum—a level that would have permitted continued periodic cuts for income. But her heir, daughter Ann Veldy, decided, after considering various options, to use timber to pay estate taxes. After most of the sawtimber was cut to raise the estate taxes, the volume of timber on the property dropped to near the 1947 level; Veldy must rebuild timber stocks. In the meantime, she has built a house on the main tract and plans to sell outlying tracts of land to produce income.

Veldy is managing the land to produce periodic income, build the timber volume, and improve timber quality by emphasizing the management of longleaf pine in natural stands and working for natural regeneration. After the heavy cutting to pay estate taxes, little or no income from timber sales can be expected until about 2006. Eventually the young stands will need thinning, and in other stands the residual seed trees or shelterwood overstory trees may have to be removed. Primary products will be longleaf pine sawtimber and poles, with pulpwood as a by-product only if market conditions are favorable. Fortunately, during the period of regrowth the owner will be able to carry out many management functions herself, which should minimize costs.

Veldy was fortunate that she was able to use timber to pay taxes. If the timber volume had not been high when her mother died, Ann Veldy might have been forced to sell most or all of her land, and the land might then have been converted out of forest use. However, as the Cary property indicates, land and timber investment, long term, can provide appreciation, and a good volume of standing timber can afford landowner liquidity.

The Frederick Property: An Investment Corporation

The 12,768-acre Frederick property indicates that careful estate planning can help maintain the productivity of private forestland to provide income and appreciation while preserving environmental values. The owner originally purchased 14,597 acres in southern Alabama in 1967 with proceeds from the sale of a family sawmill to provide long-term financial security for himself and his heirs. He had spent his career from the 1920s to the 1960s managing the woodlands and forestry operations for a large family-owned lumber company and had faith in the inherent value of land and timber as an investment and income producer.

When the owner purchased the land, timber volumes were near optimum. The new owner wished to leave this, his primary asset, to provide for his family. So he began a planned program of cutting most of the merchantable timber. He hoped this would allow the property to pass through his estate at more favorable cut-over land values. But only a small part of the property had been cut when the owner died at age 69 in 1973. The property then became part of a trust, which is managed by a bank as the corporate trustee. The bank is charged with managing the land to provide income for his widow and two daughters as primary beneficiaries, and for grandchildren and other heirs as residual beneficiaries. This trust will last until the death of the owner's wife and daughters, which could be 60 years or more. Although not an absolute restriction, the trust directs the trustee to follow sound forestry practices, so that the timber volume will be increased and land and timber retained as a long-term investment.

The corporate executor applied and received a special-use 10-year payout of estate taxes as provided by IRS tax regulations. For the next 10 years cutting was designed primarily to pay estate taxes. The land has since been reduced to 12,768 acres. Some tracts were sold to provide the trust with some diversity, block ownership, and, in recent years, to build liquidity for estate taxes due on death of the owner's widow. The widow has also given some lands as gifts to daughters. All timber sale proceeds by the terms of the owner's will are allocated to income for the family. When timberland is sold, 80 percent of the value is distributed as income and the balance retained as principal for the trust.

The trustee's overall objective is to meet the beneficiaries' needs for current income, while bearing in mind the rights of and obligations to future beneficiaries. The majority of the property, 20 separate parcels that range from 40 acres to about 7000 acres, is classified as a loblolly-shortleaf pine forest and has a site index from 75 to 90. The bank is managing the land to build the stocks of pine sawtimber to a target level of about 4000 feet (Doyle log scale) per acre of pine sawtimber, so that eventually the property will produce higher value products of pine poles and pine and hardwood sawtimber. Until the stocks are rebuilt, the trustee will cut less than the annual growth. Even-aged and multiaged stands are managed with prescribed fire and selective cutting. Trees are harvested when they reach an optimum size, usually by age 60. Both natural regeneration and planting are used to establish new stands, with emphasis on the former. If conditions are too severe for good natural regeneration, site preparation and planting may be prescribed.

The property has provided satisfactory income and appreciation for the owners' heirs. Timber sales have been conducted every year since 1967, although the size has varied considerably. All sales are designed to meet income needs and to maintain the land in as well stocked a condition as possible. Current removals, forest condition, and estimated volume and growth are continuously tracked to measure progress toward the long-term goal of cutting annual growth once the target volume is reached.

In spite of the sale of over 2000 acres (14 percent of the original purchase), there has been a 700 percent gain in total asset value over the 30 years since 1967. Some of this is attributable to a 22 percent increase in timber volume, but most is due to land and timber appreciation. Over the 30 years, pine saw-timber prices have increased from about $50/mbf Doyle to over $400, pine pulpwood from $6.50 to $34 per cord, and hardwood pulpwood from $2.50 to $20 per cord. The average value of land in the area of the property also changed from about $50 per acre to $400 per acre.

Net income has averaged less than 3 percent of current asset value, excluding income from land sales. Annual timber sales produce over 98 percent of the income; since 1967 sales have been primarily to one buyer. When acquired, the property was subject to a forest products company's right of "first offer" on the sale of any land or timber. The restriction expired in 1997. Subsequent timber sales should produce greater income through the use of competitive sealed bids. Historically, the costs of timber management (sales preparation and inspecting, monitoring, and tracking) have been between 4 and 8 percent of sale proceeds. The bank trustee is also entitled to a management fee, in this case calculated as a percentage of asset value and a percentage of timber income, similar to that of an investment manager.

The Frederick property was able to counteract the negative effect of estate taxes on the land, but not all landowners are so fortunate. NIPF owners may consider tax laws a disincentive for sustainable forestry, but inversely, they also view effective tax policies as potentially the greatest incentive for sustainable practices on small woodlots. Reduction of estate taxes on family-owned businesses would allow family tree farmers to make management decisions based on maintaining a healthy forest, rather than on tax management needs, according to K.C. VanNatta. The NIPF owners here also support further reductions in the capital gains tax and allowances for writing off reforestation and forest maintenance as expenses. Doing so, they insist, would encourage additional investments in sustainable practices. George Freeman would like to see incentive programs implemented that would allow reductions in property taxes in exchange for commitments to sustainable forestry. The development of forest-friendly capital tax programs would also help promote sustainable forestry on nonindustrial parcels.

New Models of Certification for NIPF Lands

The disincentives for sustainable forestry on small woodlots caused by the tax laws, as well as the varying circumstances and objectives of small landowners, make the practice of sustainable forestry difficult. Even though the certification of forest products has increasingly become a tool with the potential to provide forest owners with benefits for sound forest management practices, NIPF owners face a number of barriers to certified sustainable forestry. Cost, which is traditionally skewed to favor large land holdings, is foremost. As the size of acreage increases, certification becomes more and more cost-effective: In parcels over several hundred thousand acres, it may cost just a few cents per acre. But the owner of a very small parcel may pay up to $5,000 for a single assessment, according to the Forest Stewardship Council (FSC).

The requirements for certification can be equally imposing, especially for woodlot owners with fewer than 50 acres. In 1996 the FSC, the international accrediting body of forest management certifiers, required the following documentation from a forest operation: (1) environmental impact assessment; (2) guidelines for erosion control, reduced stand damage, and water resource protection; (3) a comprehensive management plan; and (4) research on yields, growth rates, regeneration, environmental and social impacts, and cost and productivity of the operation. Such requirements are a burden for small landowners, particularly when only a fraction has any management plan at all. Even if the owner of a small parcel can afford the cost and can meet the requirements of certification, still other barriers hinder the effective marketing of certified forest products. Irregular and uneven flow of species and volume and lack of access to "certified chain-of-custody" processors, manufacturers, and distributors are likely to prevent the small landowner from realizing any significant financial benefits from certification.

In the mid-1990s, certifying groups started to adapt certification to the needs of small woodlot owners. Traditional certification involves a single forest operation. New initiatives for small landowners involve creating an umbrella structure for several forest operations and landowners. This type of structure relies on a single forest manager to manage multiple small forest holdings. The forest manager may be associated with a forest owner's cooperative, a land trust, or a chain-of-custody certified buyer or processor, according to 1996 FSC guidelines for such a structure. Under an umbrella structure, each forest owner may have a separate management plan, or the entire umbrella organization can have a single, shared management plan. In either case, certification assessments are conducted by a random sampling method, rather than on every individual parcel. Traditionally, the certifier has had a contract directly with the forest owner. In the new model, the certifier has a contract with the resource manager, who, in turn, may have numerous contracts with landowner clients. However, in both cases, certification still relies on a commitment by landowners and managers to long-term management, not just a single harvest.

Umbrella certification offers advantages to small woodlot owners. Many of the written requirements of certifying bodies and the FSC, such as management plans and environmental impact assessments, could be spread out among all participants in the program. Costs of the certification assessment, program fees, and audits could be similarly shared, drastically reducing the cost to any one landowner. In addition, a cooperative approach to marketing is likely to ensure an even flow of volume and species of certified forest products and thus will have a better chance of getting the product into a certified chain of custody, which potentially would allow the small owners to command a market premium.

Innovations in NIPF Certification

Three different ways in which umbrella structures have been created to make certification more efficient and affordable for small timber owners are explored below: (1) certification of forest managers, who warrant that lands they manage are sustainable; (2) chain-of-custody certification by manufacturers, who warrant that the wood they use is from sustainable forestland; and (3) group certification of multiple tracts, which are managed cooperatively.

Certification Through a Resource Manager

To New Hampshire forester Chip Chapman, the chance to bring a group of small woodlot owners—those who own fewer than 500 acres—together for certification looked like a business opportunity. In 1994 he founded the Northeast Ecologically Sustainable Timber Company (NEST) to develop a system based on a single certified forest resource manager who provides certification services for small, private forest landowners. He signed up two dozen landowners to cooperate on a joint sustainable management plan.

In January 1997, NEST won Rainforest Alliance certification as the first New England sustainable forest resource management company. Timber harvested from lots under its supervision can display Rainforest Alliance's "SmartWood" seal. The 3500 acres that Chapman supervised covered 30 woodlots, ranging from 60 to 650 acres. Certification costs were down to $4 to $5 an acre for the largest holdings and $10 to $15 an acre for the smallest. The Alliance had also certified resource managers in California, Florida, Maine, Michigan, New York, and Wisconsin. In the Michigan program, which covered 10,000 acres, certification costs were expected to be less than $1 an acre.

Chapman considers it his task to help clients overcome some 300 years of "reverse natural selection" as owners logged off the healthiest and most valuable trees in the mixed pine and hardwood forests of New England.

"The really good stuff was creamed right off," he said. "Our job now is to thin it out and give the good stuff a change to grow so the forest can rehabilitate itself." Eventually, Chapman hopes that the arrangement will enable him to pay a premium of 10 to 20 percent directly to the landowners. He also expects certification under NEST to help small landowners link up more easily with certified chain-of-custody dealers and manufacturers.

A Manufacturer Willing to Share Certification Costs

The biggest challenge for certified manufacturers, according to Eric Bloomquist, president of Colonial Craft, a St. Paul, Minnesota, maker of milled products, is "finding an adequate and reliable supply of certified forest products." Bloomquist has issued an offer to small woodlot owners: He would share the direct costs of certification in return for access to the supply of certified wood.

Manufacturers face a twofold problem in establishing a certified line of products: (1) finding forest owners interested in participating in certification and (2) finding traders interested in purchasing certified products. In 1996, a barbecue grill manufacturer approached Colonial Craft with an offer to purchase wooden grill handles. Its customer, a major British retail chain, is a member of the 1995+ Buyers Group and is committed to trading in FSC-certified products. The deal had the potential for six-figure sales and expanded products. The buyer expected to pay a premium for certified wood. Bloomquist hoped that his offer would help Colonial Craft expand its line of certified products by persuading landowners to switch to certification, which would increase the number of certified landowners and the size of the certified land base. However, by 1997 Bloomquist still had few takers.

Cooperative Certification

Mill owners who are interested in processing certified products face the same supply problems as manufacturers. Jim and Margaret Drescher own the 140-acre Windhorse Farm in Nova Scotia, Canada, a property that has been managed by the principles of ecoforestry for 150 years. The farm's operations—a board sawmill, dry kiln, planer, and custom molding shop—produce about 150,000 board feet of lumber annually that the Dreschers sell locally. Drescher also manages about 2000 acres of timberland, owned by seven landowners, according to ecoforestry principles.

In 1997 Drescher began to explore certification, to improve forest management, and to validate the farm's ecoforestry. That year a Belgian buyer approached Drescher with an offer to buy a product that represented slightly more than the total annual yield from Windhorse Farm, as long as the products had an FSC-accredited certification label. Drescher decided to pursue certification as a SmartWood resource manager, so that he and neighboring landowners could apply for joint certification to reduce costs and to increase the supply of certified products. He was convinced that the arrangement could leverage their position in the certified marketplace. In the meantime, Drescher pays a 10 percent premium for logs from several well-managed woodlots and receives a 10 percent premium on "ecoforestry wood." With additional certified wood, Drescher hopes to double lumber production over the next three years.

Future of Certification Schemes for Small Landowners

Alternative certification schemes for small landowners are still untested and raise a number of issues—which potential models for small landowner certification make the most sense; whether certification will be economically feasible on NIPF lands; and how changes in ownership, governance, chain-of-custody certification, and annual monitoring are to be handled. Several existing programs for small landowners might serve as appropriate umbrella certification groups. The Tree Farm Program, a national program that requires landowners to have a management plan and comply with some harvesting restrictions or cooperative management programs, or the Connecticut/Vermont Project, in which landowners share management resources for shared management objectives, might be adapted to NIPF certification. State and local land trusts that hold forest conservation easements are still other candidates for umbrella organizations.

Creating an umbrella structure for certification poses challenges for certifying bodies. Finding resource managers with the skills to fulfill the certification requirements and to perform all of the administrative chores may prove difficult. Costs are likely to remain high, even with the economies of scale afforded by multiple landowners. Certifiers will need to find ways to economize on assessment and monitoring costs while still providing a rigorous and reliable service.

It is premature to predict whether certification of small forest lands will be a boom or a bust. More likely it will fall somewhere in between. If there is an adequate organizational infrastructure, the benefits of certification are likely to be comparable for any size forest ownership—receiving a premium for certified wood, better access to markets, enhanced competitiveness, improved quality of product over time, and improved overall efficiency. Similarly, as the economies of scale become more equitable for small landowners, the direct cost of certification should become relatively equivalent for most land holdings.

Knitting the Cleavage

The seven properties profiled here suggest the complexity surrounding the motivations of NIPF landowners for sustainable forestry and the conditions that foster or hamper sound, long-term forest management on small properties. Perhaps the overriding conclusion that one can draw from these profiles is that the interests of NIPF owners cannot be captured in the explicit arithmetic of business or in the implicit heuristic of "love of the land." The objective function for NIPF management incorporates the apples of hard financial logic, both short and long term, and the oranges of aesthetics, legacy, and land health. It is, in essence, the cleavage in attitudes toward the land that Aldo Leopold made famous in *A Sand County Almanac*:

> In each field one group (A) regards the land as soil, and its function as commodity-production; another group (B) regards the land as a biota, and its function as something broader. . . . In my own field, forestry, group A is quite content to grow trees like cabbages, with cellulose as the basic forest commodity. It feels no inhibition against violence; its ideology is agronomic. Group B, on the other hand, sees forestry as fundamentally different from agronomy because it employs natural species, and manages a natural environment rather than creating an artificial one. (p. 221)

Under certain circumstances, NIPF owners are able to knit that cleavage, which is the essence of sustainability.

NOTES

This chapter is adapted from *Non-Industrial Private Forest Landowners: Building the Business Case for Sustainable Forestry,* edited by Michael P. Washburn, Stephen B. Jones, and Larry A. Neilsen for the Sustainable Forestry Working Group, 1997.

The cases were prepared as follows:

The VanNatta, the Trappist Abbey, and the Brent cases were prepared by Mark Miller, Two Trees Forestry, Cooper Mills, Maine.

The Freeman case was prepared by Michael Washburn, Penn State University, School of Forest Resources, University Park, Pennsylvania.

The Cary and the Frederick cases were prepared by Keville Larson, Larson and McGowin, Inc., Montgomery, Alabama.

The section on new approaches to certification for small woodlot owners was prepared by Ervin Jamison, Forest Stewardship Council, U.S., Richmond, Vermont.

REFERENCES

Birch, Thomas W. (1996). "Private forest landowners of the United States, 1994." Proceedings of the Symposium on Non-Industrial Private Forests, Washington, D.C., Jan. 18–20, 1996, 10–18.

Egan, A.F., and S.B. Jones (1993). "Do landowner practices reflect beliefs? Implications of an extension-research partnership." *Journal of Forestry* vol. 91, no. 10, 39–45.

Ervin, J. (1996). "Forest Certification and Conservation Easements: A Case Study in Richmond Vermont." Discussion paper. Forest Stewardship Council, U.S. Initiative. R.D. 1, Box 182, Waterbury, VT 05676.

Forest Stewardship Council (1996a). "Group Certification of Small Landowners: A Discussion Paper." Forest Stewardship Council. Oaxaca, Mexico.

Forest Stewardship Council (1996b). "Principles and Criteria of Forest Management." Forest Stewardship Council. Oaxaca, Mexico.

Jones, Stephen B., A.E. Luloff, and J.C. Finley (1995). "Another look at NIPFs, facing our 'myths.'" *Journal of Forestry* vol. 93, no. 9, 41–44.

Leopold, Aldo (1949). *A Sand County Almanac.* Oxford University Press, 221.

Nyland, R.D. (1992). "Exploitation and greed in eastern hardwood forests." *Journal of Forestry* vol. 90, no. 1, 33–37.

Simula, M. (1996). "Economics of Certification." In *Certification of Forest Products: Issues and Perspectives.* V. Viana et al., eds. Washington, D.C.: Island Press.

U.S. Forest Service (1992). *Forest Resources of the United States, General Technical Report, RM 234.* Fort Collins, Colorado.

Weyerhaeuser Company. Data provided by the Strategic Planning Director, Tacoma, Washington.

Chapter 8

Rich Niches

Adding value. In the wood products industry, these two words have become a mantra for manufacturers facing globalization, tighter wood supplies, shifting markets, and the need to make their products stand out from the competition. In a commodity industry, adding value can be a potent strategy to gain access to new markets, build market share, and enhance profits. Adding value in wood products manufacturing, moreover, may be as essential for practicing sustainable forestry as is implementing more ecologically based forest management. Making more with less, maximizing the use of sustainably harvested resources, converting wood waste into profits, and creating more jobs than commodity lumber processing are the fundamental components of value-added wood processing: They also provide the framework for sustainable forestry and sustainable communities.

Wood products manufacturers add value in a variety of ways, as the activities of the three companies discussed here indicate. Each manufacturer faces its own set of challenges in the coming years to adapt to shifting business realities. All, however, have carved out potentially lucrative niches by building on the tenets of sustainable forest management. Parsons Pine Products produces an assortment of component parts and consumer products made from wood that is normally considered trash. Colonial Craft, Inc., one of the first manufacturers to gain third-party certification for its opera-

189

tions, has opened up new market opportunities for its hardwood molding and millwork products through environmental certification. And in Costa Rica, Portico S.A. practices certified SFM to ensure a supply of high-quality wood to produce its line of doors—a combination that helped it build a stronghold in the U.S. market.

Parsons Pine Products—Trash to Cash

In 1946, James Parsons, founder of Parsons Pine, decided to do something with the piles of wood scrap and waste that sawmills in the Pacific Northwest churned out from lumber processing. Subscribing to the old adage "make a better mousetrap," Parsons determined that the wood blanks used to produce mousetraps were in short supply and that the scrap wood from the sawmills could be converted into those blanks. When new employees started at the company, Parsons would show them a wood pyramid created from 250 mousetrap blanks. He would then knock the first 100 blanks off the top of the pyramid. "That's $10,000 in sales lost if thrown away or wasted in our processing line," he would say. By operating on the basis that every piece of wood scrap processed affected the bottom line, the Ashland, Oregon, company captured a large share of the mousetrap blank market.

Fifty years later, in 1996, Parsons Pine had parlayed James Parsons' philosophy of turning trash into cash, making more products with less wood, and maximizing the resources available for production into a portfolio of value-added wood products for niche markets worldwide. In its 85,000 square foot facility with 100 employees, Parsons Pine processed about 10 million board feet of wood annually—70 percent softwoods, 30 percent hardwoods—and had revenues of $7 million. The company produced components for houses, including door louvers and slats, fingerjointed and edge-glued furniture parts, and consumer products to organize space, such as knife blocks, wine racks, and compact disc box components. All of these products can be made with wood pieces less than 24 inches long, the typical size of sawmill wood scrap.

The commodity lumber producers that process large volumes of "real" wood in the Pacific Northwest until recently overshadowed value-added manufacturers. During the 1990s, however, when restrictions on logging, mill closures, and lay-offs battered commodity producers, value-added manufacturers like Parsons Pine that had growing or stable labor forces took on new importance for the region. Employment in the secondary wood products industry in Oregon had increased from 27 percent in the late 1980s to 40 percent of the total forest products workforce by 1995, according to the Oregon Employment Division. Total employment in Oregon for logging operations, sawmills, and veneer and plywood operations dropped some 13,000 jobs between 1990 and 1995, or about 30.5 percent, while employment in the value-added and secondary wood products industry—furniture, millwork, cabinetry, and the like—generated 11 percent more jobs during the same period.

Equally important, for every foot of wood processed value-added manufacturers generate more jobs than commodity lumber processing. Every one million board feet of wood processed into commodity lumber, according to The Oregon Wood Products Competitiveness Corp., on average creates only three full-time, family-wage jobs—year round positions that provide industry-competitive wages and benefits. Process that same one million board feet of lumber into components such as furniture blanks or table turnings, and 20 more jobs are created. If that same one million board feet of wood represented in component parts is processed into quality furniture for consumers, another 80 jobs will materialize.

For manufacturers, the rewards of converting waste wood into more valuable uses can be equally dramatic. Softwood waste processed into chips for board production captures about $125/mbf. If that wood is converted into fingerjoint block, it is worth between $480 and $1,000 per mbf. For hardwood, the difference is even more pronounced. Hardwood chips sell for about $50/mbf: They are worth up to 10 times that value if they are converted to hardwood fingerjoint blocks.

The Challenges of Using Wood Waste

Parsons Pine was able to use wood waste only because it circumvented the traditional industry practices and mind-set that make it difficult to buy wood waste, process it, and market products made from it. Between 1946 and 1996, over 80 percent of all products sold by Parsons Pine were produced from wood considered defective, scrap, or waste material. Producing those products using character wood (also known as wood with defects) and short pieces, however, is fraught with difficulties. Not every product manufactured into wood can be made from character wood, since the "character" often denotes some form of deterioration or structural problem in the wood. Further, the equipment used to manufacture wood products typically processes longer pieces of wood. Working with shorter pieces can increase production costs because the waste wood requires more manual handling during manufacturing. Technologies that can automate the handling of short pieces exist, but until recently many of those technologies were designed for large, high-volume production with price tags to match. For companies with small-scale production needs, adapting those technologies can be difficult.

Nor can all wood waste be used cost-effectively in value-added product development. Wood trim ends and shorts that are less than 6 inches long can be substantially less valuable to recover because of current limitations in processing equipment that size scrap pieces of wood to uniform blanks for use in multiple product applications. The process of separating a wood waste stream on the basis of piece length can require the installation of additional machinery, will often be labor-intensive, and can create an added bottleneck in the vertical processing line that could affect daily production rates in processing stations upstream. One way Parsons Pine solved the problem was to subcontract to waste recovery contractors who set up on-site to evaluate and implement wood recovery without creating additional process concerns for the sawmill.

Buying sufficient supplies of lower-grade wood, wood scraps, trim ends, and shorts also poses difficulties. Even if mills are interested in selling their waste, collecting can require inconvenient or expensive changes in processing. Typically sawmills use conveyor systems in the processing line to convey the wood scraps and trim ends directly to a fuel bin or a chipping machine called a "hog." Retrieving those pieces from the waste conveyor system can be inconvenient for the mill and can pose material handling challenges. Conveyor systems are often installed below standard production equipment in dangerous and difficult-to-reach areas, and chipping hogs can be installed in small areas where added material handling for short piece extraction can also be difficult and dangerous. Any modifications to accommodate waste recovery under these circumstances can be costly. Softwood and hardwood lumber grading rules that govern what happens to a log from the moment it is cut also work against a manufacturer like Parsons that sees profit in short pieces or the value of defect or character in the wood.

U.S. mills can be reluctant to part with their waste wood for other reasons as well. Some mills use their wood waste in cogeneration plants that produce steam, which is then converted to electricity to power the mill. Converting wood waste to electricity can cut costs since electricity costs can run as high as 25 percent of a processing facility's total operating costs—although 5 to 8 percent is more the norm for a primary wood products operation in North America. The mill can sell any excess electricity to regional power companies. To overcome that mind-set, Parsons Pine initiated in the early 1990s what Bernard Jerry Sivin, then the president of Parsons Pine, called the 80:20 lumber purchase policy. The policy requested that the sawmill consider selling 80 percent of the desired volume of lower-grade material to Parsons before it would negotiate a 20-percent purchase of high-grade material. In many cases, according to Sivin, it was the only way to get the sawmills to recognize how serious Parsons was about buying lower-grade material.

Historically, some 70 percent of Parsons Pine's wood supply came from the harvest off sawmills that processed lumber from national forests in western Oregon. During the 1990s those supplies dropped off, and Sivin expanded his search for recycled wood to Canada. There he found that the mills were "throwing away everything that wouldn't make a window part—stuff under 24 inches in length," he recalled. The Canadians told Sivin they would sell him wood waste for approximately US$9,000 per truckload. Instead, Sivin offered them US$22,000 a truckload as long as they could guarantee that he would be able to buy 30 truckloads a month. Even at that price, the cost of the wood was substantially lower than traditional lumber prices. For white pine, the price for the waste and trim ends ranged from $175 per thousand board feet (mbf) to $275/mbf compared to the normal lumber purchase price of seven times that value. For hardwoods, the price differences were closer to tenfold.

Make What You Can Sell

Marketing products made from wood waste presents an additional set of challenges. In some instances, consumers prefer that wood products have a noncharacter look. Sports floors, for instance, have always required clean, clear, straight hardwood lumber, usually maple. Working with wood waste has required Parsons to accurately track consumer trends and preferences to find niches where consumers will accept products made from short pieces of wood waste—a reversal of the conventional marketing approach. Typically, manufacturers know what species they have traditionally used to make products. The typical producer makes the product, then figures out how to sell it. Parsons Pine, however, first figures out what consumers want, then produces the product to match consumer demand.

The company's product line in 1997 reflected that approach to product development. Beginning in 1995, Parsons added a line of products that helps sconsumers organize space better, such as a roll-top counter organizer for kitchens, wine racks, a shoe organizer, and disk and CD storage shelves for home office storage—all aimed at a strong niche market. In 1993, according to research conducted by the Peachtree Group in Atlanta, consumers ranked their closets, kitchens, and home offices as areas with the highest need for organizational products. Although classified as a niche market, in 1993 U.S. consumers spent almost $1.5 billion on such organizational products for every part of the home, according to the Peachtree Group. And certain segments, particularly products for the home office, continue to grow. Parsons products made of wood louver slats are also well grounded in consumer preference. The value of louver door shipments in the United States was $46.5 million and is expected to reach almost $50 million by the year 2000, according to data from the U.S. Department of Commerce.

An All Out War on Waste

Parsons Pine has continued James Parson's war on waste in its manufacturing process, continually rethinking the process to make it more efficient. For instance, typically wood product producers cut defects out of the raw lumber first before further processing the material into product as a way to cut the cost of processing waste material. Parsons Pine, however, has reversed that standard in manufacturing for its lower-grade material and the products produced from small pieces. In the late 1980s, Sivin realized he could produce more product with less waste by training personnel to slightly vary the cutting pattern on processing defect and to pull the waste material at the end of the process line. By reworking this single production step, production volume rose an estimated 40 percent while waste dropped some 12 percent. Parsons Pine also makes money on its own waste. Some of the wood that the company buys falls below the quality that it can use. Rather than throw that by-product away, Parsons sells the material to companies that can use it. The activity generated $100,000 a year in sales, on average, from the late 1980s through 1997.

Wringing the Most Value from Wood

Fundamentally, though, Parsons Pine's success since its founding has depended on its ability to convert each piece of wood waste that it buys into products that are worth many times more. In 1995, Parsons had a good year (Figure 8.1). Wood waste originally purchased from Canada, for instance,

Item	Species	Lbr. Purchase $/mbf	Retail $/mbf
Knife Block	Oak	$2,200	$25,000
Wine Rack	Pine, White Spruce	900	8,000
Roll-Top Counter Organizer	Pine	900	10,500
Door Louvers	Pine - Western, White Spruce	900	1,700
Toy Blocks	Pine	600	3,500
Logs for Play Cabins	Pine	600	3,500
Computer Organizer	Pine	600	1,500
Shoe Organizer Anywhere Drawer CD Rack	Alder, Maple, Pine	900 to 2,200	5,500

Figure 8.1.

Converted Value of Wood per mbf.

Source: Mater Engineering

Note: Even with substantially increased raw material cost, Parsons Pine adds notable value to every mbf of wood it uses in product manufacturing.

for $2,200/mbf was converted to $25,000/mbf when processed into a value-added knife block. Parsons, however, cannot realize those dramatic increases in value unless it can buy wood waste for much less than the price of traditional lumber.

Threats to Parsons Pine

Recently, business conditions changed significantly enough to threaten the company's ability to keep its customers and realize those dramatic increases in value. In 1996, Parsons lost its supply of Canadian wood. U.S. lumber producers had long claimed that Canada was selling its lumber in the United States at unfairly low prices. Between 1991 and 1995, Canada's share of the U.S. softwood lumber market had grown 36 percent to $5.6 billion in 1995, which cost some 30,000 jobs, according to the office of U.S. trade representative Mickey Kantor. To reduce government-subsidized Canadian softwood lumber imports to the United States, the Canada–U.S. Softwood Lumber Agreement took effect in April 1996. This agreement mainly affected the Canadian provinces of Quebec and British Columbia—the provinces where Parsons Pine had its waste wood contract. The agreement stipulated that exports that exceeded 14.7 billion board feet a year would be subject to a $50/mbf border fee for the first 650 million board feet (mmbf) and a $100/mbf border fee for quantities over that. The 14.7 billion board feet number was almost 10 percent lower than the 16.2 billion board feet imported to the United States in 1995.

The duties and allotments involved in the treaty cut Parsons off from its Canadian wood waste supply and forced the company to purchase standard lumber loads at competitive lumber prices. In 1995, 70 percent of the company's wood supplies came from recycled wood and wood waste; by year-end 1997, it had dropped to 50 percent. Parsons Pine had by then begun to buy recycled and traditional lumber from Mexico and offshore countries including New Zealand, Chile, Brazil, and South Africa—countries that may use harvesting and forest management practices that are more environmentally harmful than would currently be allowed in the Pacific Northwest.

Meanwhile, low-cost competitors from Chile, Brazil, and New Zealand were underselling U.S. makers of millwork products, the components in a finished house. During 1996 and 1997, "profits in the whole industry were just about nonexistent," said Sivin, " and many companies closed up because of the disastrously low prices from the competition." Wood louver slats, which accounted for 80 percent of Parsons Pine's business in 1994, plunged to less than 50 percent in 1997.

The shifting business conditions had sobering financial repercussions. The company's costs for raw material as a percentage of total product price increased over fivefold for many of its products (Figure 8.2). Profits, which

Product	(a) Board Feet Per Unit	(b) Raw Material $/mbf Paid	(c) Total Raw Material Cost [a × (b ÷ 1,000)]	(d) Retail Sales Price (Product)	(e)* Manufacturer Sales Price (Product) (d ÷ 2)	(f) Raw Material Costs as a % of Total Manufacturer Sales Price (c ÷ e)
Roll Top Counter Organizer:						
Current	6	$900/mbf	$5.40	$60.00	$30.00	18%
Prior	6	$225/mbf	$1.35	$60.00	$30.00	5%
Wine Rack:						
Current	7	$900/mbf	$6.30	$56.00	$28.00	23%
Prior	7	$225/mbf	$1.57	$56.00	$28.00	6%
Knife Block:						
Current	2	$2,200/mbf	$4.40	$50.00	$25.00	18%
Prior	2	$400/mbf	$0.80	$50.00	$25.00	3%

Figure 8.2.

Parsons Pine Raw Material Cost as a Percentage of Sales Price.

Source: Mater Engineering

Notes: * Manufacturer Sales Price = 50% of Retail Sales Price

Legislative actions in the U.S. and the 1996 U.S.–Canadian trade agreement increased raw material costs as a percentage of product sales price for Parsons Pine.

had averaged $1 million on about $7 million in sales between 1990 and 1995, eroded. By 1996 the company was operating at a little better than break even.

New Strategies to Counter Off-Shore Competition

Starting in late 1995, Parsons began to modify its product designs, establish new customer bases, and identify new distribution channels to counter the effects of offshore competition. Small manufacturers cannot afford to remain low-tech in their production, but many find that high-tech is not necessarily cost-effective. The key is to be competitive in product offering while being flexible in production. In 1996, Parsons bought a new mid-tech router that allowed the company to convert a percentage of the louver wood slats to CD wood bookmarks. The new line increased annual sales by approximately 6 percent in 1996 and earned a return on its investment within two months.

The flexible manufacturing options that Parsons had developed over the years, in response to changing technology and markets, enabled it to respond quickly with new product developments. These production capabilities include the production of profile molding lines, radio-frequency edge-glue and end-trim lines, long and short board cut lines, a fingerjoint line, short block equalizers, product assembly lines, and stretch and shrink wrap lines. Using this flexibility, the company moved into the production of oak furniture products. It also started to supplement its line of oak furniture parts with those made from lower-grade species, such as Hemlock and alder, and stepped up marketing its space-organizing products.

Parsons started to distribute some of its new consumer-oriented, space-organizing products through the fast-growing mail-order business instead of working exclusively through brokers and wholesalers that manufacturers typically use for distribution. The shift seemed logical: The U.S. mail-order business has increased on the average of 10 percent in sales annually since 1983 to exceed more than $220 billion in 1997. It has also outperformed the retail business in both sales growth and net income as a percentage of sales. Moreover, of the total sales for mail-order catalogs, about 51 percent are for consumer products and 25 percent are for business products, segments that fit several Parsons Pine products, according to the Peachtree Group. And since 1991, the top selling product areas in mail-order catalogs have included functional home furnishings, home office, and school supplies, product categories that fit with Parsons. In addition, internationally, the fastest-growing markets for mail-order product sales are in Germany and Japan, two countries where Parsons Pine sold products through other distribution channels.

As a result of the new product offerings and the new mail-order distribution channels created in 1995/1996, Parsons Pine quickly established a new base of customers: Over 70 percent of the company's 1996 customers were new from just one year earlier. Typically in wood products operations about 80 percent of the customer base consists of long-standing customers. The shift, however, exposed the company to both risks and rewards. Long-term relationships can be a stabilizing factor for wood products manufacturers, whose markets undergo dramatic fluctuations about every seven years. In market slumps, long-standing customers often sustain wood products companies. By the same token, manufacturers may lose market opportunities because they are unwilling to look in new directions. In 1997 the mail-order effort continued to do well.

In September 1997, the Parsons family and Sivin sold the company for an undisclosed sum to three investors, led by Michael J. Rasmussen, who took a 70 percent share of the company. Initially, the investors had no plans to change the basic wood waste strategy set by James Parsons in 1946. Rasmussen, however, intended to regain the company's basic louver and slats business "louver by louver" as he put it by winning back old customers from South American competitors. Rasmussen was counting on Parsons Pine to level the playing field with these lower-cost competitors by providing better service—primarily through up-to-the-minute delivery—and lower costs by making louvers out of lower-cost species of wood. In the campaign to win back the business Rasmussen said that Parsons Pine's efficiencies—the company can produce two more louvers than competitors out of a block of wood—would be a significant factor. Management also intended to increase the company's use of wood waste by "searching for any niche we can find in raw material supply that isn't being exploited by anyone else," according to Rasmussen. The company was negotiating to buy a sawmill in California that processed small-diameter pine, which could supply inexpensive logs for louvers, and has had found new sources of trim ends and falldown for louvers. Rasmussen predicted that in 1998 the company would bring in $10 million in sales—and earn a profit.

The Importance of Adding Value

Parsons Pine Products' for decades has profitably converted wood waste into more valuable consumer products, indicating that this type of value-added manufacturing can be a profitable part of a sustainable forest products manufacturing operation. Yet, the conversion of wood waste to highest and best use applications continues to be an often overlooked option in the industry. Both U.S. and Canadian lumber producers have been reluctant, even adverse, to recognizing the value of their wood waste. As Parsons activities indicate, purchasing policies, such as limiting the purchase of high-grade lumber without the sale of lower-grade lumber and wood waste, can be effective ways to change those mind-sets. In working with wood waste, acumen in product design and marketing is critical. And as Parsons Pine's experience suggests, to be successful a sustainable forest products company may have to shift business-as-usual practices to rethink the production process, buy appropriate equipment, establish a new customer base, and engage new distribution channels.

The U.S. trade agreements with Canada to govern natural resource management and lumber flow were intended to level the playing field for primary wood product producers in the United States, but they had the opposite effect on smaller value-added wood product manufacturers. The results may be counter-productive to achieving sustainable forestry practices on a global scale. To be effective, future policies need to consider the differences in impact that any restrictions will have on different types of wood flow, whether it is commodity lumber or value-added wood products.

Colonial Craft—An Early Certification Venture

When Eric Bloomquist, president of Colonial Craft, Inc., first heard about environmental certification, in 1992, no one needed to persuade him that the idea made sense. Certification answered one of Bloomquist's long-time concerns—how to assure that the company bought wood only from well-managed forests. "Until then, we had to take the word of our vendors that they were doing a good job, and that wasn't adequate," recalled Bloomquist. "Certification gave us, as a purchaser of lumber, the only valid tool we knew of that we could use to make those kinds of judgments." Bloomquist then became one of a handful of industry executives to support certification and the creation of the Forest Stewardship Council (FSC), which certifies the four key independent certification programs. Under his direction, Colonial Craft became one of the first manufacturers to become certified.

Colonial Craft, established in 1965, is a leading producer of hardwood molding and millwork products. It is the largest U.S. producer of window and door grilles and small moldings used for pane designs in windows and doors. The company grew from just 20 employees in 1984 to more than 200 by 1996. Revenues almost doubled from $12.64 million in 1991 to $24 million in 1996. In 1997, Colonial Craft was the only molding and millwork operation to hold chain-of-custody (COC) certification from the Forest Stewardship Council-approved SmartWood program in the United States and Canada. It produced a number of certified products, including window and door grilles, picture frame molding, and wood parts for barbecue grills.

The willingness of manufacturers like Colonial Craft to work with certified wood is as important for the acceptance of certification as timber producers' willingness to certify their land. Manufacturers are a pivotal step in tracking the flow of certified wood from the forest to the consumers. Many manufacturers, however, still question whether certification is practical for their day-to-day operations; whether there is enough demand for certified products to justify changing business practices; and whether certification can deliver a competitive advantage.

Colonial Craft has found some answers. Since 1992 the St. Paul, Minnesota-based company has helped to shape the direction of sustainable forest management (SFM) and certification. In the process, Colonial Craft's experience indicates that for manufacturers, certification combined with otherwise sound business practices can deliver tangible benefits—new business opportunities, increased visibility, and better management practices.

Sound Business Management First

Bloomquist is the first to emphasize that the certified status of a product cannot substitute for the hallmarks of good management—product quality, customer service, intelligent product design, marketing acumen. These are attributes that Colonial Craft is recognized for in the industry. *Wood and Wood Products* magazine, a leading industry publication, has listed the company in its elite ranking of the top 100 value-added wood products producers in North America on the basis of financial and product performance for five years since 1990. Approximately 10 percent of the 100 companies making the list for 1995 also produced molding and millwork products. Colonial Craft was the second largest molding and millwork operation in annual sales and total number of employees making that list. The magazine credited the company for being competitive in implementing industry standards of practice that are traditionally viewed as good business strategy. Colonial Craft's management attributed its success in production to purchasing new equipment, improving quality, and listening to and implementing employee suggestions in management and facility operations.

Several key strategies are responsible for Colonial Craft's recent growth spurt—an affiliation with Andersen Windows, flexible manufacturing, product focus, and environmental leadership. Colonial Craft became the leading U.S. supplier of wood window and door grilles after 1982, when it formed a business relationship with Andersen Corp., the producer of Andersen Windows and Andersen Doors. Before the partnership, Colonial Craft's grille sales were less than $400,000 annually. By 1990 sales exceeded $5 million, and by 1997 they totaled $26 million.

The affiliation with Andersen is advantageous for Colonial Craft for several reasons. Andersen's status as a major player in the wood window and door markets in the United States and Canada assures substantial sales for

Colonial Craft, one of its major suppliers. *Professional Builder* magazine's 1996 Brand Use study documented Andersen's position as a top brand product used by a substantial percentage of U.S. builders. According to survey results, over 50 percent of builders install Andersen windows in their projects. The same survey also found that a number of builders used certain Andersen products exclusively—nearly 12 percent used just the line of Frenchwood hinged patio doors; 18.8 percent installed only Andersen's wood windows; and 35.1 percent used only the Andersen clad windows.

Colonial Craft and Andersen are also a good match because both companies are concerned with producing environmentally appropriate wood products. During the 1970s Andersen, which produced large amounts of wood waste in the form of sawdust and vinyl waste from its manufacturing, discovered a way to combine wood fiber and PVC into a material whose structural properties were similar to wood. Andersen produced a prototype window of the wood and PVC waste in 1972 and installed it in test homes in Minnesota. The project, however, was shelved because wood was inexpensive and plentiful then.

In 1991 when the wood products industry faced problems with both the supply and quality of wood, Andersen revisited those test windows to see how they held up. Management was so impressed that in 1991 Andersen introduced a new composite material called Fibrex, which combined 40 percent wood weight with 60 percent PVC weight. New independent tests found that Fibrex had a stiffness, resistance to heat, and thermal expansion superior to vinyl. It also had insulating properties on a par with pine and vinyl and was impervious to decay.

In 1992 Andersen rolled out a new window product line called Renewal made from Fibrex, the same year that Colonial Craft initiated the SmartWood certification process. By 1994, Colonial Craft was supplying certified wood grilles to Andersen's Renewal line, but not by Andersen's request. Independently, Bloomquist had decided that all Colonial Craft wood grilles that went into the Renewal line would be made from certified wood. In 1995 Green Seal certified the Andersen windows and doors for their energy efficiency. Green Seal stated that the door and window products produced by the company "are among the most energy-efficient doors and windows available." Andersen does not actively promote the certified grilles in its Renewal windows, but it does identify the grilles as being certified in the product display.

Flexible manufacturing has also played a role in Colonial Craft's growth. The company, through flexible manufacturing, can quickly adapt to varying design modifications in product development. This ability has helped it win and keep customers. It is particularly important for the relationship with Andersen Corp. Andersen has, at times, requested product design changes that need final production results within a week, a request that Colonial Craft can cost-effectively and efficiently meet.

Colonial Craft's focus on Andersen Corp. has also been an asset in product development. The company has been able to concentrate its production and service to a few products and a few customer types, which has yielded substantial benefit given Andersen's market share. The dependence on so few products and customers, however, does expose the company to risk, should market conditions change quickly or its major customer develop difficulties. In recent years, Colonial Craft has lowered the risks of its product focus on Andersen by identifying product opportunities that solve bottlenecks in its own processing while addressing sustainability concerns. The company's picture frame molding line, introduced in 1993, for instance, arose out of efforts to figure out what to do with the thin longer wood waste pieces left after lumber ripsawing—the process that cuts raw boards into smaller strips that fit into molding machinery.

The Beneficial Fall-Out of Certification

Bloomquist's decision to make Colonial Craft a certified manufacturer may be the company's most significant move since its affiliation with Andersen. Colonial Craft for more than two decades has pursued sustainable resource use and sustainable economic development through a variety of programs. Certification, according to Bloomquist, can help develop sustainable wood supplies for the industry, steer debate to how sustainable forest management should be implemented, and help the public to make informed choices about the products it buys. He also expected that as an early mover in certification, Colonial Craft would gain a market advantage. By 1997, with less than four years as a certified manufacturer, Colonial Craft could already count improved businesses practices, a higher profile in the industry, and new business as the beneficial fall-out from certification.

As part of certification, manufacturers are required to track certified wood from suppliers through the production process to document that its vendors are following the criteria required by the certifying body. To do that, Colonial Craft asked its lumber suppliers to verify where their logs came from, who harvested the logs, how the logs were harvested, and whether the forests where the logs were cut were sustainably managed. In 1994, on the basis of its survey of customers and the company's capability to separate certified wood from noncertified wood in its operations, Colonial Craft received certification from SmartWood. Colonial Craft maintains separate inventory space for certified and noncertified wood; it does separate "menu" production runs for customers requiring certified wood to make sure that noncertified wood does not get mixed in with certified supplies.

The cost of keeping the certified wood separate is negligible, according to the company.

Critics have contended that the logistics involved in chain of custody impose too great a burden on manufacturers—a view contrary to Colonial Craft's experience. "We are a better run business because of some of the things that certification made us do," Bloomquist said. Chain-of-custody certification required Colonial Craft to develop inventory systems precise enough to trace a board from a vendor, a specific load of wood, or a forest site right through manufacturing to customers. That information allows the company to better determine yields from the wood it processes and to improve quality. Before certification, wood discovered during manufacturing that fell outside specifications, for instance, often could not be traced back to the supplier. With the chain-of-custody system, the company can identify the supplier and get credit for any faulty wood.

Colonial Craft's early status as a certified manufacturer enhanced the company's visibility and helped generate new business. Bloomquist credits certification for giving the company its single largest new product opportunity in recent years. In 1997, a U.S. barbecue manufacturer identified Colonial Craft by its high profile in certified products. The manufacturer had tried, without success, to market noncertified barbecue units in European markets. The U.S. maker was then forced to find a U.S. supplier of certified parts. Colonial Craft won the order to supply certified handles. But more important to Bloomquist, Colonial Craft's quality and delivery times were so good that in 1997 the barbecue maker awarded the company a far larger contract for noncertified handles. The added business was expected to increase annual sales over $1 million and add more than 2 million board feet to annual production output.

Although difficult to verify, Home Depot's decision to stock Colonial Craft's line of architectural molding line in 1996 may also have been due to the visibility of Colonial Craft's certification efforts. Home Depot has been a leader in moving certified wood products into the North American do-it-yourself (DIY) markets.

But those orders for certified products have not fetched the premiums that are so often touted as an advantage of certification. In 1995, less than 1 percent of Colonial Craft's total sales of about $24 million resulted from certified wood sales that had premiums attached. The failure to realize premiums for certified products, however, has not been a concern. As Bloomquist has stated, "We're not in certification for the premium, we're in it for the market share." Bloomquist did expect—and has earned—the same profit margins on certified as noncertified products, which means getting compensated for any extra costs involved in producing certified products that may come from having to pay more for certified wood. The barbecue grill maker, for instance, paid Colonial Craft the additional 15 percent it cost to produce certified handles.

The greatest hurdle that Colonial Craft faced in certified wood products manufacturing as of 1997 was a lack of certified wood. Bloomquist predicted that for the next three years the company would not be able to get enough certified wood in the species it needed to meet orders. In 1996, Colonial Craft offered a full line of 100-percent certified picture frames moldings (called Warm Woods), and more than 17 percent of its window and door grille volume were produced from certified wood. But only 8 percent of the total volume of 10 million board feet of wood processed annually for all products combined was certified. The company expected to increase that amount to 13 percent by year-end 1997. For the majority of its certified wood, the company has paid a slight premium.

Colonial Craft, however, has a goal of offering 100-percent certified wood products by 1999. To increase supplies of certified wood and to make certification more of a mainstream concern in the industry, Bloomquist has tried to jump-start the certification of public forestlands. In 1997 Colonial Craft was instrumental in creating a joint venture between Minnesota State foresters and forest managers for Aitkin County to participate in the nation's first pilot project involving an FSC-certification assessment of 614,000 acres of state and county forestlands. By year end, the land had received the stamp of approval from the FSC-accredited SmartWood.

Strategies for the Future

Colonial Craft continued to look for new growth opportunities that would increase business while improving the quality of the environment through increased production efficiencies, new product offerings, identification of new product distribution channels, and a commitment to certified products.

Increased Production Efficiency

Colonial Craft estimated that about 50% of the total annual volume of raw wood that comes into its operations for production ends up as waste material, sawdust, chips, and short pieces of wood. The company was implementing a number of actions to convert more of the solidwood waste into more valuable wood components—which would not only increase the company's yields but also provide environmental benefits by using the resource more efficiently.

In mid-1996, Colonial Craft installed color optimization systems to more efficiently and effectively separate material based on grade/appearance. The optimization system, installed in midsummer of 1996, generated the first readings on yield in August. As Figure 8.3 indicates, in August about 55,000

Species (lineal feet)	Volume % Yield	Average (lineal feet)	Volume Waste
Certified Basswood	78,984	93%	5,528
Hard Maple	390,670	90%	39,067
Red Oak	337,837	97%	10,135
Total	807,491	93%	54,730

Figure 8.3.

Colonial Craft Production at Luck Plant for August 1996.
Source: Mater Engineering

lineal feet of solidwood material went to the waste bins destined for the chipper or set aside for giveaway wood scraps to local residents. A spot check of the waste bin by Mater Engineering, Ltd., indicated that about 80 percent of the waste material was six inches or longer and had color variation as its only defect.

In 1997 the company was investigating the costs and benefits of buying equipment to process those wood scraps into good quality fingerjoint blocks or cabinet parts. Since fingerjoint blocks sell at approximately $500 per million board feet, Colonial Craft might realize an annual increase in gross sales of up to 10 times the gross sales that would be generated through selling chips of the same volume.

New Product Offerings

New product offerings that capture the use of custom grade material and increase new products and product lines—especially those that take advantage of development opportunities in nontraditional products and do not rely on the cyclical housing industry—were also on Colonial Craft's radar screen. According to the company, about 30 to 40 percent of the hard maple it processes has color variation rather than the preferred clear look desired by the company's existing customers. The company had started to work with customers to market wood with color variations as certified custom-grades, which can sometimes command higher prices than regular maple.

New Distribution Channels

As long as supplies of certified wood are limited, manufacturers face a set of conditions that can affect the movement of certified wood. Lumber brokers and wholesalers—often the middlemen in the forest owners' sales of logs to the primary sawmill and the sawmill's sales to the value-added producer—may try to "lock-up" certified wood supply. Suppliers may be reluctant to sell too large a percentage of their certified resource to one buyer and may do so only if they are able to increase the base price plus any premium they may charge for certified wood. Such a situation could make the price of certified wood noncompetitive and stop the flow of wood. Customers that supply products for new building construction have product needs that rise and fall with the number of new building permits issued, so their suppliers follow the same cycle of rising and falling orders. Colonial Craft had identified diversification away from the new housing construction industry into other product distribution channels as an intelligent way to even the flow of product demand and supply over time.

Certification for a Business Advantage

Colonial Craft's experience offers several important guideposts for wood product manufacturers evaluating the business of green certification. Certification is not well established, but it is growing. As Colonial Craft has demonstrated, offering certified products can strengthen the bottom line. Certification as a product differential, however, cannot substitute for the other requirements of good business—quality production, customer service, on-time delivery, and competitive pricing. If those other attributes of good business are in place, certification can serve as a clear product differentiation for the wood product producer.

Even though some certified products appeared to command premiums in the market long term, realizing premiums may not be as important for manufacturers as achieving consistent cost-plus-normal-profit recovery matched with increased market share for certified products. Moreover, companies that invest in certification may find, like Colonial Craft, that the increased business exposure resulting from offering certified products can generate new business in noncertified product arenas. And in contrast to conventional wisdom, the cost of working with certified wood appears to be less a constraint to the wood product manufacturer than might be expected. It is more important for the producer to understand that the demand for targeted wood products currently exceeds supply.

Finally, as Colonial Craft has demonstrated, one company can influence the development of sustainable forestry and certification. Colonial Craft's leadership in the private sector has increased the awareness of and opportunity to evaluate SFM practices for use on public forestlands, which could significantly help to increase the growth of independent, third-party certification of forests and wood products.

Portico S.A.—The Power of Vertical Integration

Wood products companies operate in some tropical countries under what could be described as hardship conditions. Poverty leads to multiple, short-term demands on forests that governments in poor countries often cannot balance with broader concerns of forest sustainability. The resulting defor-estation combined with underdeveloped wood markets can make it difficult for manufacturers to secure supplies of high-quality tropical hardwoods. Too often governments in tropical countries, under pressures at home and abroad to stem rampant deforestation, have enacted regulations that are not only ineffective at slowing deforestation but are also detrimental to the industry.

Under such conditions, in the past 15 years Portico S.A. of Costa Rica has evolved from a small workshop into a forestry and manufacturing operation that has captured a significant share of the U.S. market for high-end resi-dential mahogany doors. Key strategic decisions—the acquisition of forest-land, the integration of its forestry and manufacturing operations, the adop-tion of sustainable forest management, and the environmental certification of its forests—enabled Portico to avoid much of the controversy over log-ging in Costa Rica, shape the nation's forest regulations, and gain entry into major U.S. chains. As Portico's experience demonstrates, companies can combine SFM with successful businesses in a setting characterized by heightened government control of forests, and can benefit by increasing product quality and by prevailing in competitive international markets.

The Path to Vertical Integration

Leopoldo Torres and a group of business associates founded Portico 1982 when they purchased Puertas y Ventanas de Costa Rica (Costa Rica Doors and Windows). The original company produced 70 to 80 wooden doors and window frames a month for the Costa Rican market and for export to Panama and El Salvador. But Torres and the other investors had more ambitious goals: They wanted to convert the company into a manufacturer and exporter of premium exterior mahogany doors for the U.S. market. They chose mahogany because it was available in Costa Rica, popular with U.S. customers, and a strong and attractive wood.

To compete in the more demanding U.S. market, Portico needed to reor-ganize production processes, factory layout, and product design. The doors the workshop made for the Costa Rican market in 1982 were custom-made, since doorframes were not standard. The U.S. market required doors of standardized dimensions and much higher, consistent quality. The large accounts and retail chains that sold the doors in the United States purchased them in large volumes—a 450-door order was routine. To sell the doors at a profit Portico needed a stable, year-round supply of high-quality wood to meet volume orders and produce doors that did not crack in transit or after installation.

It proved impossible for Portico to buy a consistent, year-round supply of the mahogany at reasonable prices, even though the species grew abundantly in Costa Rica. Overexploitation of forests was part of the problem. In addition, logging was impossible during Costa Rica's six-month rainy season because heavy equipment could not be transported through the mud. Logging practices were poor, with loggers often allowing the cut logs to receive either too much or too little humidity before they arrived at the mill, which often damaged them. Costa Rica also had no formal lumber market. Few standards for lumber quality existed, and prices had to be negotiated individually for every purchase. Nor did the lumber mills generally use high-quality equipment, such as drying kilns, circular saws, or band saws, and thus could not supply high-quality lumber with consistency.

In 1982, management decided to use vertical integration to address the problems of price, supply, and quality it faced in obtaining a wood supply. Owning forestland that would supply its own manufacturing operations would, management reasoned, give the company "total quality control" over its products from the Costa Rican forest through distribution to U.S. retail outlets. Portico then began to acquire forestland to supply its wood, trucks for transportation, and saw mills, kiln dryers, and other equipment for processing lumber.

Portico had no choice but to purchase natural forestland. The mahogany used in Portico's doors does not grow well on plantations because the trees need the surrounding biodiversity to protect them from ants and other wood-destroying pests. From the outset the company decided to adopt sustainable forest management. By the late 1980s, it was clear to Torres and the other investors that not only were consumers in the United States and Europe developing a "green attitude" toward tropical hardwoods but that also increasingly U.S. distributors wanted assurances that the tropical hardwood products they carried were produced without decimating tropical forests.

In 1987, Portico S.A. acquired over 5000 hectares (12,355 acres) of natural forest in the northeast coastal region of Costa Rica, near Nicaragua. The deal was a debt-for-nature swap agreement with the government of Costa Rica and Norwest Corporation, a U.S. financial institution based in Minneapolis, Minnesota. Norwest, which held a large amount of Costa Rican government debt, financed the deal in exchange for a stake in the company. Later Portico bought additional hectares of native Costa Rican rainforest and by 1997 managed slightly more than 20,000 acres.

A Less Damaging System of Forestry

Portico created a subsidiary, Tecnoforest del Norte, to manage its primary and secondary-growth natural forest. Tecnoforest is guided by two basic principles to ensure long-term natural forest management for its commercial purposes: (1) to guarantee the stability of the species used to manufacture the doors (*carapa guianensis*), which Portico labeled as "Royal mahogany," so that it would have consistent long-term yields of high-quality wood and (2) to do less damage to the forest during logging and extraction than traditional industry practices.

Traditional commercial logging practices in Costa Rica, which are selective, do more damage to the surrounding forests than selective cutting in temperate forests. In Costa Rica, selective logging is typically a one-time operation, where the loggers try to reap the maximum high-value lumber in one harvest. Because the logger never intends to return to the site, sustainable resource management is not a consideration. All valuable trees on the site greater than 60 cm (24 inches) in diameter at breast height (DBH) are marked for cutting or preservation. When the trees are cut, the networks of vines in the forest bring down trees other than those being cut. The cut trees then destroy more trees and vegetation by falling on them. Removing the logs from the forest further damages vegetation and tears up the soil. Skid trails, for instance, created by dragging the trees out of the forest, damage seedlings needed for regeneration, cause erosion, and can dry out the soil and reduce drainage. And finally, the loggers have no economic incentive to minimize damage after the one-time harvest.

To minimize damage to the forest, Tecnoforest adopted different practices for preharvest planning, harvesting, transportation, and maintenance than those typically used in Costa Rican forestry. First, Tecnoforest took a prelogging inventory. Foresters marked and mapped all trees greater than 70 cm (28 inches) DBH for the management plan that was submitted to the Direccion General Forestal (DGF), the government agency that enforced forestry laws. The best specimens of royal mahogany trees were left as seed trees, others were left as shade trees to protect Royal Mahogany seedlings from excessive sunlight, and the remaining trees (about three to four trees per hectare) were marked for cutting. Only trees greater than 70 cm DBH were marked for felling because they were believed to be "overmature," even though trees greater than 60 cm DBH could legally be cut.

The average Tecnoforest crew was twice as large as a traditional crew and always included an accredited forester. All Tecnoforest loggers were trained in techniques to minimize damage to the forest, and the subsidiary employed five full-time researchers to study its forest holdings and oversee maintenance. Tecnoforest researchers were further helped by programs developed with U.S. universities. Preselected (marked) trees were cut to fall in a specific direction and land in a predetermined place to minimize damage to other trees and to allow the skid trails to be run along the same paths. Running the skid trails in confluence reduced damage to the forest floor by minimizing the area over which the heavy trees were dragged. The practice destroyed fewer saplings, damaged less undergrowth, and disturbed fewer animal dwellings. Fifteen years after logging, the area would have regenerated sufficiently to support another harvest.

The new practices accomplished their goal. In 1990 one study found that Tecnoforest's procedures did less damage to the forest than did traditional logging. After comparing a traditional site with a Tecnoforest site after logging, the study found that the traditional site contained only 61 percent of the trees from before the cut, while the Tecnoforest stand contained 95 percent of the original trees. The forest composition, or the relative numbers of each species present, was relatively unchanged at the Tecnoforest site; at the traditional site, it was significantly altered. The forest canopy opening was computed to be 46 percent at the traditional site after logging, but only 29 percent at the Tecnoforest site. Excessive sunlight coming through large openings in the canopy can damage saplings and interrupt wildlife corridors that are crucial to maintaining biodiversity. Finally, while roads and skid trails covered 9 percent of the traditional site, they covered only 3 percent of the Tecnoforest site. The study concluded that "Portico's logging operation seems much more viable than traditional logging in terms of long-term sustainability."

An Easy Decision for Certification

In the early 1990s, as the controversy over logging in tropical forests climbed higher on the public agenda with the approach of the 1992 environmental summit at Rio de Janeiro, Portico's management decided to seek certification for the forestry operations. They thought certification would help the company land accounts with key U.S. distributors. The decision was a fairly easy one, since Portico's sustainable forestry practices were already consistent with the standards required for certification, and the cost of certification was roughly $1/door.

In 1992, Tecnoforest received certification of 7000 hectares (17,300 acres) as a "state-of-the-art, well managed forest" by Scientific Certification Systems (SCS) of Oakland, California (formerly known as Green Cross), an independent environmental certification organization. The SCS process took six months and cost Portico more than $40,000. SCS scored the operation in three standard forestry categories: sustaining timber yields, maintaining the ecosystem, and providing socioeconomic benefits to the surrounding community. A score of 60 or better in each area won certification.

Tecnoforest scored 82, 79, and 73, respectively, to become the first SCS-certified "forest manager" in Costa Rica and the third in the world.

By the mid-1990s, Portico's certified forestry supplied the company with most of the wood it needs—about 85 percent in most years. The rest Portico buys on the open markets. But forest management costs are also higher. Although the company does not release those figures, according to some Costa Rican analysts Portico has invested more than $1 million in forestry research, and its sustainable forestry costs about 20 percent more than traditional Costa Rican forestry.

High-Quality Manufacturing for Export

Portico's primary export is to the U.S. market for exterior hardwood doors, which each year fluctuates from 500,000 to 1.5 million doors, depending on the health of the economy, housing starts, home renovations, and general construction. Most exterior wood doors are used in homes and a select group of commercial properties. The market consists of low-end ($50) plywood composites, through mid-range ($150–$500) softwoods and hardwoods, and up to premium ($500 and up) 100-percent hardwood doors made from mahogany, other hardwoods, and oak. Oak doors fetch anywhere from $50 to $150 more than mahogany. In the United States, home center chains, such as Lowe's and The Home Depot, and small independent housing supply and hardware stores that cater to high-end markets are the primary distributors of high-end doors. The highest quality wood doors are generally stile and rail, 1-3/4 inches thick, made out of premium grade hardwood that use mortise and tenon type joinery. Premium grade hardwood refers to mahogany or oak that is free of defects such as knots, kiln burn, or bright sap spots, which might affect its performance.

Portico started selling single-piece, premium grade, solid mahogany doors priced from $500 to $2,500. By 1994, Portico sold about 60,000 units per year and held about 60 percent of the U.S. market in the southeast. The company later broadened its product line to include doors made from American red oak in the same range of styles and prices as the mahogany line and a line of low-end doors made with another species of tropical hardwood, priced at about 60 percent of the mahogany doors. In 1997 Portico's line of royal mahogany doors accounted for 50 percent of sales; American red oak doors made up 30 percent; and the low-end doors were 20 percent of the total.

Portico broadened its product offerings as it made inroads into large national accounts. By 1992 most of the company's business was with national chains such as Lowe's and Addison Corp. Portico shipped its doors directly to their central warehouses. By 1993 Portico offered high-end American red oak doors, because buyers wanted to carry lines from premium door manufacturers that offered both oak and mahogany to meet the regional preferences of customers—consumers in the Midwest and Northeast preferred oak; in the Southeast the preference was mahogany. The oak doors cost the company about 15 percent more to produce than mahogany doors, but it also received $20 to $50 more for each one. In retail stores the oak doors each sold for $50 to $150 more than mahogany doors.

In 1993 Portico shifted from solidwood doors to engineered doors. The engineered doors look like their single-piece ancestors but are constructed using the very best wood for a 2-mm veneer that covers the surfaces of the stiles and rails. Fitted and glued pieces of mahogany called butcher block are underneath the veneer. Portico started offering engineered products primarily to stretch supplies of its valuable mahogany. But engineered products have other advantages. The use of smaller pieces reduces internal stresses that cause larger pieces of wood to warp and allows scrap mahogany to be used in the cores.

Torres credited SCS certification with enhancing the company's image among its key accounts and allowing the company to gain greater market share between 1992 and 1996. Initially, said Torres, certification played a role in the company's doing business with The Home Depot, which has tried to carry environmentally certified products. A buyer for The Home Depot confirmed Torres' opinions and explained, "We have to buy out of managed rainforests because everything we buy has to be certified if it's imported. There's a lot of bad wood out there, but Portico does well-managed forestry, and they've got the highest quality doors too."

A Volatile Regulatory Environment

Throughout its history Portico has operated in a volatile regulatory environment. Before 1977, the Costa Rican government freely handed out logging and forestland grants. The 1977 Reforestation Law limited extraction on logging sites to 25 percent of the standing trees. A stipulation that a tax be paid on the projected harvest before cutting, however, stimulated illegal logging. By 1981, Costa Rica still had the highest rate of deforestation among tropical countries, at 4 percent of forested lands. In 1987 and again in 1992, the government established new guidelines requiring forest management plans and agreements to reduce logging to 60 percent of saleable trees with diameters of 24 inches or wider. As of 1990, though, an estimated 60 percent of ongoing logging continued to be illegal, according to the DGF. And by 1992, only 5 percent of Costa Rica's remaining forest cover had logging potential—the rest was in protected areas.

Although environmentalists generally agreed that Costa Rica has had tough legislation to control logging and to keep forests intact, enforcement under the Direccion-General Forestal (DGF) had been ineffective during the 1990s. Moreover, permits for logging, which the agency also issued, were a sore point with the country's wood products industry. They were issued each year, which made them difficult to get through the government bureaucracy and resulted in long delays. In 1994 Portico's permits were delayed so long that the company was forced to buy more than the usual 15 percent of its mahogany from outside.

The situation in Costa Rica was so difficult for logging that in 1994 Torres decided not to expand Portico's forest holdings there. Management did not feel that it could assure growth in the door business if it depended on Costa Rican holdings. Not only were permits a problem, but environmentalists opposed anyone cutting mahogany, regardless of their records as forest managers and even though Costa Rica still had large stands of lowland coastal forest that contained the royal mahogany Portico used. Under the circumstances Portico did not want to risk transforming its "small" forest management undertaking into a large-scale commitment.

In response to the uncertain regulatory climate, more than 600 private businesses founded the Costa Rican Chamber of Forest Producers with the goal of passing enforceable laws that would make forestry as profitable as other land uses. Portico, as one of the leading exporters and certified forest managers, spearheaded a legislative initiative. The new law, enacted as Forest Act No. 7575 in May 1994, contained several major provisions. It required that all wood harvested in Costa Rica come from environmentally certified sources, a measure designed to lower supply and raise wood prices. The government agreed to compensate forest owners for certification costs and for the increased costs of maintaining a certified or sustainable forest, which would make the value of forestland competitive with other uses and would gradually open up the export market for logs and lumber. The law also addressed the ineffective enforcement of government forest management regulations and, at the instigation of the Chamber of Forest Producers, removed DGF from issuing logging permits and collecting taxes—and logging permits did not expire every year. In 1996 the government adopted another law that banned further land-use changes in natural forests, deregulated forestry plantations, and gave municipalities and private forestry engineers the power to issue logging permits for farmlands and nonforested areas.

By 1997 Torres was so encouraged by changes in the laws and the new attitudes that he had encountered in government agencies that he reversed his earlier decision not to expand Portico's land holding in Costa Rica. He was in the market for more forestland in Costa Rica. "With the new legislation and controls everything that's being cut legally is from sustainably managed, selectively harvested lands," Torres commented. "The government has . . . removed many of the problems we used to experience."

A Turning Point

In 1997 Portico was at a turning point. Door sales were flat at the 1994 level of 60,000 units, when Portico held 60 percent of the southeastern U.S. market. But Portico's share of market had declined. An influx of competitors in the high-end residential door market from the lower-cost regions of Southeast Asia and from South American markets—such as Bolivia, Peru, Brazil, Guatemala, and Chile—had undercut its prices by as much as 20 percent. Portico, which exported 10 to 15 percent of its high-end line to countries other than the United States, planned to expand to those export markets. Torres expected certification to help the company win accounts in northern European countries such as Germany.

Portico, however, did not plan to seek broader environmental certifications for its total operations or for other types of wood. Torres continued to credit certification with helping the company open up opportunities to sell into accounts such as Home Depot that has wanted to insulate themselves from controversy or criticism over tropical deforestation. However, he was convinced that Portico kept its customers, not with certification, but for the package of services Portico offered, including price, quality, delivery, consistency, and service. Moreover, Torres had no clear evidence that retail customers would pay more for a door simply because it was made out of certified wood. Customers, he maintained, still based their buying decisions primarily on the price-to-quality ratio.

Portico's certified forestry, however, had positioned it for another strategic decision: In 1997 management planned to separate the forestry operations from the manufacturing and distribution division. The company made the decision in hopes of capitalizing on a growing interest in forestry investments among pension funds and other investors looking for diversification. Portico planned to raise capital to finance the acquisition of more forestland through the Environmental Enterprises Assistance Fund (EEAF) in Virginia, a nonprofit venture capital fund that provides long-term capital and management assistance to environmentally oriented businesses in developing countries. Portico's decision to adopt SFM, according to Torres, was responsible for creating the new opportunity to finance the acquisition of more forestland. "Investors are interested in us because we are a proven forest manager that can provide a reasonable long-term yield."

The Synergy of Forestry and Manufacturing

Portico confronted a business challenge—to assure supplies of raw material in an environment of diminishing supplies, controversy over logging in tropical forests, and antagonistic government policies toward the industry. Vertical integration and environmentally certified forestry proved to be linchpins in Portico's success. Portico's market position depends on product quality and reliability of delivery. Vertical integration allowed Portico to efficiently meet its need for high-quality wood with a long-term guarantee of quantity and quality. The company's investment in forestlands was basic to ensuring its wood supply and hence its ability to produce doors for export. The company's investment in environmentally certified forestry provided an advantage for entry into key distribution channels. Conversely, the value-added manufacturing enabled the company to absorb the added costs of sustainable forestry. In Portico's case, the synergy between the forestry operations and value-added manufacturing proved to be a potent benefit.

Despite Portico's success, however, it is unclear whether certified, high-value-added products actually command premium prices or whether the key benefit, at least in the United States, is simply gaining access to distribution channels. If environmental pressures on retailers moderate in the United States, there might be a concurrent reduction in any interest in certification. In addition, if the prices of certified products include high margins, those products may become vulnerable to undercutting by competitors offering products produced from traditionally harvested wood.

NOTES

The section on Parsons Pine is adapted from "Parsons Pine: Trash to Cash," prepared by Catherine L. Mater for the Sustainable Forestry Working Group, 1997.

The section on Colonial Craft is adapted from "Colonial Craft: A Rich Niche," prepared by Catherine L. Mater for the Sustainable Forestry Working Group, 1997.

The section on Portico S.A. is adapted from "Portico S.A.: The Power of Vertical Integration," prepared by Betty Diener for the Sustainable Forestry Working Group, 1997.

Chapter 9
_____

The Difficulties of Sustainable
Forest Management in the Tropics

Few forestry issues are as contentious or as ecologically significant as the destruction of the world's tropical forests. Between 1981 and 1990, about 154,000 square km of tropical rainforest were destroyed each year, an area slightly larger than the state of Georgia. In Latin America, 27 percent of the region's forests were destroyed by 1980; another 13 percent vanished by 1990. Tropical forests, which harbor 50 percent of all plants and animal species, are not only repositories of great ecological, medicinal, aesthetic, and economic value but they are also a defense against rising levels of carbon dioxide in the atmosphere that threaten to alter the earth's climate.

Public outcry and environmental activism led to international treaties and national government reforms to stem the destruction during the 1990s. But the deforestation continued—in some cases, it accelerated. In Brazil deforestation actually rose 34 percent between 1991 and 1994 during the annual burning season.[1] Although clearing forest for livestock and agriculture is a principal cause of Brazilian deforestation, the growth of the timber industry has played an increasingly important role. Not only does the burgeoning industry extract more timber but in the process, logging operations compact fragile soil and open the forest canopy. The disturbances damage plant and animal life and increase the risk of fire. Typically, for every tree

216

cut, 27 others are severely injured, 40 meters of road are built, and 600 square meters of canopy are opened. It can take 70 years for the forest to recover from the intrusion.

The Intricacies of Supply and Demand for Tropical Timber

Tropical hardwood producers compete in an industry with low barriers to entry, little product differentiation except in value-added activities, and relatively low capital requirements.[2] Only species and quality distinguish tropical sawnwood, which is a commodity product. Producers have little or no leverage to influence prices for their products in domestic or international markets. Quality is measured by the presence of defects in the wood (ranging from knots and holes to twisting or warping in dried wood), and wood is categorized according to its level of defects. Import markets such as those in Europe, Japan, and the United States typically buy only the highest-quality, lowest-defect woods, while domestic markets in Brazil and other developing countries tend to be less demanding. From 1950 to 1992, international tropical timber prices drifted up a modest average of 1.2 percent in real terms. During the 1990s, prices were stable due chiefly to competition from high-quality temperate hardwoods, which are good substitutes for many tropical wood products.

Demand for tropical wood varies by species. The properties and characteristics of each one differ considerably and determine how a particular wood is used. Some of the heavy, dense Brazilian species, for instance, are unsuitable for light construction woods, but their strength and resistance to decay make them popular in marine construction. Typically, buyers judge wood by characteristics of density, strength, appearance, and durability—but they need to know and be confident in a wood's suitability for a particular application before they will buy it.

Just a fraction of all the tropical wood produced enters international trade—about 6 percent. Tropical Asia, led by Malaysia and Indonesia, produces about 80 percent of the total exports. During the 1990s, the production of tropical roundwood slowed.[3] In some tropical countries demand for timber rose sharply. But because their own forest reserves are declining, exports from some tropical producers, including Indonesia and Malaysia, fell while imports rose. Bans on log exports enacted to stimulate domestic wood products manufacturing also contributed to the decline of Asian log exports. Although exports of value-added products, such as plywood and veneer, rose during the same period, it was not enough to offset the drop in log exports. Brazil, the dominant Latin American producer of tropical wood products, accounted for 8 percent of international exports in 1995. However, an influx of Asian firms to the Amazon in search of timber supplies that started in the early 1990s is an indication that Brazil's forests are considered increasingly valuable.[4]

The Volatile Amazon Timber Industry

Logging in the Amazon took off after the mid-1970s. The Brazilian government promoted cattle ranching in the hundreds of miles of Amazon forest opened up in the 1960s by the Brasilia-Belem highway that led to the country's new capital, Brasilia. The government offered subsidies for the costs of starting a ranch—as high as 75 percent—and property to those who could demonstrate they had cleared land. But cattle ranching began to lose its appeal in the 1980s as the nutrient-poor Amazon soil failed to produce enough grass for grazing and as fiscal incentives dried up. Faced with the collapse of their livelihood, ranchers turned to selling logging rights to their properties. Logging began selectively with just a few high-value species such as mahogany but grew steadily as hardwood stocks declined in southern Brazil. Prices rose in domestic and international markets. From 1976 to 1988, hardwood production in the north Amazon skyrocketed from just 7 million cubic meters a year to 25 million cubic meters. This transformed the wood industry into a dominant economic force in the northeastern Amazon. By 1990 loggers were extracting more than 100 species.

The Amazon timber industry developed most rapidly in Pará, a state of 22,000 square km in northeastern Brazil. In the mid-1990s, almost 90 percent of all roundwood in the Amazon was produced in Pará.[5] The majority of the growth in logging and sawmills took place between the late 1970s and the late 1990s (Figure 9.1). By 1990, the industry was composed of three principal types of firms: loggers, who extracted wood from the forest; processors, who cut and processed the logs; and value-added firms, includ-

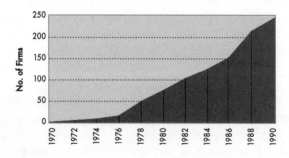

Figure 9.1.

Wood Processing Firms in Paragominas (1970–1990).

ing veneer and plywood producers. About 80 percent of the roughly 240 sawmills in Paragominas were small and produced 4000–5000 cubic meters per year. The remaining 20 percent were large mills that had annual production of 10,000–14,000 cubic meters. Almost all of these mills operated with used capital equipment of relatively low quality, and about 60 percent of the sawmills harvested and processed logs. Until the late 1980s, the mills sold most of their sawnwood domestically because demand was strong. But most firms were also too small and did not produce products of high enough quality to export.

During the 1990s, economic conditions for the industry deteriorated. Land ownership stabilized, and the forest became less accessible. As a result, the fees charged for cutting logs jumped more than 30 percent in real terms. During the same period, high *real* Brazilian interest rates slowed civil construction, depressed a principal source of demand for wood products, and caused real domestic prices for end products, principally sawnwood, to decline an average of 20 percent (Figure 9.2).

Even though international prices for export-quality tropical woods remained stable, Brazilian producers' margins were slashed by higher log costs, inflation, and the currency's appreciation.[6] The costs of electricity and diesel fuel also leaped almost 70 percent between 1990 and 1996. Although

Grade	Product	Average Price		%Change
		1990	1995	
High	Log	$60	$82	37%
	Sawnwood	$336	$291	-13%
Medium	Log	$38	$43	13%
	Sawnwood	$216	$174	-19%
Low	Log	$24	$30	25%
	Sawnwood	$168	$98	-42%
Very Low	Log	$18	$27	50%
	Sawnwood	$96	$89	-7%
Average	Log	$35	$46	30%
	Sawnwood	$204	$163	-20%

Figure 9.2.

Wood Prices in Paragominas, 1990 and 1995 (in constant 1995 US$ per m³)

Source: Steven Stone, "Economic Trends in the Timber Industry," p. 22

the cost of labor declined in real terms, firms continued to pay social security and other taxes on labor that amounted to an additional 50 percent of
salaries.[7] Profits for those firms that merely processed wood took the hard-
est hit from the combination of higher log prices and energy costs and lower
market prices. To enhance margins, many firms tried to vertically integrate
and capture economies of scale. Consolidation, particularly among smaller
firms, rippled through the region. Some firms expanded into value-added
processing, including the manufacturing of veneer and plywood. Between
1992 and 1997, consortia of six to seven of the largest processors created
"export clearing houses" that packaged wood for export from various suppliers. Each handled an estimated 40,000–50,000 cubic meters a year.

More Stringent Government Policies

Just as economic conditions were eroding the forestry industry's profits, the
government increased its oversight. The basic legal framework for regulation of the Brazilian forests, established in the Forest Code of 1965, set up
permanent preservation areas, created biological reserves, and established a
public forestry agency, IBAMA, responsible for the oversight of the forest
products industry. Subsequent decrees and regulations created additional
conditions that must be met before Amazon forests can be tapped. These
include management plans that must be approved by IBAMA before logging, prohibitions against felling trees smaller than 45 cm in diameter,
exporting logs and sawnwood thicker than 76 cm, and the cutting of certain
species such as the Brazil nut tree and the rubber tree.

But many of these regulations were not enforced. Management turnover
and inadequate financial support made IBAMA ineffective. In the mid-
1990s, the Brazilian government, stung by international criticism and lambasted at home by activists, toughened its regulatory efforts. New federal
protocols gave IBAMA broader powers, more staff, and greater financial
support. In 1996 IBAMA rejected 70 percent of all forestry projects submitted to it for approval on the basis of inadequate forest management plans.
Although the added government control raised costs for logging firms that
comply with regulations—forest management plans cost $10,000–$20,000
and could take six months to process—for those companies with well-struc-
tured sustainable forestry projects, the likelihood of government intervention or shutdown fell significantly.

An Alternative to Destruction

Sustainable forest management (SFM) has emerged as a set of less damaging practices for harvesting natural forests that may have the potential to

enhance the long-term economic value and ecological condition of tropical forests. But few commercially successful tropical SFM operations exist. In the early 1990, less than 1 percent of the world's tropical forests were managed even on a sustained-yield basis, an approach far less environmentally rigorous than SFM. By the mid-1990s, most experts believed the figure was much lower.[8]

Precious Woods, Ltd., one of the few companies attempting to practice SFM in the tropics, was launched by Swiss entrepreneur Roman Jann. Jann's interest in tropical reforestation was sparked in 1988 when, as a financial advisor to Swiss investor Dr. Anton E. Schrafl, he evaluated a Costa Rican teak reforestation project. Encouraged by the economic potential and ecological benefits of reforestation, the former corporate lawyer developed a proposal for an independent company. With contributions from Dr. Schrafl, Jann founded Precious Woods in 1990 to invest in tropical reforestation and hardwood plantation management in Costa Rica. Later he expanded operations to include an SFM operation in Brazil.

By year-end 1997, the company had raised $32.68 million in equity capital from European investors, hired leading forestry engineers and experts to craft and implement a management plan, and launched production and sales of tropical hardwood products to high-paying European markets. During the start-up phase in Brazil, however, Precious Woods suffered setbacks and a financial crisis, which prevented the company from meeting its revenue projections in 1996. The board instituted a turn-around strategy in late 1996 built around management changes and cost cutting. Despite this troubled beginning, a financial analysis suggests that Precious Woods' approach to sustainable forestry can, under the right conditions, meet its investors' expectations for financial returns and become a commercially viable alternative to conventional tropical timber operations.

The company's strategy hinges on harvesting multiple species of trees and finding appropriate niches for both sawnwood and semi-finished products made from those species in domestic and export markets. Both Precious Woods' financial success and the success of its SFM program, according to board member Daniel Heuer, depend on "having the best possible product mix by using all the resources harvested from the forest, not just the high-quality, export-grade portion of a few well-known species."

An Imported System of SFM

Shortly after launching its Costa Rican operations in 1990, Precious Woods considered expanding its activities into natural forest management in Brazil to widen the company's operations in tropical forestry and to enhance its medium-term cash flow. In late 1991, the company commissioned Hans

Peter Aeberhard, a Swiss economist with business management experience in Latin America, to prepare a feasibility study for expansion. A forest management plan and a capital investment program were developed. Precious Woods' Swiss board approved the Brazilian project in January 1994 and began further fund raising. The company then conducted due diligence for the purchase of appropriate forestland in the Amazon region. A forest management plan was completed and submitted for approval to IBAMA. In May 1994, the company bought 80,000 hectares of forest approximately 230 km east of Manaus along the Amazon River, 90 percent of which was undeveloped, for $4.3 million. The deal included some road infrastructure, buildings, and sawmilling equipment.[9]

From the outset, Precious Woods was organized to practice SFM in Brazil's primary tropical forests. The company's forest management program is based on the Celos model, a polycyclic (or multiple-cycle) forestry system developed in Suriname between 1965 and 1983 by Dr. Reitze de Graaf of the Wageningen Agricultural University in The Netherlands.[10] The Celos management system differs from conventional logging methods in key ways. The Celos system—through its organization of labor; use of low-impact logging techniques, controlled, regulated harvesting; and use of forest inventory as a planning tool—emphasizes protecting and sustaining the forest ecosystem while increasing the efficiency of timber extraction.[11] The silvicultural practices used in the Celos model stress active management to enhance the regeneration of trees between cutting cycles, especially the commercially desirable species.

Scientists at INPA, a scientific research institute based in Manaus, and Dr. de Graaf, who was retained as a scientific advisor, adapted the Celos SFM system to Amazon conditions. In the process, the company conducted a preliminary review of the topography and natural characteristics of its property. On the basis of those results, it set aside about 20 to 30 percent of its land, which included ecologically sensitive areas, in reserves. The remaining 70 to 80 percent of the forest was divided into 25 parcels of roughly equal size, about 2100 hectares. The division reflected the company's decision to adopt a period between each successive harvest of 25 years.[12]

As prescribed by the Celos system, Precious Wood's management plan called for the completion of a comprehensive forest inventory, or preliminary survey of the forest's characteristics, followed by a thorough prospecting of the entire forest. The prospecting review, conducted on each parcel—ideally at least six months to one year before that parcel is harvested—identifies each tree over 50 cm in diameter by species and location and documents topographical features such as gullies, steep hills, and other natural obstacles that can influence logging. The data from the inventory are then computerized to yield a map of each compartment (Figure 9.3). Foresters

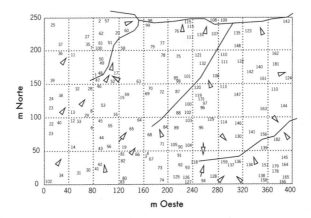

Figure 9.3.

Sample of Prospecting Printout Precious Woods.

Note: Chart of botanical and topographical data gathered in the field.

use the inventory to locate all of commercially desirable trees, determine which ones will be extracted, and map out roads and skid trails that provide the most direct, least damaging access to the trees. Loggers may use directional felling, cut vines, and use winch skidding—measures that minimize damage to nearby trees, the forest canopy, and the soil while the targeted trees are cut down and removed from the forest.

Silvicultural treatments are designed to increase the number of commercially desirable species and to enhance natural regeneration. These practices include cutting vines and other plants in the forest canopy to allow sunlight to reach the desired species, selectively killing undesirable species by cutting around the base of the trees, and applying an arboricide. In Suriname, these post-harvest steps increased growth in commercial timber volume by a factor of at least four, from about 0.5 to 2 cubic meters per hectare per year.

A Marketing Strategy Based on Quality

The many small competing tropical wood producers often have irregular access to supply, use low-quality capital equipment, and, it is often alleged, conduct business with a low level of professionalism. In this environment, the company that can consistently produce high-quality products, pay attention to customers' specifications, and provide professional service in orders and delivery has the potential to gain a competitive advantage. Precious Woods intended to market two forms of high-quality, sustainably harvested tropical wood products—sawnwood and semi-finished products—to high-paying international markets including Europe, the United States, and Asia.

In 1997, the company expected to sell approximately 24,000 cubic meters of sawnwood and approximately 2500 cubic meters of semi-finished products produced from roughly 30 different species. The company hired an international salesman in 1996 with lengthy international experience in tropical woods and strong sales contacts in Europe. Precious Woods bought capital equipment of sufficiently high quality to produce for European and other international market standards. The company's European sawmilling machines, for example, are capable of variable speeds; they have stelite-edged saw blades and a hydraulic log-rotator. With such equipment, Precious Woods opened its doors as one of the more technologically advanced tropical producers in Latin America.

The Importance of Lesser Known Species

Typically, loggers target one, two, or maybe a handful of commercially prized species in the rich tropical forests. Successful sustainable forest management, however, requires producers to selectively log as many of the diverse species that grow in tropical forests as possible, even those that may be unfamiliar in international markets. In this way, producers avoid over-harvesting and depleting the genetic stock of the few commercially popular species, such as mahogany.

But logging a variety of species is also essential to make an acceptable financial return. Producers need to generate sufficient harvest volumes—and thus sales—to be commercially viable. But SFM limits the volume of any one species that can be logged at a given time to maintain sustainable supplies of that species. Some 60 species with commercial potential grow on Precious Woods' land. But the better known species, such as Ipe, Amapa, and Jatoba, under SFM yield volumes that are just a fraction of the volumes the company needs to be commercially viable. In 1997 Precious Woods planned to produce 70,000 cubic meters of wood from 38 species. Just 1 percent of that volume was Jatoba, 2.6 percent Amapa, and 0.03 percent Ipe.

The Advantages of Semi-Finished Goods

A strategy based on the production of semi-finished goods, such as moldings and door parts, in combination with the production of sawnwood was designed to give Precious Woods advantages and efficiencies as it develops markets for multiple species. The company needs to find a suitable niche for each of the 30 or so species that it plans to harvest. The task "is easier if you offer processed goods" along with sawnwood, according to Heuer. Value-added production also enhances the company's margins. Semi-finished goods, which can be created from the scrap and odd-size pieces left over from the production of standard-size sawnwood, enable the company to prepackage less known tropical woods for end users. By doing so, Precious Woods accelerates market acceptance by predetermining which uses are suitable for each species. Finally, value-added production contributes to sustainability by creating jobs near the company's forests. "Sustainability is not only a question of ecology, but [is] also a social issue," commented Heuer. In 1997, the company planned to offer semi-finished products that included kiln-dried lumber, molding, door parts, parquet flooring, and certain furniture items.

The Essential European Market

Precious Woods' competitive strategy depends heavily on success in the premium European markets that pay strong prices for high-quality tropical woods. In 1995 those markets represented 21 percent of the total imports for tropical wood products. In June 1997, Precious Woods was certified by the SmartWood Program, which is accredited by the Forest Stewardship Council (FSC). Management had no expectations that FSC certification would necessarily bring the company a "green premium" for its products in Europe. But senior managers did anticipate that sustainable forestry would give it an edge in gaining access to those markets and forging strong relationships with European buyers, who are sensitive to tropical deforestation issues. According to World Bank studies, in the future certified sustainable producers like Precious Woods could capture as much as 20 percent of the European tropical timber import market, up from an estimated 1 to 3 percent in 1995.

Precious Woods' central challenge in competing for European sales will be supplying sufficient volumes of its diverse tropical species. By the standards of typical European buyers, Precious Woods' total production levels are low: Projected annual sales levels are 25,000–30,000 cubic meters, or roughly 3 percent of only one of Europe's smaller markets—The Netherlands. For most buyers of tropical hardwoods, a company's ability to supply minimum, regular volumes of a wood species is as important as the quality of the wood itself. Constant, ongoing supply becomes even more important in winning market acceptance for lesser-known species.

An Inauspicious Beginning

The management plan for Precious Woods' Brazilian SFM operation projected total start-up costs of $15 million, which included the purchase of the land, the purchase of equipment for harvesting and transporting logs, machinery for sawmilling, working capital, and project development expenses. By mid-1996, Precious Woods had completed an inventory of the first 4000 hectares, had logged roughly 1500 hectares in Compartment A—its designation for the first parcel of land scheduled for harvest—and had started to log Compartment B. The company held an inventory of more than 35 tropical wood species in sufficient volumes to begin sales and a scheduled ongoing monthly harvest of roughly 5500 cubic meters. The first sawmill line began operations midyear; the second was operational in early October 1996. In November 1996, the company had six kilns installed and functioning, with a total annual capacity of about 5000 cubic meters.[13] The company had placed orders for approximately $350,000 in semi-finished production machinery and expected to take delivery in February 1997. It had also opened discussions with a nearby Brazilian firm that manufactured parquet flooring to supply some 500 cubic meters per month of high-quality parquet flooring from its sustainably harvested wood for European markets.

The company, however, suffered a series of setbacks in 1995 and 1996 that caused costs to balloon and delayed the start-up of manufacturing operations. The chief difficulties were (1) inadequate administration and management of the capital investment program, which delayed production and caused losses of wood inventory; (2) expenses prior to the start-up of operations—about $9.95 million—which were higher than expected; (3) excessive operational costs, caused by the difficulties inherent in new business development in the Amazon; (4) inadequate financial controls and high administrative costs; and (5) a weak sales record that resulted in excess production (and further at-risk inventory) and ongoing cash shortages. High inflation and an appreciating exchange rate further increased costs.

In the summer of 1996 the board, faced with cost overruns and slower-than-expected sales, demanded stricter financial controls and more discipline over expenses. The board's concerns triggered operational and management changes in Brazil and resulted in the initiation of a program to cut costs and improve production efficiencies. Despite these changes, by November 1996 Precious Woods did not have enough cash left to finish out the year. In December 1996, the board intervened again and made sweeping changes. The president and chief executive officer of Precious Woods stepped down from both of his positions. The Executive Committee of the Board of Directors assumed the functions of CEO and president. The board hired a Swiss-Brazilian dual national with forest operations experience in the Amazon to take over responsibilities as General Manager of the Brazilian operations. A former chief financial officer of Champion International Brazil was hired to revamp the financial systems and controls of the Brazilian operation to improve financial reporting, cut costs, and increase operational efficiency. The company also revised its compensation for management to link pay to the company's financial performance.

With these changes in place, the company reported several improvements and successes in 1997. Swiss investors registered their confidence in the board of directors' intervention by investing an additional $3 million in Precious Woods in January 1997. These funds helped carry the company through early 1997 until sales and revenues picked up. The new management team put a third sawmill line into service, which increased overall production at the mill close to 100 cubic meters a day, the level needed to operate profitably. The average price for sales made in early 1997 was slightly below $300 per cubic meter. In May, the sales contracts secured prices of between $350 and $400 per cubic meter, which the company hoped to maintain during the remainder of 1997. By year end, the company had export orders for 25 different types of wood and four times more orders than it had wood to fill, which suggested that Precious Woods was gaining ground in marketing lesser-known species. In the case of acaricuara, a heavy, dense wood, a test contract to make reinforcement piles to protect the coastline of the Baltic Sea produced further orders for the spring and summer of 1998. On the domestic side, the company opened a sales outlet in Manaus as a showcase for its products and a testing ground for marketing lesser known species. The company completed the first model for a do-it-yourself wood house to be marketed in cooperation with partners in Brazil. For the year, Precious Woods processed about 46,000 cubic meters of roundwood, produced 16,758 cubic meters of sawnwood, packaged 12,107 cubic meters for delivery, and produced 212 cubic meters of finished and semi-finished products. Finally, the company's efforts to cut administrative and supervisory costs by decentralizing the management structure lowered those costs from about $2 million in 1995 to a projected $750,000 in 1998.

Precious Woods ended 1997 with a $5.38 million operating loss and an extraordinary write-off of over $6 million, most of it related to the excessive preoperational expenses of the Brazilian operation. In its first full year, the Brazilian operation had total costs of $1.5 million a quarter and a higher-than-expected operating loss. The overall quality of timber from company lands was disappointing—too many trees had holes in them or the diameter was small—and the company was forced to write off the wood that did not meet its quality requirements. The variety of wood handled at the sawmill—over 30 types of trees with various characteristics and dimensions—presented a complex logistics problem for temporary storage: Some wood was damaged and written off as well. And a Philippine customer that ordered some 3300 cubic meters of wood was unable to pay for it and did not collect the merchandise, which is the chief reason the company produced more wood that it sold.

To address those issues, management restructured operations at the sawmill, changed wood storage, and increased the minimum diameter of harvested timber to 60 cm late in the year. Heuer predicted that Precious Woods would produce some 15,000 cubic meters of product in 1998, ring up sales between $4.5 million and $5.8 million, achieve a 45- to 50-percent yield in the sawmill, and break even in 1998.

Profits in the Future?

Precious Woods' start-up troubles and nascent sales record complicate any evaluation of its future profitability. Charles A. Webster and Diana Propper de Callejon of Environmental Advantage, Inc., however, concluded that the company's income statement could indeed improve by year-end 1999. Their analysis showed, using a set of conservative assumptions, that the company had the potential to generate a net margin of 28 percent within two years under the following circumstances: (1) Sales would rise to 23,800 cubic meters; (2) sales of semi-finished products would grow from zero in 1996 to 5500 cubic meters; (3) the company's overall rate of converting timber to product would rise from 35 percent in 1996 to 47 percent; and (4) the company would cut operational costs 30 percent.

If that were the case, even if prices for its wood products did not rise significantly from their 1996 levels, Precious Woods could generate a net margin of 28 percent. Total equity capital allocated to Brazil was approximately $23 million in early 1997. Assuming that an additional $5 million to $6 million in shareholders' funds would be required until the company broke even in 1998, the company's net margin of 28 percent could produce a return on equity (ROE) of about 11 to 12 percent in 1999.[14] What is significant about these numbers is that the company's projected returns would meet the range of returns (11 to 16 percent) that the investors expected at the time they made their long-term investment in Precious Woods.

The Risks for Precious Woods

But Precious Woods faced several risks to its eventual profitability in 1997. Inadequate sales were the principal risk factor. The two most important variables for Precious Woods' profitability will be the *volume* of product it is able to sell and the *price* at which it sells this product. In late 1996, Precious Woods tallied its conversion rate of natural logs to European-quality sawnwood at an average of 35 percent, which is fairly low.[15] But the company expected that this yield would improve with the addition of semi-finished products—to as high as 50 to 55 percent.

The importance of these variables becomes clear in a sensitivity analysis.

If, for instance, Precious Woods had a conversion rate of 50 percent for the 70,000 cubic meters of wood it expected to harvest and sold that wood at $350 per cubic meter, it might realize a net profit as high as $4.9 million and a net margin of 40 percent. If the company had a conversion rate of slightly more than 35 percent—about its 1996 rate—net profits would drop to about $1.9 million and net margins to 22 percent. But if Precious Woods' conversion rate stayed at slightly more than 35 percent and its wood sold for $250 per cubic meter, the company would lose about $220,000 and have a net margin of − 4 percent. The company, however, did not begin implementing its semi-finished goods strategy until 1997. The program's success hinged not only on adequate operational management of semi-finished goods but also on market acceptance of those products.

Other factors posed risks for the company as well. Inadequate management in Brazil was still a potential problem. Many of the company's financial and operational start-up difficulties stemmed from its corporate structure, which split the CEO's time and efforts between activities in Costa Rica, Switzerland, and Brazil and did not place management experienced in Brazilian forestry on-the-ground in Brazil. The results of the changes made by the board in late 1996 to rectify the situation were still not in by year-end 1997.

The company's cost-cutting program is also critical to its future profitability. But given the economic environment in Brazil, Precious Woods may find it difficult to cut costs sufficiently to ensure profitability. Precious Woods had a higher cost structure than other firms in the area, which leaves the company less flexibility than competitors might have in a difficult economic environment. The company's management team, former managers at Champion International, were expensive to hire. The company's logging costs, due in part to SFM, are higher. So are sales and marketing costs to develop markets for lesser known species, which is essential for commercially viable SFM. Sales to export markets also cost more since those markets demand higher-quality products. Although some of these costs will come down over time, projections for profits assume that the company cuts expenses in real terms. Inflationary pressures as well as continued appreciation of the Brazilian currency will make it more difficult for the company to reach its expense reduction goals of 20 percent in 1997 and an additional 10 percent in 1998.

The Potential to Improve Profits

Precious Woods' distinctive position as a sustainable harvester of high-quality tropical woods gives it advantages that it can leverage toward profitable growth. The company's industry-leading SFM program limits its exposure to environmental liabilities and creates opportunities to differentiate its products. High-quality machinery will enable Precious Woods to produce products that meet the standards of high-paying European and international markets. And as a vertically integrated producer that owns its own land, Precious Woods is protected from the rising costs of wood.

The company has other opportunities to improve profits. Greater familiarity with the characteristics and properties of its many species will make Precious Woods more efficient in both harvest and production. The company has already discovered, for example, that certain species have extremely high levels of defects and yield low output in sawnwood; it will stop harvests of these species in favor of lower-defect woods. Other woods have proven suitable for end products such as parquet flooring; so the company will increase their use. The expansion into semi-finished products should cut wood waste while boosting revenues. Using wood chips to generate electricity could cut the costs of diesel fuel. And as the company's investment program nears completion, senior management will be able to improve operational efficiency in a number of areas ranging from divesting unnecessary assets and regulating production to minimizing wood damage in inventory, reducing administrative personnel, and streamlining sales and order procedures.

In 1997, Precious Woods also had additional opportunities to capture economies of scale in wood processing that would enhance profitability, including an option to purchase an additional 270,000 hectares of neighboring primary forest. The company was in no financial position to consider the expansion at the time. But in the future, processing larger volumes of wood could improve the company's efficiencies and help the company gain market acceptance for its lesser-known species.

Success: An Issue of Business Management Not Sustainable Forestry

Precious Woods' start-up difficulties and inadequate sales record make an analysis of the company's profitability imprecise and broad conclusions about the viability of its SFM model premature. In 1997 the company had not yet demonstrated that it could successfully sell lesser-known species, which is essential for profitable SFM. The financial control systems needed further improvements to accurately measure how much SFM will add to costs. It was still unclear whether the company's higher cost structure was caused by its SFM program or the expense of establishing high-quality operations that meet European standards.

Nevertheless, even at an early juncture in Precious Woods history, the company's experience is an encouraging one for the future of SFM in tropical forests. Certainly, some aspects of Precious Woods' experience are unusual. Most important of these was the company's ability to raise almost $32 million in private equity capital, a feat that may be difficult to replicate. But in Precious Woods' short history, issues of business management, exacerbated by the economic climate of the region, emerged as the most formidable challenges for the company—not issues of forest management. For the next several years, the importance of maintaining a stable, experienced management team, implementing efficiencies throughout operations, instituting sophisticated financial systems, and successful marketing will continue to be the linchpins for the company's success. If Precious Woods can meet those challenges, it should operate at an attractive profit. The projected net margin of roughly 28 percent compares favorably with other wood processing businesses in the Amazon. Companies that follow Precious Woods into tropical sustainable forestry may fare even better: They should be able to save investment and even operational expenses, create a leaner capital structure, and generate more attractive returns on shareholder funds. Such a performance would demonstrate that SFM has the potential for commercial success in tropical forests.

NOTES

This chapter is adapted from "Precious Woods: Meeting the Challenge of Tropical Forestry," prepared by Diane Propper de Callejon and Charles A. Webster, Environmental Advantage, Inc., for the Sustainable Forestry Working Group, 1997.

1. "Amazon Deforestation Has Accelerated," *The New York Times,* September 12, 1996.
2. Capital investment requirements rise with the level of value-added products. In developing countries such as Brazil, where interest rates have been extremely high by industrialized standards, the cost of capital represents an important barrier to entry.
3. See Joanne C. Burgess (1993), "Timber Production, Timber Trade and Tropical Deforestation," *Ambio* vol. 22, May, 136–137. See also International Tropical Timber Organization (ITTO), *1995 Annual Review,* production tables, 46–53.
4. Several large Asian wood products firms recently purchased land and began logging and processing operations in the Amazon, as most of the high-quality, easily accessible hardwood forests in their home countries have been depleted. See *The Wall Street Journal* November 11, 1996, 1.
5. In 1992, logging and sawmill companies in Pará generated sales that totaled roughly $190 million and earned approximately $62 million in profits (or an impressive net margin of almost 33 percent). The industry also generated almost 6000 jobs—in Paragominas, a city in the center of Pará's forestry activ-

ity, roughly 56 percent of the urban population depended directly on the wood industry for its livelihood. The next biggest industry, ranching, was only 10 to 20 percent of this size.

6. In the 1990s, Brazilian macroeconomic policy was characterized by monetary policies to control inflation, which resulted in high real interest rates. The government also used the *real* as a tool to control inflationary pressure, allowing the currency's value to adjust, but at a slower rate than that of inflation. As a result, the currency appreciated in real terms, which squeezed profit margins further for export firms whose costs are principally denominated in *reals* and whose revenue is denominated in hard currencies such as the U.S. dollar.

7. Real interest rates in Brazil remained above 25 percent per year for most of the 1990s, falling briefly in 1994 but rising sharply again under the Plano Real.

8. Statistics come from Richard Rice, Raymond E. Gullison, and John W. Reid (1996). "Can Sustainable Management Save Tropical Forests," unpublished manuscript, September 24, 4; current situation from author's personal research.

9. A total of $4.3 million was paid for 80,900 hectares of land. This land, however, contained roughly $565,000 in houses, buildings, sawmill equipment, and other infrastructure. In addition, approximately 5500 hectares of the land area were already deforested, which is more expensive in the Brazilian land market. Thus, the company asserts it paid an equivalent of $35 per hectare for 75,400 hectares of primary tropical forest. The company considered this a fair price.

10. The four principal SFM systems include the Malayan Uniform System, the Tropical Shelterwood System, the Strip-Clearcut System, and the Suriname Celos System. For further detail, see Reid and Rice (note 8 above), "Natural Forest Management," 3–6.

11. For additional detail, see *The Celos Management System: A Polycyclic Method for Sustained Timber Production in South American Rain Forest,* by N.R. de Graaf and R.L.H. Poels.

12. This time period was derived from an estimate of the average volume of wood that could be harvested per hectare of natural rainforest (approximately 35 cubic meters) and from the estimated growth rate of the forest (1.5 to 1.8 cubic meters per hectare per year). Thus over 25 years, the forest would regenerate at least 37.5 cubic meters of wood (25 × 1.5), more than replacing what had been harvested. For more detail on the scientific studies contributing to these statistics, see writings by Silva et al. in *Forest Ecology and Management,* 1995.

13. The company believed annual capacity could be as high as 7000 to 8000 cubic meters per year, depending on the type of wood being dried (less-dense wood generally dries faster) and the dimensions of the wood (smaller pieces meant greater overall surface area and faster drying).

14. Total equity capital allocated to Brazil to date sums to approximately $23 million (with the remainder of monies raised allocated to Costa Rica). Adding $5 million in equity capital gives a base of shareholders funds of $28 million. Earnings of roughly $2 million are projected for 1998, producing an ROE of 7 percent. Note that original profitability projections used by management to raise financing for Precious Woods cited a constant dollar-based IRR for the project of 11 to 12 percent, indicating that investor expectations for profitability have been quite modest (particularly in light of the many risks of the project).

15. The forestry industry is renowned for its low efficiency rates in converting raw material into product. U.S. manufacturers have yields of 40 to 60 percent; Brazilian tropical wood processors average 30 to 35 percent conversion rates.

REFERENCES

Almeida, Oriana Trindade de, and Christopher Uhl (1995). "Brazil's Rural Land Tax." *Land Use Policy* vol. 12, no. 2, 105–114.

Barreto, Paulo, Paulo Amoral, Edson Vidal, and Christopher Uhl (1996). "Impacts of Forest Management on the Economics of Timber Extraction in Eastern Amazonia." Unpublished manuscript.

Barros, Ana Cristina, and Christopher Uhl (1995). "Logging Along the Amazon River and Estuary: Patterns, Problems and Potential." *Forest Ecology and Management* April.

Browder, John O., Eraldo Aparecido Trondoli Matricardi, and Wilson Soares Abdala (1996). "Is Sustainable Tropical Timber Financially Viable?" *Ecological Economics* 16, 147–159.

Burgess, Joanne C. (1993). "Timber Production, Timber Trade and Tropical Deforestation." *Ambio* vol. 22, no. 2-3, May.

de Graaf, N.R., and R.L.H. Poels. *The Celos Management System: A Polycyclic Method for Sustained Timber Production in South American Rain Forest.*

Hardner, Jared J., and Richard E. Rice (1994). "Financial Constraints to Sustainable Selective Harvesting of Forests in the Eastern Amazon." DESFIL paper prepared under AID contract, June.

Holloway, Marguerite (1993). "Sustaining the Amazon." *Scientific American* July, 91–99.

Howard, Andrew F., and Juvenal Valerio (1996). "Financial Returns from Sustainable Forest Management and Selected Agricultural Land-Use Options in Costa Rica." *Forest Ecology and Management* 81, 35–49.

Kishor, Nalin M., and Luis F. Constantino (1993). "Forest Management and Competing Land Uses: An Economic Analysis for Costa Rica." LATEN Dissemination Note #7, the World Bank, October.

Pancel, L., ed. (1993). *The Tropical Forestry Handbook* vol. 2. Germany: Springer-Verlag.

Reid, John W., and Richard E. Rice. "Natural Forest Management as a Tool for Tropical Forest Conservation: Does It Work?" Forthcoming in *Ambio,* submitted September 1996.

Rice, Richard E., Raymond E. Gullison, and John W. Reid (1996). "Can Sustainable Management Save Tropical Forests." Unpublished manuscript, September 24.

Stone, Steven W. (1996). "Economic Trends in the Timber Industry of the Brazilian Amazon: Evidence from Paragominas." CREED Working Paper Series No. 6, July.

Verissimo, Adalberto, et al. (1992). "Logging Impacts and Prospects for Sustainable Forest Management in an old Amazonian Frontier: The Case of Paragominas." *Forest Ecology and Management* 55, 169–199.

Verissimo, Adalberto, Paulo Barreto, Ricardo Tarifa, and Christopher Uhl (1995). "Extraction of a High-Value Natural Resource in Amazonia: The Case of Mahogany." *Forest Ecology and Management* 72, 39–60.

Vincent, Jeffrey R. (1992). "The Tropical Timber Trade and Sustainable Development." *Science* vol. 256, June 19, 1651–1656.

Chapter 10

The Socioeconomics of Brazilian Pulp Plantations

Fast-growth plantations in the Southern Hemisphere seem to be tailor-made for an industry facing tight wood supplies. In the mild climate, these plantations—rows of trees, often clones—can produce from five to 10 times more fiber per hectare than a natural forest in a similar locale.[1] That fiber is also more uniform and of better quality for pulp than fiber produced by natural forests, making these plantations particularly suitable for an industry marching toward greater standardization. Increasingly, environmentalists acknowledge that plantations can take pressure off natural forests and protect land and wildlife. The Forest Stewardship Council, the largest independent arbiter of sustainable forest management, accepts plantations as a "sustainable alternative to natural forests that can contribute to satisfying the world's need for forest products."

However, critics—and even supporters—will reel off a litany of hazards that these monoculture agricultural systems can present for sustainability. The efficiencies of plantation forestry can challenge the environmental, economic, and social sustainability of neighboring communities. Plantations, with their more simplified ecosystems, do not offer all of the environmental services—everything from water storage to soil regeneration—of "full-service" forests, nor do they support as rich a diversity of plant, microbial,

235

and animal life. Large-scale corporate plantations also have the potential to alter nearby economies and disrupt the social fabric of communities. In Indonesia, which will be discussed later, fast-growth plantations have been a government priority that conflicts with social and ecological sustainability. It is the tension between efficiency and sustainability that makes plantations controversial.

In Brazil, a center of fast-growth plantations for the pulp and paper industry, two producers, Aracruz Celulose S.A. and Riocell S.A., take different approaches to the challenges of balancing efficiency and sustainability on fast-growth plantations. Aracruz, the world's largest and lowest-cost pulp producer, has transformed environmental issues into opportunities for efficiencies and new markets. Riocell, however, considers environmental issues primarily as an added cost for its manufacturing operations. Although each company improved its environmental performance during the 1990s in response to pressures, both still come under fire for a variety of environmental and social issues. Their experiences illustrate the promises and pitfalls inherent in the sustainable management of fast-growth plantations.

Aracruz Celulose S.A.

In 1992, while the world's attention was riveted on Brazil during the United Nations Conference on Environment and Development (UNCED), the Greenpeace vessel Rainbow Warrior II, a veteran of high-profile environmental confrontations, sailed into Portocel, blockading the private terminal of Aracruz Celulose. The international environmental group and local activists lambasted the pulp producer for its plantations that they blamed for destroying 10,000 hectares of endangered Atlantic rainforest.

The action was one more graphic illustration of the continuing tension and contradiction between Aracruz's business success and its role as a lightning rod for the environmental and social controversies that surround the Brazilian forest products industry. As the world's largest lowest-cost producer of bleached eucalyptus pulp, Aracruz Celulose is one of Brazil's great business successes. Brazil's favorable climate and Aracruz's technological superiority in cloning and growing trees, have given the company a competitive advantage in its industry that financial analysts estimate would take a well-financed competitor over two decades to match. Aracruz consistently generates positive cash flow and high margins, no easy feat in a commodity market plagued by rapid price fluctuations. Through sustainable yield forestry and top quality manufacturing, Aracruz has become a global leader and a model for industrialization within Brazil. Supporters even credit the pulp giant with reclaiming eroded, abandoned agricultural lands and improving local environmental and social conditions.

Yet the company's public image has not matched its business achievements. For years, Aracruz has fended off critics that routinely target it for some of the very actions that its supporters credit. Detractors claim that Aracruz's plantation forestry is insensitive to the rights and needs of local communities, impoverishes and dries up the soil, destroys wildlife, and causes unhealthy levels of air pollution.

The Volatile Pulp Market

Both Aracruz and Riocell compete in a growing market. In 1995, the global production of wood pulp was 168 million metric tons, according to the 1995 FAO Forest Products Yearbook; the pulp and paper industry's total sales were estimated to be between $500 billion and $650 billion.[2] The FAO projects that global pulp consumption will rise 2.9 percent annually between 1991 and 2010. Meanwhile, wood pulp production is projected to increase just 1.1 percent a year.[3] The pressures on demand are expected to bring higher prices for the pulp producers fortunate enough to have a stable fiber supply.

But the pulp industry is cyclical, driven by high capital intensity, uneven supply, and wide fluctuations in price year to year. In recent years, improvements in technology, transportation, and communication that enable producers to sell their pulp worldwide have only accentuated the traditional pulp cycle. Now price swings are steeper and more frequent, and average margins throughout the industry are generally low. To mitigate some of the cyclical effects on earnings, most pulp producers combine pulp production with paper product manufacturing.

A Market Heavyweight

Aracruz, headquartered in Rio de Janeiro, was founded in 1967 by Erling Lorentzen, a Norwegian-born businessman. Lorentzen and his partners saw the advantages of Brazilian tree plantations for fiber export and decided that domestic conversion to pulp made good business sense. With help from favorable national government policies, the company bought 30,000 hectares in southeastern Brazil from the state government at a low price, planted eucalyptus seeds, and constructed a pulp mill.

Thirty years later Aracruz holds a commanding share in its markets. In 1996, the company's Espirito Santo manufacturing plant had a capacity to produce some 1,025,000 tons of bleached hardwood kraft eucalyptus pulp (BHK). Aracruz pulp has earned a reputation of "premium" grade, according to analysts, and is used in the production of fine writing papers, tissue, and other paper products. In 1996 Aracruz supplied a full 19 percent of the world's bleached eucalyptus pulp, making it the industry leader. In the United States in first-quarter 1997, Aracruz held 64 percent of the market for bleached eucalyptus kraft market pulp (BEKP). Overall, Aracruz supplied 3 percent of the total world capacity for chemical market pulp. In the

1990s, Aracruz set its sights on expanding sales in Europe and Asia, and in first-quarter 1997 the company held market shares of 13 percent and 25 percent, respectively, in the regional BEKP markets.

The Golden Triangle for Eucalyptus

Aracruz's lead in the pulp market rests in part on advantages of geography and the species of genetically improved eucalyptus trees that it uses. Eucalyptus, a native of Australia and some parts of Indonesia, is a highly efficient fiber producer, which is why it is a favorite worldwide for plantation forestry. But nowhere does eucalyptus grow as well as in southeastern Brazil, Uruguay, and Paraguay. In this golden triangle for eucalyptus, the trees produce over twice as much wood per hectare as they do in South Africa, the next most productive region (Figure 10.1). This hyper productivity gives Brazil an enormous leg-up in production costs. In 1995, for instance, Brazil produced fiber for pulp at a low $97/ton. The southern United States produced pulp at $112/ton, and in the world's most expensive growing area, Finland, the cost was $276/ton (Figure 10.2).

The Pursuit of Efficiency

Even though Aracruz markets a premium pulp, which attracts long-term customers, its success in a commodity market is tied primarily to its ability to control costs. In the late 1980s, Aracruz adopted an aggressive cost-cutting strategy that has made it the world's low-cost producer of bleached eucalyptus pulp. Aracruz's emphasis on eco-efficiency—the efficient use of raw materials through waste minimization—in the forest and mill is a major contributor to its low costs.

As company documents make clear, Aracruz Celulose is in the business

Region	Species	Rotation (yrs)	m³/ha/yr
Brazil (Aracruz)	hybrid	7	44.4
South Africa	E. grandis	8-10	20.0
Chile	E. globulus	10-12	20.0
Portugal	E. globulus	12-15	12.0
Spain	E. globulus	12-15	10.0

Figure 10.1.

Rotation Age and Average Productivity of Major Eucalyptus Growing Regions.
Source: Aracruz, Facts & Figures, January 1996

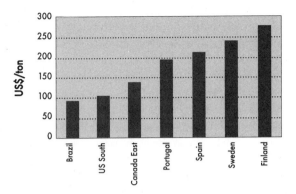

Figure 10.2.

Regional Fiber Costs per Ton.
Source: Oppenheimer & Co., Inc., International
Research Latin America, Klabim, 4/9/96

of agriculture to produce a crop of eucalyptus as efficiently as possible on its 203,000 hectares of forestlands. Of those, 132,000 hectares are sustained-yield eucalyptus plantations, mostly on land previously degraded by agriculture, and about 57,000 hectares are set aside as natural forests, according to the law.

Aracruz has developed an intensive management system for growing eucalyptus that produces top yields in the industry and fibers of unusually high quality. Seedlings are produced primarily through coppiced shoot cloning, known as "vegetative propagation." This way, the company selects individual trees with superior characteristics and makes many identical copies for planting. Aracruz foresters then match individual varieties with the sites best suited for them, which reduces the need for fertilizer, pesticides, or other activities that have a negative environmental impact. Aracruz uses three major species of eucalyptus: *E. grandis, E. urophylla,* and, most often, a hybrid of the two. The short maturation period of Brazilian eucalyptus—tree height can reach 100 feet after seven years—has enabled Aracruz to genetically improve its trees quickly. Aracruz's plantations deliver an average yield of 45 cubic meters per hectare each year, up from 30 cubic meters in 1990—well above the national average yield of 20 cubic meters per hectare annually in Chile, the closest competing region.

The cloned seedlings are first grown in the company's nurseries, then moved to a prepared site on either Aracruz Celulose land or occasionally to land owned by a local farmer. For the first few years, the company supplies the nutrients that the seedlings need, including pesticides and herbicides that may be applied to protect the seedlings from undergrowth. When the seedlings are tall enough, natural undergrowth is allowed to develop. When the trees are ready for harvest after seven years, they are felled and trimmed, and the undergrowth is knocked down. Recently, Aracruz adopted high-tech harvesters to replace the chain saws formerly used in logging, thus improving yields and efficiencies. The logs are then sent to the mill. The crowns and smaller branches are left on site, and larger branches may be sent to charcoal manufacturing operations. After harvesting, the plots are recovered by natural regeneration through sprouting (coppicing) and anoth-er production cycle is initiated without having to plant. After the second harvest, the site is prepared for the next planting. The branches and other plant material left behind decompose to return nutrients to the soil—fertil-izer is also added. Then a new batch of eucalyptus seedlings is planted, and the cycle begins anew.

Eco-Efficiencies in the Mill

Aracruz has pursued efficiency as relentlessly in its pulping operations as it has in forestry through technology improvements and waste minimiza-tion during the 1990s. These efficiencies have not only improved the com-pany's bottom line but they have also reduced the company's impact on the environment.

Pulp mills typically place serious burdens on the local environment. The logs that enter the plant must be cleaned, chipped, chemically treated to remove lignin, dried, and often bleached. These operations use massive amounts of water and energy and produce large amounts of chemical and biological waste, which is emitted into the air, water, and soil. Over the years, pulp producers have increased their efficiency to lower costs, result-ing in the environmental benefit of lowering emissions.

Few pulp producers are as efficient as Aracruz. In 1997 the company met 87 percent of its own energy requirements by using waste and recycling 94 percent of the chemicals used in the digestion phases of pulp production. Bark from the logs, for instance, becomes fuel for boilers and dryers. And when lignin is extracted through the company's kraft process, the wastes are again used as fuel. Between 1990 and 1997, biological oxygen demand (BOD) effluent levels created by organic wastes plunged 90 percent, and lev-els of toxic absorbable organic halogenated compounds (AOX) plunged even more. Those effluents that remain are piped more than one mile out to sea before being released.

All of Aracruz's pulp is bleached. In the past, most pulp bleaching processes used elemental chlorine as the main active agent, a practice that has been linked to the creation of toxic dioxin. Since the discovery of its tox-icity, pressure has mounted for producers to move away from elemental chlorine, and a market has developed, primarily in Europe, for totally chlorine-free (TCF) pulp. Aracruz reacted quickly. In 1997, 56 percent of the company's production was elemental chlorine-free (ECF) and another 11 percent was TCF. The company was phasing out all elemental chlorine-based production, so that eventually it will produce only ECF or TCF.

Impressive Productivity Gains

Aracruz's efficient forestry and successful waste minimization at the mill have helped increase productivity. But other, nonenvironmentally related efficiency gains are also responsible for a rise in productivity since 1989. Through down-sizing and outsourcing, the number of Aracruz employees has dropped from 8301 to 2600 in 1996. In that period Aracruz's productivity leaped from 67 tons per employee in 1989 to 407 tons in 1996. Within a low-cost region, Aracruz is a more efficient producer of pulp per ton than two local competitors (Figure 10.3).

Aracruz's programs for eco-efficiency have been an integral part of a cost-cutting strategy that has improved financial performance. Annual production costs, according to the company, range from $260 to $310 per ton. Between 1990 and 1996, prices in the cyclical market for Aracruz pulp fluctuated from $641 per ton in 1990 to a low of $366 per ton in 1993, back to a high of $810 per ton in 1995, then plunged to $469 per ton in 1996. But as shown in Figure 10.4, which compares Aracruz's average price (FOB) of pulp to its production costs per ton from 1990 through 1996, the company's pulp has consistently fetched prices comfortably above production costs.

In 1995 sales totaled $796 million, then dropped to $516 million in 1996, reflecting a downturn in the pulp market. But since Aracruz's price for pulp has not recently dipped below $364 per ton, the company has maintained a consistent positive net cash flow—from a low of $74 million in 1993 to a

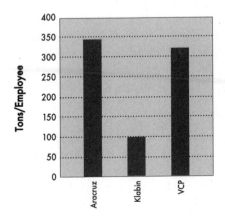

Figure 10.3.

Brazilian Pulp and Paper Industry: Metric Tons Produced per Employee, 1995.
Source: Oppenheimer & Co., Inc., International Research Latin America, Klabin, 4/9/96

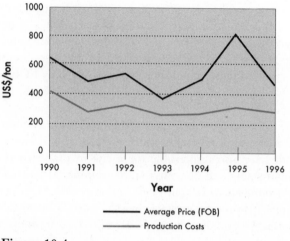

Figure 10.4.

Aracruz Average Price (FOB) vs. Production Costs.

high of $524 million in 1995. Even in the downturn year of 1996, Aracruz achieved a net cash flow of $193 million. Aracruz also earns hefty pretax margins. In 1995, for instance, Aracruz earned 52.3 percent, a performance that compared favorably to top North American pulp and paper companies whose margins that year ranged from 8 to 24 percent and to Brazilian euca-lyptus-growing competitors such as Klabin, who achieved a 39.8 percent margin.

For about 10 years Aracruz enjoyed a significant tax holiday that ended in May 1997. The terms exempted Aracruz from taxes on all export revenues, over 90 percent of sales, as long as the company met a number of criteria. These criteria included performing within established environmental limits and maintaining a positive foreign-trade balance. By the end of the period, the company had met all of the conditions. These tax concessions gave Aracruz a tremendous advantage as it expanded into the global market. In the future, analysts estimate that Aracruz will have a 10 percent effective tax rate.

A Controversial Social Performance

The company, however, has had far greater difficulty building a reputation for social responsibility than for business acumen. From the rights of native peoples to investment in local communities, there is a wide gap between management's perceptions of company actions and those of its critics.

Indigenous Peoples' Land Claims

When Aracruz originally purchased its land from the legally recognized owners in the late 1960s, it inherited land ownership disputes. An indigenous tribe, the Tupinikim, claimed that it had had rights to some 13,000 hectares of Aracruz's land since the 1600s and brought a complaint before the National Indian Foundation (FUNAI). Critics contended that as many as 32 indigenous communities may have been forced to relocate during Aracruz's start-up period. Aracruz management contended that when the company acquired the land, no indigenous peoples lived there, so relocation was unnecessary. The company, however, donated 1819 hectares of land, as well as materials and equipment, to help establish a permanent reservation for the communities. Aracruz supports a government-NGO-private sector partnership program to provide assistance to indigenous peoples called the Interinstitutional Indigenous Health Nucleus (NISI). In 1998, the company finally settled the dispute. As part of a 20-year agreement, Aracruz agreed to pay a total of $10 million to support the collective needs of the indigenous community.

Working Conditions

Another gap between management's and critics' perceptions of the company is the characterization that Aracruz is unfair to workers. Labor unions have brought several lawsuits that have alleged poor working conditions and have pressed for greater benefits. In some cases they have been successful. Critics contend that the company often lays off experienced workers without severance benefits, then replaces them with younger, cheaper laborers; and they fault working conditions. Some activists have even suggested that the percentage of workers who retire with disabilities is unnecessarily high.

Aracruz managers express surprise over the criticism. According to the company, it pays the highest salaries in the Brazilian pulp and paper industry, has a variable compensation system for employees that links corporate performance to bonuses, and has a competitive fringe benefits package that includes vacation bonuses, medical and dental assistance, and more. "Furthermore," Aracruz management wrote, "the company has always been open to dialogue with labor unions."

The company also exerts a powerful—albeit sometimes controversial—influence on the local economy. Between 1990 and 1995 as part of its cost-cutting, Aracruz downsized by some 4440 employees while it doubled production. The move disrupted local communities and ignited a firestorm of criticism. At the same time, Aracruz contracted out many of its operations to local firms to soften the economic blow. In 1997 Aracruz increased capacity by 20 percent, which generated more local employment. As part of the outsourcing and expansion, Aracruz helps local farmers grow eucalyptus for its mill under its Forestry Farmers Program. The company has also invested more than $3 billion in the local community, more than $125 million of that for social programs. Aracruz also supplies its 3000 employees and their 8000 dependents with housing, a school, a hospital, and other infrastructure improvements.

Without Aracruz, the weak economies of the communities of Espirito Santo would undoubtedly be in worse shape. The company's investments have provided services for employees, such as schooling and health care, that might not otherwise be available. But the extent of the criticism that still rains down on the company suggests that Aracruz has not successfully engaged local communities to build strong relationships between the company and the community.

Contentious Environmental Issues

The company's forestry operations and its pulp mill create a significant impact on the surrounding areas. Although plantation forestry has become more acceptable to environmentalists and Aracruz practices progressive plantation management, here, too, the company is not shielded from criticism.

The Criticism of Aracruz's Environmental Performance

Although the Aracruz system produces enviable efficiencies, critics argue that the intensive forestry methods pose hazards for ecological sustainability because they deplete the soils, damage the water supply, and fail to preserve biological diversity. In 1995, The United Nations Environment Program (UNEP) warned of the dangers of altering or simplifying ecosystems to maximize yields.[4] The resulting losses in biodiversity can reduce the ability of ecosystems to adjust to environmental changes; damage their capacity to provide clean soil, water, and air; and deny people the future use of species of as yet undiscovered value. According to that report, simplified ecosystems, which have a diminished capacity to provide free services such as controlled nutrient delivery and pest control, need to be subsidized by the use of fertilizers and pesticides. The introduction of alien species for cultivation can have a negative impact on ecosystems because there are no local biological controls to help keep the invaders in check.

Certainly, the Aracruz plantations come close to monocultures. Although the company uses different varieties and species of eucalyptus, it maintains large tracts of land for the intensive management of eucalyptus, which do not harbor the wide variety of plants and wildlife that natural forests do.

Some environmentalists consider eucalyptus, an exotic, a threat to the integrity of Brazil's native biodiversity.

The impressive yields achieved by Aracruz Celulose are also a bone of contention. Plants through photosynthesis capture sunlight and nutrients from the air for growth, but they also require nutrients from the soil. Aracruz's trees are not self-sufficient. Although the company leaves much of the trees' nutrients on site after logging, it also adds fertilizer, which critics contend indicates that the net nutrient flow on the plantations would otherwise be negative. Critics also maintain that the use of fertilizers, pesticides, and other chemicals pollute nearby water supplies, and local NGOs monitoring the effects of Aracruz blame the fast-growing trees for consuming massive amounts of water.

Although the company's eco-efficiency program at the mill has cut pollution, the company remains a target of critics concerning air pollution. Critics accuse Aracruz of neglecting necessary, but costly, air pollution abatement, even though the company invested more than $2 million in gas treatment equipment and processes in the mid-1990s and put in place a more effective air emissions control system. Between 1994 and 1997, for instance, particulate emissions from one of the plant's major boilers plunged almost 80 percent, and sulfur dioxide emission dropped over 50 percent. Nevertheless, local complaints about air quality near the plant continued, and in the late 1990s the Pulp Industry Workers Union accused Aracruz of "incorrect eco-management of both waterborne and airborne emissions."

The Aracruz Response to Critics

Do these criticisms accurately characterize Aracruz Celulose's impact on the environment? Neither the company's management nor its supporters think so. The company's plantations grow on land that was degraded by agriculture when Aracruz bought it, so the biodiversity of the land was already compromised. Now that land is covered with trees and harbors, by the company's count, more than 1500 species of wildlife, including 18 that are endangered. Aracruz maintains 57,000 hectares of natural forest reserves, intermixed with the eucalyptus plantation sites, to preserve biodiversity where once there was none. As management points out, by the time Aracruz started, the Atlantic rainforest, which once covered 90 percent of Espirito Santo, was one-tenth its original size. Aracruz's practices may be intensive forestry, but they are also relatively environmentally friendly agriculture.

The company clearly tries to minimize the environmental impacts of its plantations and operations. Those efforts include minimizing disturbance to the soil and using biological, instead of chemical, controls for pests when possible. "In terms of biodiversity," Lorentzen stated, "eucalyptus plantations should be regarded as being like any other crop, such as corn, soybeans, wheat, sugar cane, and coffee." When compared to these crops, eucalyptus plantations require less fertilizer and chemicals. Company research—begun in 1994 and filed with the state environmental agency—found that chemicals used in the forestry operations do not contaminate the local watershed. Several studies presented at a 1997 meeting on eucalyptus forestry in Brazil indicate that the region's annual rainfall of 1400 mm is sufficient to meet the demands of any crop, including eucalyptus, with a surplus for ground storage or to replenish springs.

Finally, Aracruz managers defend fast-growth plantation forestry on the basis of its efficiency at meeting the world's growing demand for fiber. Plantations, according to company literature, "reduce the world's—and Brazil's—wood deficit by alleviating the main pressures on native forests and consequently helping to preserve them." The fiber supplied by Aracruz would have to come from somewhere, and the company's negative environmental impacts are still less than many other sources.

Well Positioned for the Future

Aracruz may not have mollified its critics, but the company is well positioned for continued success. The company has built a competitive advantage based on a variety of factors—geography, technology, tax relief, product mix, and fiber supply—and is focused on defending this advantage. Aracruz has forged a significant cost advantage based on its eco-efficiency programs and the productivity of eucalyptus growing in southeastern Brazil, Uruguay, and Paraguay. Among competitors in the region, such as Klabin's Riocell subsidiary, CENIBRA, Jari (Monte Dourado), and Bahia del Sul, Aracruz produces the superior quality product at larger volumes with lower costs. And the productivity of other eucalyptus-growing regions, including Australia, Chile, and Portugal, does not match that of Espirito Santo. In a low-cost eucalyptus-growing region, Aracruz is the low-cost producer. Success has made Aracruz cash rich, and in 1997 the company was considering expanding its forestlands into neighboring states or even into Uruguay or Paraguay.

As a world leader in eucalyptus plantation forestry, Aracruz is generations ahead of competitors in the genetic refinement of the eucalyptus varieties it uses. The company has developed knowledge of the best forestry techniques both for matching specific genetic variations to particular sites and for silvicultural activities throughout the growth cycle. Aracruz will continue to improve the genetics of its tree stock and its silvicultural methods, ensuring that it remains farther up the learning curve than any potential competitor.

An Early Mover on Environmental Trends

Aracruz management has a history of correctly anticipating environmental trends. It adopted less polluting pulp-producing technologies such as TCF early and is poised to take advantage of these expanding markets. Demand for totally chlorine-free (TCF) pulp, which has established a small, stable market in western Europe, is projected to rise from an annual global consumption of 3 million tons in 1993 to between 5 million and 6 million tons in 2000.[5] Demand for "environmentally friendly" pulp and paper products such as elemental chlorine-free (ECF) and effluent-free pulp is also expected to grow. In 1997 the company was considering forest certification, either under Forest Stewardship Council (FSC) guidelines or ISO-14000 environmental management guidelines, or possibly both. Thus, if regulatory restrictions or consumer demand mandate a shift in the industry, Aracruz should be ready to move in this direction with less disruption than some competitors might experience.

Management also intends to parlay it technological expertise into solid-wood and engineered board production from eucalyptus. In 1997, Aracruz forged a joint venture, Techflor Industrial S.A., with U.S.-based Gutchess International Group to produce solidwood products. Plans called for a new sawmill in southern Bahia to produce 75,000 cubic meters a year of structural and decorative millwork—half for the Brazilian market, the rest for international markets—by 1999. The venture will put Aracruz in direct competition with Amazonian solidwood, which helps to relieve pressure on endangered tropical hardwoods. The company planned to supply the mill entirely with wood from its plantations and use all waste as chips for the pulp operations or as fuel for the sawmill.

Nothing Is Ever "Good Enough"

Aracruz Celulose is a strong eco-efficiency success story. The company's success through a sustained yield, technology-driven, eco-efficient strategy demonstrates the potential for better business and environmental performance throughout the industry and specifically the environmental and business advantages of plantation forestry for pulp production. Sustainable forestry, however, requires companies to move beyond short-term efficiency in their operations and investment choices. Aracruz management claims that it is engaging these challenges in a responsible manner, and the company is looking into forest certification.

Can any forestry operation fully achieve the moving target that is sustainable forestry? Aracruz is a global leader based in an economically and environmentally depressed region and has environmental and social responsibility as explicit goals. The company touts its environmental performance and is often considered an example of the future of fiber forestry. Yet despite this positive attention—or perhaps because of it—Aracruz is often singled out for criticism. One lesson that Aracruz's experience suggests is that when it comes to environmental performance, nothing is ever "good enough."

Riocell S.A.

Like Aracruz, Riocell S.A. is a Brazilian pulp producer and exporter that depends on plantations, primarily eucalyptus, for its fiber. Over the years, it too has repeatedly clashed with environmental activists, local communities, and government officials over its environmental and social performance. In 1996, Riocell, a subsidiary of the forest products conglomerate Industrias Klabin, or the IKPC Group, produced 283,000 tons of pulp and generated some $187.8 million in sales, which accounted for about 15 percent of Klabin's annual sales. The company, located about 30 km from Porto Alegre in Brazil's southernmost state of Rio Grande do Sul, is one of the most important in the region.

Riocell, with an annual production capacity of 300,000 tons of short-fiber pulp, in 1995 ranked as the fourth largest producer of bleached short-fiber pulp in the region, with Aracruz as number one. The company holds about 1 percent of the world market in short-fiber pulp, but Riocell's share of pulp made from eucalyptus wood accounted for 5.5 percent of world production in 1996. Riocell's best known product, in terms of both size of production and billing, is paper grade bleached pulp (72 percent of total sales earnings in 1996). The company also markets pulp filler, dissolvable pulp used in textile manufacturing, and fine writing paper. Of all the pulp sold by the company in 1996, 69 percent was exported—29 percent to European markets, which are the most demanding environmentally. The remainder went to Asian and Latin American markets and to the United States.

A Turbulent History

During the 1960s and 1970s, a booming economy and government incentives made Brazil a haven for forestry investments. Riocell was founded in 1972 as the pulp mill and plantation of a Norwegian company, Borregaard, and opened with the promise of 2500 new jobs for the region. During its first years of operation, Riocell became infamous for its high levels of air and water pollution. Public outrage and pressure reached such a pitch that in late 1973 the government suspended factory operations, then took over the company. After passing through another owner and government control, the company was purchased by Klabin in 1982.

Riocell's management has not forgotten its past as a public enemy. Although the pollution issues that led to the closing of the pulp mill have long since been addressed, the history continues to influence the outlook of Riocell's management. It is reflected in the company's emphasis on environmental protection and maintaining good relations with the communities surrounding its operations. To maintain those relations, the company has altered its forestry strategy.

An Evolving Forestry Strategy

Riocell has a total forest area of 71,717 hectares, 53,216 hectares of which are eucalyptus with the remaining 18,501 hectares (25.7 percent) set aside as conservation areas. The company operates with three types of ownership: fully owned forests make up about 83.5 percent of its land; leased forests, from which the company owns all production, total 15.6 percent; and jointly owned forests, where Riocell plants trees in partnership with other owners and divides production with them, account for 0.9 percent of its land base.

Riocell's forest management has evolved through four distinct phases. Initially, while the company began to plant its own eucalyptus forests, it bought most fiber from third parties—primarily acacia wood. During the 1980s when its own plantations came on line, Riocell took most production from its own lands. In the late 1980s, Riocell began farming out its production system. In this era, even on its own lands, all of the forestry operations were handled by third parties, including inventories, planting, maintenance, harvesting, and transportation.

By the mid-1990s, Riocell had adjusted its forestry strategy once again. It was farming out or subcontracting a large portion of the production of its raw material, following the "Scandinavian Model" in which the production of raw materials is carried out in areas owned by third parties. By 1997 the company had only 70 employees at work in forestry. Besides activities related directly to forestry, outside contractors also received logs at the factory and chipped the wood, activities that are usually carried out by the industrial unit in similar companies. Even though it had more than enough raw material to sustain the factory's current production, the company still bought about 20 percent of its raw materials from third parties in 1996—18 percent from acacia forests and 2 percent from jointly owned forests. The outside and joint holdings of eucalyptus are dispersed mosaic-fashion among 162 tree farms in 23 municipalities, averaging 3 percent of the total land area of each of the municipalities.

This rather unusual agrarian structure for the region is grounded in environmental and social considerations. Riocell claims it does not concentrate wood production in large plantations because they can cause social tension and environmental disturbances. By using jointly owned plots and outside contractors, the company keeps the region's timber market competitive and also gives farmers an incentive to plant more trees. That is part of Riocell's strategy to keep prices of its raw materials attractive while simultaneously reducing competition with other eucalyptus wood consumers. The surplus decentralized forest holdings give Riocell a competitive advantage in terms of social and environmental sustainability.

Conserving an Asset

Riocell considers the soil a vital asset that needs protection. As part of that effort, Riocell runs a Technology Center, which has a laboratory dedicated to analyzing the fertility and the rejuvenation of soils and vegetation. Those activities have led the company to adopt a variety of progressive practices to protect its soil. In 1988, Riocell stopped burning forest debris before preparing the soil for planting. Although burning the debris makes it easier to prepare the soil for subsequent planting, it also destroys nutrients, especially nitrogen, and microorganisms that contribute to the health of the soil. If burning is not used, the nutrients remain in the soil and the layer of debris acts as a protective cover to guard against erosion and secondary growth. This decreases the need to use herbicides to control the secondary growth during the first six months after planting. To offset the additional costs of these practices, the company adopted minimal cultivation. In this technique, rows of seedlings are planted by hand in minimally prepared soil. Herbicides are applied only to the rows, which are planted about three meters apart, to keep down secondary growth.

Riocell is the only large Brazilian pulp and paper producer that carries out 100 percent of its bark removal mechanically in the field, and it has done so since 1972. Removing the bark in the field has a number of benefits. The bark, branches, and leaves of the trees contain 70 percent of the nutrients the plants absorb from the soil. Removing the bark in the field helps to conserve the soil's productive potential. The bark shavings also form a protective carpet for machinery as it moves over the ground, which helps to minimize erosion. Finally, by stripping bark in the field, the company transports less volume of wood to the pulp mill, thus cutting transportation costs and fuel consumption.

Industrial Production

Riocell's industrial unit is made up of three basic processes: (1) the fiber line, where the wood is processed; (2) the line that recovers the chemical reagents used in the process of wood digestion, for subsequent reuse; and (3) the utilities system, which provides treated water, compressed air, steam and electric power, the chemical products used in the whitening process, and the treatment of industrial wastes.

The company, which has the capacity to produce 300,000 tons of pulp a year, consumes about 1.7 million cubic meters of wood, or about 5500 cubic meters of wood each day. At its 1997 fiber production rate, an average of 35 cubic meters of solidwood per hectare, Riocell needs 34,000 hectares of planted forests to produce 1.7 million cubic meters a year, or about 65 percent of the company's current land base. Riocell clearly owns enough forestland to maintain its current production. Any wood surplus is sold in the form of timber to saw mills for civil construction, power line poles, and fuel. If the company needs added production capacity, it can be found through efficiency gains at other stages of production instead of simply using more wood.

Riocell has always conducted research to develop new products and techniques that can be adapted to a changing marketplace. Riocell had the technology to produce less-polluting TCF pulp in the mid-1990s but had not introduced it, according to management, because the market was not yet ready to pay more for products that used TCF technologies.

The company, since its closure in 1973, has worked to reduce pollution. Some 99.86 percent of the solid wastes generated during the industrial process are recycled into fertilizers and correctives for the soil or are sold as by-products, mainly to be added to cement. In March 1990, Riocell installed a new oxygen delignification unit, which recovered 30 percent of the pollution from the bleaching process. And in 1993 the company slashed the sulfur used in the kraft pulping process by 50 percent.

The Pursuit of Quality

Riocell practices total quality management (TQM). As part of that effort , it was one of the first pulp manufacturers to receive certification under the ISO 9000 standards in 1993 and ISO 14000 in 1996. The company has implemented a variety of policies to meet total quality management standards, which include rewarding team work, providing incentives for TQM, stressing transparency, and recognizing employee contributions.

Under its quality program, Riocell will develop products with specialized characteristics for customers and provide post-sales service. The company maintains a pilot manufacturing facility that tests ways to produce pulp with optimum characteristics for particular customer needs. The potential to provide paper with differing characteristics is one of the company's comparative competitive advantages. Riocell is an international supplier of world renown in various markets where it has carved out a place for itself among large international paper groups such as STORA, Arjo Wiggins, Kymmene, Courtaulds, and Kimberly-Clark.

The Price of Environmental Protection

Riocell's production costs for 1995 and 1996 reached US$320 per ton and US$354 per ton respectively, according to the company. The largest cost component of the business was industrial processing, which represented 73 percent in 1995 and 74 percent in 1996. Transportation costs were relatively low (2.6 percent of the total cost in 1996), which indicates that Riocell's "mosaic strategy" of dispersed forest holdings does not necessarily entail significantly higher transportation costs. Even though Riocell's forests are spread out, the average distance to the mill is not great—about 70 kilometers.

Riocell's efforts to address environmental issues, however, have entailed significant investments and costs, according to management. Since it was founded Riocell has plowed some US$135.1 million into environmental protection. Investments to control air emissions and liquid waste were the two most expensive aspects of the company's environmental protection plan (Figure 10.5).

The cost of stripping the bark from wood in the field—compared to removing bark at the factory—represents a 5.66 percent increase in the cost of forestry production. In Riocell's case that leads to an annual additional cost of about US$1.5 million, or an increase of 1.23 percent in total production cost. Figure 10.6, which shows the breakdown of environmental costs in overall forestry and industrial production costs, including fixed and variable costs as well as depreciation reserves, indicates that environmental costs accounted for 9.3 percent of costs in 1995 and 9.0 percent in 1996.

Figure 10.6 takes into consideration not only the financial costs of these activities but also any quantifiable benefits gained. An example of this eco-efficiency gain would be the reduction in fertilizer use that comes from debarking the trees in the forest. These activities actually do result in net costs for the company. Among the various costs that are incurred to treat

Item	Description	Value in US$1000 since establishment
1	Emissions into the air	28,397
2	Effluvia or liquid wastes	85,340
3	Solid industrial wastes	4,987
4	Environmental quality control	1,352
5	Forest areas	14,700
6	Environmental research	390
Total		135,166

Figure 10.5.

Investments in Environmental Protection.
Source: Riocell

contaminants, the highest cost is that of treating liquid wastes, which makes up about 50 percent of the total costs incurred by Riocell's efforts to achieve environmentally responsible production.

As Figures 10.6 and 10.7 indicate, Riocell's adoption of more environmentally sound business practices reduced the company's gross profits an average of 15 percent in 1995 and 1994. That average, between 11.2 and 18.9

Item	Share of Company's Total Production Cost
Forest management	1.2%
Solid wastes	0.7%
Treatment of liquid wastes	4.7%
Emissions into the air	2.7%
Total	9.3%

Figure 10.6.

Share of Environmental Protection and Control Costs within Riocell's Total Production Cost.
Source: Riocell

	(US$1.000.000)		
Item	Value in 1994	Value in 1995	Value in 1996
1. Net sales income	196,228	229,794	150.000
2. Cost of goods sold	(131,510)	(125,516)	(123.300)
3. Gross profit	64,718	104,278	26.700
4. Cost of environmental protection and control	12,230 (9.3% of item 2)	11,673 (9.3% of item 2)	11.100 (9.0% of item 2)
5. Gross profit excluding costs of environmental protection and control	76,949	115,951	37.800
6. Relative increase in gross profit without environmental protection and control	18.9%	11.2%	41.6%

Figure 10.7.

Impact of the Costs of Environmental Control and Protection on Riocell's Gross Profits.
Source: Riocell
Note: Based on 1995 exchange rates US$1 = 0.973 reals.

percent, corresponds respectively to the estimated value of gross profit reduction in 1995 and 1994. In 1996, the reduction was a far greater 41.6 percent.

Sustainability—A Cost or an Opportunity?

In an era of constrained natural resources, no business can afford to ignore its impact on communities or the environment. However, by treating socioenvironmental concerns as merely a cost and not an opportunity, a business risks adding inefficiency through higher production costs, greater investments in pollution control, and lower worker productivity, without gaining any compensatory benefits. The evidence from several decades of corporate adaptation to environmental pressures indicates that those companies that integrate socioenvironmental concerns into a business strategy can enhance efficiency and the quality of production and can even develop sustainable competitive advantages. But the full integration of these issues into business strategy requires major changes in thinking by management and stakeholders. In the forest products industry, some of the issues raised by the integration of socioenvironmental issues into the business strategy become evident when comparing the approaches of Riocell S.A. and Aracruz Celulose S.A. to environmental issues.

Comparing Aracruz to Riocell

In the global pulp market, Aracruz and Riocell share common advantages. In a commodity market characterized by a general lack of product differentiation, low-cost producers have an advantage. Brazil's pulp and paper industry enjoys a global low-cost advantage, due principally to low energy costs, the high productivity of the soil and climate in the southeastern states, and the use of fast-growing eucalyptus trees. As a result, it costs 18.5 percent more to produce a ton of bleached hardwood pulp in Sweden than in Brazil, and over 15 percent more in the southern United States. The growth of similar plantations in southeast Asia has begun to challenge Brazil's advantage, but only to a limited extent.

Riocell and Aracruz stand to benefit from common advantages in the current market. Both Brazilian pulp producers depend primarily on eucalyptus fiber and export to markets worldwide. Both were started in the late 1960s and early 1970s, during a period of Brazilian government incentives and economic growth. Both had Scandinavian backing, which brought in expertise from that region's well-established forest products industry. Both companies strive for environmental and social responsibility and believe, with some justification, that they contribute to sustainable development.

Strategic and Financial Differences

The two companies view the environment and surrounding community, in light of their business strategies, quite differently. Riocell considers the environment as a cost for the manufacturing process. Company documents primarily address the costs incurred through its efforts to comply with environmental regulations. They give little attention to the potential cost savings of preventing waste. Indeed, the company states that adoption of better environmental practices cost it 15 percent of its profits annually in 1994 and 1995. Riocell has focused primarily on cutting end-of-pipe emissions rather than pollution prevention—this has contributed to inefficiency at the mill.

Aracruz, by contrast, views the environment as an opportunity for both efficiency and new markets. The company has emphasized eco-efficiency in its manufacturing to drive down costs by minimizing waste. It has extended this effort to the plantations, by working to ensure that its eucalyptus fiber has consistent yield and quality. The company has achieved these goals through research that matches superior genetic strains of a particular eucalyptus hybrid with advanced silvicultural methods. As a result, Aracruz is the world's low-cost bleached eucalyptus pulp producer. The company has taken advantage of the shift in some markets, notably Germany, toward environmentally friendly products by investing in TCF technology. It has presented itself as a sustainable producer by supporting efforts like the International Institute for Environment and Development (IIED) study on sustainable paper cycles and the WBCSD.

Riocell emphasizes compliance with environmental regulations and end-of-pipe pollution control, activities that add costs but little benefit. Aracruz, on the other hand, stresses product stewardship by examining the life cycle of its products to identify efficiency gains that increase quality and lower costs. The difference in approach to the environment affects the companies' financial performance. In a commodity industry, Aracruz achieves consistently positive returns. One financial analysis of the company, by Morgan Stanley Equity Research, concluded that Aracruz has established, through its genetic improvements and land holdings, a competitive advantage that should be sustainable for at least 20 years.[6] Compared to a three-year average operating margin of 11 percent for the Latin American pulp and paper industry, Aracruz's 1993–1995 average operating margin of 17 percent was impressive.

Riocell managed to earn only $20 million in net income for the entire cyclical-high year of 1995—an operating margin of under 10 percent. Inefficiency at the mill caused by outdated equipment and waste is apparently hurting the company. Cost is also a significant driver in this industry;

Riocell's costs per ton produced have been estimated at $20 more than Aracruz's ($320/ton versus $300/ton).

Next Steps

Riocell has taken promising steps, even though it has not integrated eco-efficiency or product stewardship throughout its business. The company's efforts to preserve its soil by removing bark from logs in the forest rather than at the mill to return nutrients to the soil and cut fertilizer use is one example. Riocell has also considered the needs of local communities by contracting out many operations and services. This strategy represents socioefficiency because it lowers the company's costs while providing local entrepreneurs and farmers with more opportunities. While some may debate the value of outsourcing as a socially responsible practice, local communities appear to welcome Riocell's "mosaic approach" of small, outsourced forest holdings. The mosaic approach and dispersed debarking appear to be inefficient, yet Riocell's cost per ton of raw fiber at the mill is significantly lower than Aracruz's—about $86/ton versus $100/ton for Aracruz. If Riocell integrated environmental and social sustainability throughout its operations, it might be better able to preserve this value and capitalize on the investments.

While Aracruz has been proactive on environmental issues, on social issues its frame of thinking appears to be in line with protecting its franchise. The company acts beyond its legal requirements to provide health care for local communities, housing for its employees, and free education. But without these investments, the company could not operate. Even though the company contributes heavily to the development of the regional economy and to strengthen the local infrastructure and social system, outsiders criticize Aracruz as being insensitive to the needs of local stakeholders. These significant social investments do not provide commensurate returns for Aracruz. If Aracruz treated social sustainability as an opportunity for strategic advantage as Riocell does, rather than as a franchise protection or "social obligation," it might be better able to capitalize on the prosperity it has brought to the region, avoid controversy, and have better relations with neighbors and a more positive public image.

Total Quality Management for Forests

The experiences of Riocell and Aracruz are relevant to all forestry operations, even those not based on plantations. A company that seeks simply to protect its franchise by complying with environmental regulations will derive few benefits to outweigh the costs. But environmental and social sustainability, when integrated fully into the business, can open up opportunities for superior business performance. If these considerations, however, are only partially integrated into a company's strategy, opportunities will be lost.

In this way, sustainable forest management is comparable to total quality management. It needs to be applied throughout a company's business operations or the potential performance gains will not be fully realized.

Indonesia's Destructive Plantation Strategy[7]

By themselves the plantation strategies of individual companies such as Aracruz and Riocell may not appear to be at odds with sustainability. But conflicts with sustainability may well arise when plantation development becomes widespread in a region or nation. The Indonesian government has adopted a goal to make the country one of the world's top 10 plantation-based pulp and paper producers by the year 2004. The government plans to plant as much as 10 percent of the country's land in plantations to supply newly constructed mills. Between 1996 and 2010, some 23 pulp plantation and mill projects are expected to consume $28 billion in investment. In theory, plantations are supposed to take pressure off natural forests. In practice, Indonesia's strategy to create a pulp and paper industry based on plantations is every bit as dangerous for natural forests as conventional logging. In the 1990s the nation's plantation program was rapidly destroying natural forests. It had become a graphic reminder of the ecological and social hazards of establishing plantations and the damaging role that business practices and government policies can play.

The Indonesian government had good reasons for adopting a forestry strategy based on plantations. Since the 1960s Indonesia has depended on the timber trade to fuel economic growth. In 1995 forestry represented 10 percent of Indonesia's gross domestic product and 12 percent of its trade.[8] But the natural forests are disappearing. The future of plywood, a major Indonesian product, is uncertain—and wood product exports are declining. In 1998 the government projected that exports would drop at least 25 percent.[9] Plantation-based pulp and paper production offers a more stable future for the industry. Indonesia's climate and soil are favorable for acacia and eucalyptus plantations, and energy and labor costs are relatively low. Until the plantations mature, Indonesia's remaining large tracts of natural forest can supply the new mills. The plantations are also a solution to other social and economic problems: They are intended to relieve population pressures on Java, which supports 60 percent of the nation's 204 million people, by attracting Javans to the outer islands with the promise of employment.

Until the plantations come on line, companies can use wood from their own concessions, which include secondary and primary forests, to supply their large-scale paper mills. As of June 1997, the Ministry of Forestry had allocated 2.625 million hectares to 13 companies for pulpwood estates, but only 805,354 hectares were ready for harvesting—not enough to meet cur-

rent or future needs of the mills.[10] In fact, many analysts doubt that plantations will come on line as fast as the planners anticipate. Industry analysts
suggest that even if the government is able to develop its originally projected 4.4 million hectares of industrial plantations by 2004, pulp mills will continue to rely heavily on the 55.4 million hectares of designated natural forest
logging concessions until well after that date.[11] This will undoubtedly lead
to further dependence on natural forests, because pulp and paper companies
cannot afford to let expensive mills remain idle or operate below capacity.
Meanwhile, logging rates in Indonesia are estimated to be nearly double the
"sustainable" rate recommended by The World Bank; commonly cited figures estimate that Indonesia loses as much as 1 million hectares of forest a
year.[12]

In its effort to support the timber industry and increase the capacity in
pulp and paper, government policies exacerbate and encourage high rates of
conversion and deforestation. Some studies have concluded that programs
either sponsored or encouraged by the government account for 67 percent of
all deforestation.[13] Low timber royalties based on volume cause wood to be
undervalued and provide no incentive for companies to be efficient. The
combination of low royalties and short-term concessions encourage concessionaires to cut as much as possible as fast as possible before their time
expires and higher royalties are established.[14]

Plantations are incorporated into this pattern easily and profitably. Clear-
cutting for pulp in industrial forest estates is allowed only if there is less than
20 cubic meters of timber per hectare in a given area. To get around this,
concessionaires practice selective logging in natural forests until the stand's
concentration falls below the cutoff, then they clear-cut for pulp. Once
cleared the company can establish plantations, often taking advantage of
funds collected from logging companies for land restoration to plant their
new monocultures. The government encourages plantation entrepreneurs
further by allowing them to import pulp and paper machinery duty free and
giving them equity capital and no interest loans from the Ministry of
Forestry.[15]

The incentives have had serious economic, environmental, and social
repercussions. Burning, the least expensive method of clearing land, is
favored by the industry and routinely practiced in violation of government
guidelines. During the summer and fall of 1997 forest fires raged out of control in parts of Indonesia, blanketing portions of Southeast Asia with clouds
of smoke. In September, Indonesian President Suharto gave companies 15
days to stop any fire setting, which created a frenzy of burning as companies
rushed to complete clearing before the deadline. Satellite images confirmed
that larger firms were responsible for 80 percent of the fires.[16] Oil palm
plantations were the worst offenders, but 27 pulp makers were also listed.

The fires cost Indonesia an estimated $1 billion and Malaysia $300 million, largely from health care costs as pollution reached record levels. But tourist visits to Indonesia also plunged 26 percent, and Malaysian travel agencies lost 30 percent of their business in the late summer months.[17] The cost to the environment was equally steep: An estimated 2 million hectares of forests were destroyed.[18] Experts predicted even worse fires for 1998 caused by plantation and timber estates that were clearing land.[19]

Indonesia's conversion of natural forests to pulp plantations threatens biodiversity as well as local communities. Indonesia has only 1.3 percent of the world's land area, but its forests hold 10 percent of the world's flowering plant species; 12 percent of the world's mammals; 17 percent of all reptile and amphibian species; and 17 percent of bird species. In addition, natural forests provide a partial livelihood for as many as 65 million people.[20] The government's awards of concessions to private companies, however, conflict with the long-standing use of forest resources by local communities. Even the government has admitted the inequity. "Forest exploitation by concession holders is often to the detriment of the needs for forest resources of communities living in and adjacent to the forest."[21] The state, however, continues to help companies acquire land by discouraging the enforcement of traditional land claims.

Far from a Flawless Solution

The experiences of Aracruz and Riocell document the efficiencies and advantages that plantations have to offer the industry as productive, reliable sources of fiber for the next century. But Indonesia's efforts to establish plantations would score poorly on any rating of sustainability, given the fires, the conversion of biologically rich natural forest, and the land disputes with local populations. How those efforts fare in terms of management is less clear. Few independent scientific studies exist that have evaluated the environmental impacts of existing plantation operations. Although the growth of plantations is inevitable to satisfy global demand for wood, as the case of Indonesia illustrates, industrial timber plantations are far from a flawless solution as a source of fiber. Their establishment and management pose potentially serious environmental risks. The extent to which those risks are minimized may well depend on the commitments of governments and private corporations to environmental sustainability.

NOTES

This chapter is adapted from "Aracruz Celulose S.A. and Riocell S.A.: Efficiency and Sustainability on Brazilian Pulp Plantations," prepared by The World Resource Institute for the Sustainable Forestry Working Group, 1997. The Aracruz section was prepared by Rob Day of World Resources

Institute in 1997. The Riocell section was prepared by Isak Krugliankas, associate professor at the Universidade de Sao Paulo and Tasso Rezende de Azevedo, executive director of ImaFlora, 1997.

1. Obviously, the productivity of fast-growth plantations can vary enormously, depending on locale, cultivation techniques, and other factors. The figure cited was provided by Riocell S.A.

2. Personal communication with Gary Stanley, Paper Industry Specialist, U.S. Commerce Department Industry Trade Administration, March 1998.

3. *The Sustainable Paper Cycle,* World Business Council for Sustainable Development, 1995, A27.

4. Mooney et al., "Biodiversity and Ecosystem Functioning: Basic Principles," *Global Assessment,* UNEP, 1995.

5. *The Sustainable Paper Cycle,* World Business Council for Sustainable Development, 1995, 123.

6. "Global Investing: The Competitive Advantage," Morgan Stanley Equity Research, October 25, 1996, 169.

7. The section on Indonesia was prepared by Joseph D. Strathman, the researcher for this book project in 1998.

8. *State of the World's Forests,* Rome: FAO, 1997, 36.

9. Pleybdell, Geoffrey, ed. "Crisis Indonesia," *Tropical Timbers* vol. 13, no. 2, Feb., 1998, 1–2.

10. "Indonesia Prepares to Tap Plantations," *International Woodfiber Report,* June 1997.

11. Data Consult, Inc., (1990). *The Pulp and Paper Industry in Indonesia: Its Current State and Prospects,* Jakarta: Data Consult.

12. Thompson, Herb (1996). "Indonesia's Wood Resource: Trends and Policies," *Journal of Mineral Policy, Business and Environment* vol. 12, no. 1, 14–23.

13. "Indonesia Forest Sector Review," The World Bank, 1997, unpublished draft, 18.

14. Sunderlin, William D., and Ida Resosudarmo (1996). "Rates and Causes of Deforestation in Indonesia: Toward a Resolution of the Ambiguities," Occasional Paper No. 9. CIFOR, December, 11–12.

15. Carrere, Ricardo, and Larry Lohman (1996). *Pulping the South: Industrial Tree Plantations and the World Economy,* London: Zed Books, Ltd., 219.

16. "Timber Tycoon Sees Land Clearing as a Service to the Govt.," *Singapore Strait Times,* pp. 1, 48–50, Derwin Pereira in Jakarta, Friday, October 3, 1997.

17. Davis, Devra, and Changhua Wu (1998). "Revisiting Indonesia's Ecological Disaster and the Public Health Crisis in Asian Countries," *World Market Series: Medical Briefing* ASEAN, 24–26.

18. *Arbor Vitai,* The IUCN/The World Wildlife Fund Forest Conservation Newsletter, January 1998, 3.

19. Mydans, Seth. "Parched Borneo Catches Fire Again," *International Herald Tribune,* February 24, 1998, 4.

20. Zerner, C. (1992). *Indigenous Forest-Dwelling Communities in Indonesia's Outer Islands: Livelihood, Rights, and Environmental Management Institutions in the Era of Industrial Forest Exploitation,* Consultancy to The World Bank, Resource Planning Corporation, 4.

21. GOI, *Rencana Pembangunan Lima Tahun Keenam,* Chapter 26, 319.

Chapter 11

The Road to Certification

Management of Sweden's pulp and paper giant STORA wasted little time in adapting to the environmental issues that began to dominate the forest products industry in the 1990s. In the late 1980s, in reaction to the concerns of Sweden's conservation-minded public and its major customers, STORA overhauled forest management on its 2.3 million hectares of pine and spruce forests in favor of a more ecologically based approach to forestry. Under a new management system, Ecological Landscape Planning, the Falun, Sweden-based company implemented conservation measures to improve the health of its forest that reduced the company's annual harvest by about 10 percent and retrained its entire forestry workforce. Local environmentalists, who had been among the company's harshest critics, were invited to participate in the forest management planning process as part of a new strategy that emphasized cooperation—not conflict—with environmental groups.

STORA's management, however, soon concluded that even these changes were "not enough to assure the public that we were doing the right thing," recalled Ragnar Feiberg, STORA's chief forester. To make its sustainable forestry credible with the public, management realized it needed to do more than merely practice sustainable forestry. It needed a message to assure the public that it was doing so, and it needed a messenger that its customers and the public trusted. Independent third-party certification seemed to offer both. After considering STORA's competitive position, market dynamics, and strategic opportunities, management decided to pursue FSC certification because it believed certification would leverage the company's sources of competitive advantage and help make STORA's environmental performance credible with stakeholders. In 1996 STORA became one of the first large commercial forestry operations in the world to become certified by Scientific Certification Systems (SCS), under FSC standards. When Sweden adopted a national standard for certification, STORA committed to full certification of its productive forest.

STORA's adoption of certification ranks as a milestone for sustainable forestry. STORA, which had revenues of $5.95 billion in 1997 and was the world's fifth largest paper company, produced about 1.9 percent of the world paper and paperboard output. For so large and important an industry player to embrace certification indicated that certification had become a strategic consideration for forward-looking wood products companies in Europe by the late 1990s. The decade-long process that took STORA from conventional forestry to certified sustainable forestry is the saga of one company's adaptation to mounting environmental considerations. STORA's evolution toward FSC certified sustainable forestry demonstrates just how influential the industry's industrial and commercial customers can be in propelling companies toward certification.

From Copper Mining to Forestry[1]

The modern STORA owes its place in the industry, in part, to its medieval origins in copper mining. The company, which claims to be the world's oldest, began at the Copper Mountain in the town of Falun. A Deed of Exchange belonging to Bishop Peter of Västerås, the oldest existing document dealing with the mine at the Copper Mountain, is dated June of 1288. Early mining methods required huge volumes of wood, not only to smelt ore but also for support timbers in the mine and in the extraction process. Because it used wood extensively, the mine controlled large tracts of forest. In the late 1800s, the company started large-scale development of forest products with the construction of saw and paper mills. Through its mining experience, the company also became a producer of iron ore and steel.

By the late 1970s, the iron ore and steel part of the business was losing large sums of money. Through reorganization and help of the Swedish government, STORA divested itself of the steel component of its business. From these spin-offs, the state-owned Swedish Steel Company (SSAB) was created. After the company eliminated the steel sector, it funneled significant capital investments into forest products and developed that business.

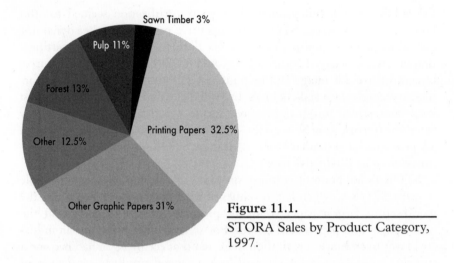

Figure 11.1.

STORA Sales by Product Category, 1997.

During the late 1980s and early 1990s, STORA purchased a number of companies, mostly in the forest products industry. The additions added a total of 78,800 employees. Between 1990 and 1995, STORA sold portions of these acquisitions to increase efficiencies and focus on the core business. After the company sold the building products business in 1996, total employment dropped to 22,716.

Even though STORA confronts formidable competitive challenges in its largely European markets, including lower forest productivity and higher labor costs, it turned in a financial performance comparable to that of many competitors' during the 1990s. In 1996 when prices for paper, which accounted for 63 percent of STORA's sales (see Figure 11.1), dipped 12 percent from prices in 1995 as production outpaced demand, STORA's revenue declined to $6 billion from $7.6 billion in 1995, while net income plunged to $209 million from $718 million during the same period.[2] The divestment of STORA Building Products also contributed to declining revenues in that year. In 1997 sales declined slightly again to $5.9 billion, but net income ticked up to $212 million. In 1996 and 1997, STORA's return on capital was 7 percent, but the company's 10-year average totaled 11 percent: Its U.S. and European counterparts averaged 12 and 11 percent for the same period, respectively.

The Transition to Sustainable Forestry

STORA's transformation from conventional forestry to certified sustainable forestry took place in two stages: First the company made the transition to an ecologically based forest management; then management opted to adopt FSC certification. Both these pivotal decisions were made within the context of STORA's competitive realities. Environmental concerns are priorities for the company because of the strong ethic of forest conservation that prevails in Sweden and because STORA's European markets are among the most environmentally sensitive.

The Preservation Ethic in Sweden

The nearly nine million Swedes who live in an area slightly smaller than California have a long history of concern about the protection of their forests. The Swedes spend a good deal of time at outdoor pastimes and many still live a farm and forest lifestyle. This closeness to the natural environment is manifest in a popular sentiment that the people own the forests—a sentiment that is codified in Swedish law, which guarantees citizens the right of access to all lands. The Swedes frequently take advantage of that right to hike, bike, or ski across privately owned property.

Public interest and involvement with the forests have limited the latitude that industrial owners have to extract economic benefits from their land. In the late 1800s and early 1900s, logging operations across Sweden degraded significant portions of the forest. The Forest Act in 1903 was Sweden's first attempt to address public concerns over the potential of these cutover forests to produce wood in the future. In 1923, the government gave forests additional protection through a law that prohibited the clear-cutting of young forests. In 1950 the General Director of the state-owned forest lands, Eric Höjer, issued Letter Circular #151, which concluded that Sweden needed to reconstruct its forests.

Clearing the land to establish softwood plantations, primarily of spruce and pine, was the most common rehabilitation method, with the goal of growing as much fiber as quickly as possible. When growth slowed, the trees were cut—so the time between cuttings, or rotations, was short. Over the next several decades companies and small landowners followed suit. By the 1970s, intense management of forests was the Swedish norm. These forests are successful timber producers—Swedish wood growth has exceeded harvest for several decades. In the late 1990s it ran about 20 percent above harvest.

During the 1970s and 1980s, two developments altered the course of STORA's forest management. Corporate restructuring and acquisitions took the company away from mining toward forest products production, and management decided to emphasize paper over other forest products. As a major papermaker, STORA focused on the most valuable species for paper production. Spruce, which produces almost double the yield of other native tree species on the best sites and is used to make high-value papers, was the clear winner. Like other major papermakers, STORA adopted a forest management strategy that maximized pulp production from intensively managed plantations of spruce, or pine where spruce was not suitable. Throughout the 1970s this meant clear-felling, burning, the use of herbicides to control weeds, and cutting other tree species. During the 1970s, under mounting public pressure to discontinue burning and use of herbicides, STORA and other Swedish companies abandoned both practices, removing more wood from the logging site and cutting weeds manually instead.

But those changes did not allay concerns about the environmental impacts of the industry's intensive forestry. Scientists studying old-growth boreal forests in Russia and northern Sweden began to identify parts of the old forest systems that were either missing in the plantations or becoming scarce. They determined that as many as 1500 species of organisms—from animals to fungi—were threatened by plantation management. The scientists also identified other detrimental impacts of the plantation forestry that included damage to wetlands and meadows; the disappearance of habitat features such as vertical structure and rotting wood; and a loss of biodiversity. Beginning in the late 1980s Greenpeace Sweden, Swedish Society for Nature Conservation, and WWF waged successful campaigns to put these concerns on the public agenda.

A New, Ecologically Based Forest Management System

In response to public concern and criticism from environmental groups, in the late 1980s the STORA forest and timber group decided to develop an ecologically based forest management on the company's Ludvika District lands (see Figure 11.2). In 1991 the company hired Börje Pettersson, a forest scientist with credibility among Swedish environmental groups, as staff ecologist to help develop a corporate forest management strategy that would maintain the company's economic viability while improving its ecological performance.

Timeline of Events—STORA's Evolution to Ecology-Based Land Management

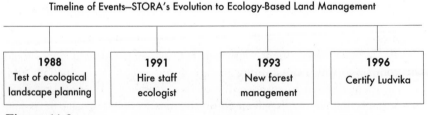

| 1988 Test of ecological landscape planning | 1991 Hire staff ecologist | 1993 New forest management | 1996 Certify Ludvika |

Figure 11.2.

STORA Forest Management: The Road to Third-Party Certification.

The land management system developed by STORA, Ecological Landscape Planning (ELP), is designed to achieve multiple goals on adjacent tracts of forest. Under the system, foresters exclude key ecological areas, especially those with endangered species, from normal management and create corridors so that populations of endangered species can travel across the landscape to improve their chances of reproduction. ELP is also intended to maintain adequate proportions of old forest, hardwoods, and other important stages of forest growth and balance the use of intensive management practices against the goal of maintaining a diversity of species in the landscape.

STORA used an 8500-hectare area of Grangärde in southern Dalarna as its test site for ELP. Foresters used maps, aerial photos, GIS technology, and field inventories to identify key ecological features and populations of endangered species. They set aside fire resistant areas along streams and bogs to develop naturally. Those areas with tree species that depend on fire to regenerate were restored with a variety of management methods that included controlled burning, increasing the proportion of hardwoods in the stands, retaining green trees in harvested units, and leaving more rotting wood to maintain the overall health of the ecosystem. In 1993 STORA adopted the goal of having all suitable lands under ELP management by 2003.

Listening to Neighbors and Environmentalists

As part of its environmentally based forest management, STORA's managers adopted comparatively innovative tactics to address public pressures on environmental issues. During the 1960s and 1970s, like U.S. companies in the 1980s, the relationship of Swedish foresters and wood products companies with environmental groups was defined by conflict. As part of its ELP program, STORA abandoned the defensive, hard-line posture toward environmental groups and adopted a collaborative approach. The decision to collaborate with environmental groups was driven, according to senior managers, by enlightened self-interest. STORA wanted to convince the public that the industry had improved its environmental performance. In their view, winning the endorsement from environmental groups was essential in that process. "The environmental nongovernmental organizations are trustworthy," explained Ake Granqvist, supervising forester, who managed the ELP program at Grangärde. "They are the only ones, really, that can reach the end consumers."

To win the support of environmental groups, STORA made them part of the ELP process and incorporated their views in planning. The company established local "reference groups" to learn how local people view and use their forests. By listening to people and groups who use their forests—

hunters, fishermen, bird watchers, berry pickers, skiers, hikers, educators—
STORA personnel not only increased their own knowledge about the
resources that exist on STORA lands but they also gained credibility in the
Grangärde area. Forestry officials reported that they are no longer treated as
part of an impersonal corporation, but instead have become more integrated with the local community. Since STORA has included locals in the ELP
process, the company's relationship with the local media has evolved from
confrontation to the point where the company had had "no negative press"
from local environmentalists in several years, according to officials at year-
end 1997.

1993—A New Companywide Forest Management Strategy

In 1993 a growing consensus over the importance of minimizing clear-cuts,
preserving biodiversity, and old-growth forests led the Swedish government
to adopt a new national forest management code. With the new code, the
priority in forest management shifted from wood production to the shared
priorities of producing wood and preserving ecosystems. In response to public support in Sweden for forest conservation and the government's new
code, STORA extended its ecologically based management to all its forests.
While the ELP experiment was underway at Grangärde, STORA forestry
executives designed a new management strategy for all company lands that
has as its goal to "contribute to the achievement of STORA's financial goals,
as well as maintain a high, valuable, and sustainable forest production, while
preserving the biological diversity." STORA's new standards are comparable to those in the national code.

The new strategy led to a number of changes. The company planned to
increase harvest levels 0.5 to 1.0 percent per year over the next 50 years from
spruce and pine plantations, which contain a high proportion of young trees.
To meet its twin goals of maintaining the productive capacity of the land
and preserving biodiversity, STORA incorporated a greater emphasis on
nature conservation in its formal forest management program, including
ELP: matching management techniques to individual sites, principles for
day-to-day nature conservation, and discussion of ways to address special
forest management questions such as the use of fertilizers.

Maintaining Ecosystems

The restoration and maintenance of forest ecosystems, an integral part of the
new forest management strategy, required STORA foresters to identify
unique ecological areas, threatened or endangered species, and areas that
needed restoration as part of their inventory process. STORA, like other

large Swedish forest companies, conducts an inventory of its lands every 10 years, as part of a national project. A range of methods, including computerized GIS technology, is used to document and track the various ecosystem components.

STORA targeted three types of areas for restoration and set asides. Critical habitats for species such as the white-backed woodpecker were reserved and protected, which included a number of wetland sites. Difficult or risky landscape features, such as rock outcrops, bogs, steep slopes, and other low productivity sites made up the second category. Areas that could serve as corridors between populations of animals, enhancing the flow of genetic diversity between nonadjacent forest tracts, were the third target. By restoring and setting aside these areas management expected to meet government requirements for preserving old forest habitat. To implement the new strategy, STORA trained its entire forestry workforce in ecologically based forestry principles and techniques and made them responsible for carrying out of day-to-day nature conservation measures. By 1997 the new practices were visible on STORA land. More green trees were left standing within clear-cut areas. A higher percentage of deciduous trees dotted the landscape, and controlled burning was used to regenerate harvested areas.

In the process, though, STORA traded some productivity to improve the health and diversity of its forests. The areas set aside for conservation reduced harvest levels about 10 percent annually. STORA forest managers estimate that meeting the requirements of Swedish law accounts for about a 5-percent set-aside, while ELP requirements make up the additional 5 percent. In addition, by 1996 the 50 employees assigned to ELP management, combined with other ELP expenses, cost the company some $10.5 million a year.

Competitive Challenges for STORA

STORA, by virtue of its location in Sweden and its dependence on the European market, confronts several strategic considerations that, along with STORA's desire to increase its market share by competing with other materials such as plastics and aluminum, were instrumental in management's decision to adopt FSC certification.

Sweden's cold climate keeps the Swedish industry's fiber yields relatively low compared to other regions. Swedish forests produce an average of 4 to 5 cubic meters of wood per hectare each year, with a high of 14 cubic meters per hectare annually. Those yields are well below yields in the Pacific Northwest, which can range between 20 to 30 cubic meters, or those of the best plantations in New Zealand, Chile, or Brazil, which rise as high as 30 to 40 cubic meters per hectare per year. Slow-growing Swedish trees, however, yield fibers of unmatched quality that are used in such products as high-quality magazine paper and lumber for windows. In these markets, STORA competes primarily with Canada, other Nordic countries, and other northern forest regions.

Generous social programs and workforce regulations add to Sweden's labor costs, among the highest in the industry worldwide. The combination of lower fiber yields per hectare, higher labor costs, and other factors keep the production costs of Swedish pulp and paper companies relatively high compared to other regions. In the mid-1990s, Sweden's cost to produce a ton of fiber was $246, according to research by Oppenheimer & Co., Inc., versus just $96 for Brazil, $112 for the southern United States, and $140 for Canada.

Higher costs put STORA and other Swedish companies at a competitive disadvantage in the global market and make STORA vulnerable to price competition from the lower-cost regions of Asia and South America. In the past, Sweden has devalued the krona to lower the costs of its products to export customers. Since Sweden has joined the European Union (EU), however, this tactic is no longer an option. In the future, STORA's competitive challenges are likely to be even more difficult. Production capacity in Southeast Asia and South America is rising, and companies in both of these regions have targeted European markets. Meanwhile, a number of substitutes threaten important wood markets—in particular, steel for lumber and plastic for paper.

The Greening of European Markets

STORA, like other large wood and paper products producers in Europe, has also taken the pulse of the European markets, which account for nearly 90 percent of its sales (see Figure 11.3). By the mid-1990s, large amounts of European customers for wood and paper products were demanding certified forest products from suppliers and expressing their opinions in a variety of ways that loomed large in STORA's decision to adopt certification, according to Bjorn Hagglund, president of STORA Forest Timber. In 1993, German publishers, important customers for STORA's high-quality paper products, began to lobby their suppliers for environmentally safe products to curry favor with the public (Box 11.1).

The actions of "Buyers' Groups," especially members of the 1995+ Group in the United Kingdom, also caught the attention of STORA's management. These heavy-purchasing customers, which include the most influential retailers in the United Kingdom, wanted to buy FSC certified products— not just sawnwood, but all manner of paper products as well. The 1995+ Group members pressed STORA and other suppliers to adopt certification,

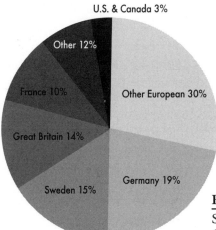

Figure 11.3.

STORA Sales by Market Category, 1997.

Box 11.1. Yielding to Environmental Pressure[3]

Until 1993 the German publishing industry paid little attention to the environmental impact of its paper-purchasing practices. Publishers assumed that the environmental issues surrounding the pulp in the paper they used were the responsibility of their suppliers, many of whom were Scandinavian and Canadian companies. But then the German publishing companies became the targets of Greenpeace activists. Several companies were faced with demonstrators questioning the source of their paper and accusing them of wasting resources and destroying forests.

The publishers quickly realized they could not afford to be insensitive to pressure from environmental groups. The German public has historically been sympathetic to actions taken by environmental activists. At about the same time that Greenpeace targeted the publishers, its also rallied the German public to protest the proposed sinking of an offshore drilling platform in the North Sea that belonged to Shell U.K. Exploration and Production. The action led to a boycott of Shell gas stations. The publishers were concerned that if they ignored the opinions of environmental activists they could become the targets of similar boycotts, public disapproval, even the loss of market share.

The role of publishing companies, as both consumers and sellers, makes them particularly vulnerable to pressure. As Dr. Furstner from the German Magazine Publishers Association (VDZ) explained it, "publishing houses . . . are part of the public opinion. They sell it. They live from it. You can't do your own business wrong in the environmental point of view, and then sell environmental positions in your magazines."

The publishers decided that they needed to restore consumers' trust in the use of paper by convincing the public that using paper is not a waste of

(continued)

resources and that the production of paper does not harm forests. The publishers chose to work together with the VDZ and the German Paper Producers Association (VDP). The group produced a series of position papers on a variety of environmental topics, including sustainability, biodiversity, and protection of old-growth forests. The group concluded that the industry wanted an early introduction of a certification system that was recognized worldwide. The group, it said, "calls upon all supplier of wood and pulp to play a part in such a certification system." The publishers then began questioning the forest practices of their suppliers. Some visited their paper suppliers to make them realize the critical nature of the sustainable forestry issue. Neither STORA nor the Swedish forest products industry could afford to ignore the actions of these important customers.

using a variety of tactics that included visits from their managers to stress their commitment to FSC-certified products and questionnaires asking suppliers to document their sources of supply. J. Sainsbury Plc., the U.K. retail giant, effectively used its supply tracking system, TimberTracker, to prod Swedish timber and pulp producers toward FSC certification (Box 11.2).

Box 11.2. The Dogged Pursuit of Certified Products[4]

The mention of certification along with the name of J. Sainsbury Plc. elicits a nod of recognition from STORA executives. Sainsbury is known throughout the executive offices of Europe's wood products industry for its tenacious campaign to acquire FSC certified products. The Sainsbury group of companies, one of Europe's largest retailers, tallied sales of $9.69 billion in 1997 (1 U.S. dollar equals 1.60 British sterling). Its chain of Sainsbury's supermarkets held about 12 percent of the U.K. grocery market, and its Homebase DIY home improvement centers claimed about 10 percent of that U.K. market. More than 60 percent of the products sold by its DIY chain and supermarket chain are its own brand. The extensive use of its own label not only helps cement customer loyalty but it also gives JS added leverage with suppliers and ensures that the company closely monitors issues of environmental responsibility and liability.

As a member of the U.K.-based 1995+ Group started by the World Wildlife Fund, management committed the Homebase and supermarket chains in 1995 to a goal of stocking 100 percent FSC-certified wood and paper products. By selling FSC-labeled products, the company is relieved of the responsibility of evaluating forests and forest practices. "If FSC certification tells me that it is a well-managed forest, I am happy to honor that," explained William Martin, senior technical manager. "If there are any questions I can refer them to the FSC or to the certifier."

The company's Technical Division (TD) and the Environmental Management Department, which are primarily responsible for the company's environmental programs, evaluate products and approve suppliers for the company's 250 centralized buyers. The TD has invested considerable resources in a computer database called TimberTracker, which monitors the "forest of origin" for all sawnwood and pulp and paper products sold by JS in the United Kingdom. The tracking system assigns suppliers a letter grade. Grade A means that the supplier is fully FSC-certified; Grade G signals the supplier has been delisted. TimberTracker accomplishes far more than data collection, however. As TimberTracker questionnaires work their way up the supply channel, they alert suppliers who may be two or three steps removed from JS that the company is an important customer. Sainsbury's size and reputation then give it leverage when it approaches timber producers to inform them about its commitment to FSC-certified wood.

JS management prefers to use persuasion and negotiation to convince suppliers to adopt FSC certification. Martin and other JS representatives have visited suppliers worldwide to deliver Sainsbury's message of commitment to FSC certification. As part of the program, suppliers are asked to create action plans that will move them toward buying their raw material from certified sources. The effort, however, has required JS personnel to spend "more time up front to explain, debate, and persuade on our motives behind certification" than management had originally anticipated, according to Martin.

The company is willing to make that effort because, as one member of the Technical Division pointed out, "It's easier to work with suppliers we know than it is to get new ones, because you don't know what other problems they bring in as well." With recalcitrant suppliers, Martin said, Sainsbury is "prepared to push a point" by delisting them or switching to suppliers who can offer certification. In 1997 Homebase dropped lumber suppliers in Finland to buy from Sweden's Assi-Doman, which had committed to certification, so that Homebase could offer FSC-labeled lumber.

Suppliers—including STORA—have responded to the Sainsbury Group's approach to acquire FSC-certified products. In late 1998, the company had 491 FSC-certified wood products on the shelves, including 10 paper products, sheet materials, and lumber. As one JS buyer explained, "The suppliers see this as a good opportunity to be the first FSC product on the shelf. They know that Sainsbury will back it."

Moving Toward FSC Certification

In 1996 STORA conducted a pilot certification effort with Scientific Certification Systems (SCS) of Oakland, California, in its Ludvika Forest District, in south-central Sweden. At the same time, STORA was a member of the Swedish FSC Working Group that was drawing up national forest certification standards for Sweden. The company wanted to gain experience with the certification process and to see how Swedish forest management would measure up to international standards.

An SCS auditing team of three experts, who customized the certification standards used by SCS to fit Swedish forestry, conducted an eight-day evaluation of the 300,000-hectare Ludvika Forest District in June 1996. Under FSC guidelines, forest certifications examine three elements—sustainable wood production, the maintenance of ecosystems, and the socioeconomic benefits to the human communities involved. A "passing score" is judged as 80 out of 100. STORA received a score of 92.7 for sustainable wood production, 80.4 for ecosystem maintenance, and 88.9 for socioeconomic benefits. The team commended the company for efficiency in harvesting, investments in timber stand improvement, and treatment of workers. STORA also received high marks for self control in managing forest inventories and cutting less than it was growing.

However, the team found fault with the STORA's forest management. It pointed out that the Ludvika forest overrelied on conifer species and had inadequate numbers of deciduous trees. The team also called for a better inventory of natural resources other than trees and more training for forest workers in ecosystem management techniques. It also asked the company to pay more attention to the economic development of communities surrounding the Ludvika District.

The FSC team imposed six conditions on STORA for continued certification. STORA was required to inventory all of its stands of trees older than the lowest age for cutting and determine which ones might be held to achieve a goal of 5 percent old forests in the district. The team asked STORA to incorporate methods into its strategic plans to assess biological diversity over the long term, to develop new policies and methods to reduce damage to deciduous trees by moose and deer, and to restore wetland spruce habitats. The FSC team also directed the company to draw up plans to minimize and counteract the effects of acid rain on the forest and to help a local community devise strategies for more diversified economic development.

In 1997 the SCS evaluated an additional 641,000 hectares of STORA forest. In June 1997 the Swedish FSC Working Group adopted national standards after the small woodlot owners associations left the discussion. As of mid-1998, five of the six major Swedish companies that participated in the discussion had agreed to certify their land, including STORA, Assi-Doman, Kornas, MoDo, and Graninge. By year-end 1998, STORA management planned to have all the company's productive forest FSC certified.

The Competitive Advantages of Certification

The challenges confronting STORA are daunting: industry overcapacity; cost and productivity disadvantages; and new, lower-cost competitors in traditional markets. Under the circumstances, the decision of STORA management to identify innovative ways to differentiate its products and/or seek preferential relationships with specific customer groups makes sense. The commitment of its customers to purchase certified wood products has the potential to help STORA hold on to its share of European markets. Fortuitously for STORA, a handful of its strongest European markets has chosen to publicly align with FSC-accredited certification. Some members of the 1995+ Group are so eager to put FSC-certified products on their shelves that they are willing, at least in the short term, to pay a price premium for these goods.

Since few suppliers provided third-party certified products in the late 1990s, by adopting FSC certification early, STORA has a strategic opportunity to sell a differentiated product that, at least in the immediate future, may give it an edge in price-sensitive markets. Sweden's geographic and political links with Europe, including EU membership, give STORA preferential access to European customers that is unavailable to U.S. and Asian producers. By moving quickly to fill demand for certified products by major customers, STORA may be able to forestall lower-cost competitors and maintain strong relations with existing customers, many of whom belong to buyers' groups. Promoting the value and importance of certification to a healthy environment may enable STORA to attract new customers as well. Relationships with members of the 1995+ Group members based on environmental performance could also help STORA cultivate a form of brand equity that would be difficult for competitors to replicate.

Third-party certification could also become a market barrier that will protect STORA's European markets. After investigating the practices of a number of Southeast Asian competitors, STORA managers concluded that those producers would have great difficulty obtaining third-party certification. Armed with FSC certification and sufficient supplies of product to satisfy the assorted buyers groups and similar customers, STORA management thinks it could effectively close the market to these competitors.

Certification may also help slow down the inroads of substitutes into wood markets. Third-party certification could provide ammunition to demonstrate that wood products are environmentally sound through life cycle analysis, which assesses the environmental impact of products from design to disposal. The documentation required for certification is likely to help generate the detailed information needed to make these life cycle computations. The upshot should be a demonstrably stronger environmental position for forest products compared to many substitutes.

Finally, STORA appears to be particularly well suited to accomplish third-party certification for its forests. Swedish culture fits well with the requirements of third-party certification. The Swedish government's social and environmental policies relating to company responsibility for social welfare and worker treatment are consistent with the underlying philosophies of FSC-accredited certification. Swedish forestry's record of cutting less than annual growth is also an advantage in earning certification. In addition, Sweden has little of the old-growth forest left that has been so contentious with environmentalists in other regions of the world. Any expectations for growing new, ancient forest on these cutover lands seem to be lower in Sweden than in other places in the world where large stands of ancient forest remain to compare with plantations.

The Value of Converting Critics to Supporters

Starting from an atmosphere of distrust and conflict, STORA's strategy of collaboration with environmentalists in the ELP process and through FSC certification proved to be an effective way to convert critics to supporters of the company's environmental performance. STORA's actions have bolstered its public image, reduced conflict with environmental groups, and even gained the endorsement of local environmentalists. The relationships have led to forest management actions that achieved mutually beneficial environmental, social, and economic objectives. From the company's perspective there is no stronger image enhancement than a former critic supporting company activities. Attaining FSC certification has, in the company's interpretation, made environmentalists messengers that can help STORA and the industry improve its image and relationships with the public and environmental groups.

The Financial Risks of Sustainable Forestry

STORA appears to have begun the successful implementation of ELP and ecologically based forest management, but the real test will come as the system is fully implemented across the company's 2.3 million hectares. Since most Swedish sites already produce less than 50 percent of the fiber that high-yield plantations produce elsewhere, the ecological set-asides made as part of the new forest management are financially riskier in the long-term for STORA than they might be for companies elsewhere. In fact, one Swedish forestry expert pointed out that the test unit near Grangärde was unique: It cost less to implement ELP there than it will cost on other company lands. In 1997 the long-term costs to fully implement ELP were still unknown, even though management thought they knew the level of total implementation costs. Companies set limits to environmental concessions to maintain economic viability. In STORA's case, the board of directors placed that limit at a 10 percent loss in potential harvest. Management considers any losses beyond that to have the potential to endanger the company's ability to compete in the worldwide pulp and paper markets. Whether STORA could fully implement ELP across all of its lands within that cost parameter was unknown in 1997.

The Competitive Risks of Certification

As an early move into certification, STORA confronted certain risks. Third-party certification may provide no more than a short-term competitive advantage. The success of certification as a competitive strategy for STORA depends on the validity of management's assumption that competitors will not be able to quickly mimic STORA's practices. A potentially even greater risk is that European customers will not have a lasting commitment to FSC accredited certification. Although these buyers say they are willing to pay a small premium for these products in the short term, they will be unwilling to do so indefinitely. The buyers take the position that third-party certification comes at little cost to the suppliers and merely reflects practices that any credible supplier should follow anyway.

STORA's management unquestionably considers environmental positioning a source of competitive value in its markets. In 1995, Stora Fors AB became the first cartonboard mill in Europe to receive Ecomanagement and Audit Scheme (EMAS) Registration, an environmental management system similar to ISO 14001, an achievement the company promoted in a video and several print pieces. But in early 1998, STORA management had not yet announced which products it would market as certified and when. Managers at the certified Ludvika forest district and the sawmill were frustrated at being unable to market certified products. Sawmill managers recognized pent up demand in specific market segments and wanted to capitalize on what they perceived as a competitive advantage if they could sell certified material. Some company personnel worried that if management did not capitalize on certification quickly, STORA might lose any potential competitive advantage.

Noncompany Lands—A Key to Sustainability Claims

STORA may also fail in its efforts to provide FSC-certified products unless other Swedish landowners that supply raw material to STORA adopt similar practices. STORA, however, has just begun the difficult task of responding to 1995+ Group members by documenting its sources of wood supplies. The raw materials for STORA's products come from five basic sources: company lands (25 percent); independently owned land that is harvested by company crews (15 percent); small woodland owner associations (25 percent); imports (15 percent); and chips from sawmills (20 percent). STORA's vertical integration gives it some advantage in keeping track of the chain of custody of its products, which is required by certification. In the production of solidwood, the chain is fairly intact. Nearly all of the wood used in the company's sawmills comes from its own lands.

The challenge lies in paper production, which can involve raw material from multiple sources that is fed simultaneously into a continuous process. In STORA's case, less than 50 percent of the fiber it uses in paper comes from its own lands. Under these circumstances, maintaining a clear chain of custody when mixing many different sources of fiber to make paper is much trickier. In 1997 the FSC allowed paper with just 70 percent certified content to carry the FSC logo—and even that content demand is "still a little high," according to Ragner Freiberg. The company, said Freiberg, preferred that the partial content limit begin at 50 percent and be raised over time.

The greatest obstacle to STORA's ability to provide sustainably produced products, however, will lie in convincing the small, independent landowners from whom it buys timber to come into environmental compliance. If other owners refuse to adopt and maintain ecologically based management practices, it will weaken STORA's claim as a sustainable operation. Where STORA personnel are supervising or conducting harvest operations, they implement these techniques. But even then, they admit that the landowners involved may not maintain the practices. The group that represents small landowners in Sweden, the Forest Owners Association, which withdrew from the Swedish FSC Working Group in 1997, planned to develop its own environmental standards to fit within the European Union version of the ISO 14001. The failure of the Forest Owners Association, which represents about 25 percent of forest area in Sweden, to endorse national standards will complicate STORA's efforts to produce certified paper.

A Degree of Uncertainty

As STORA prepared to continue certification audits for its remaining Swedish land holdings in early 1998, management acknowledged the uncertainty that dogged its certification campaign. It was questionable, Freiberg agreed, whether STORA's customers would compensate the company for the increased costs of sustainable forestry and certification. Nor was it clear whether and for how long STORA might reap a competitive advantage with FSC certified products. Said Feiberg: "Many of these questions can only be answered by the market. We have had a very positive response so far, but it is nearly impossible to predict what will happen."

While the outcome of STORA's certification efforts may be uncertain, the company's success in achieving pilot certification and its commitment to full FSC certification revealed a great deal about the state of certified sustainable forestry in the late 1990s. The forest management practices that pulp and paper producers typically use to promote rapid growth, including shorter periods of rotation than solidwood producers, would seem to make it more difficult for pulp producers to get certification for ecologically based forest management, which is based on multiple benefits and a long-term orientation. STORA's ability to achieve certification suggests that with an internal commitment to change forest management practices and allocate the necessary resources, a large pulp and paper company can certify its lands under FSC guidelines.

STORA's decision to certify its forest was based largely on demand for third-party certification by its European customers, buyers' groups, and high-purchasing paper customers. The Swedish industry's evolution toward FSC certification leaves little doubt that these commercial and industrial customers will play a decisive role in the spread of certification in Europe and that, at least initially, the European forest products market appears to be linked to FSC-certified products. STORA's adoption of certification supports the conviction of those industry observers who have long maintained that the industry's acceptance of environmental certification does not depend entirely, or even in large measure, on demand from final consumers.

In pursuing FSC certification, STORA is trying to capture a competitive advantage open to it by virtue of ecologically based management practices that it already practiced. Essentially, certification is a reward for behavior that STORA was already committed to. Even though certification was not a determining factor in STORA's adoption of ecologically based forest management, the company's changes in forest management positioned STORA to efficiently certify and reap whatever reward may come from marketing certified wood products. The impact of Sweden's national FSC standards, and the initial marketing efforts of Swedish companies, including STORA, will be a critical predictor of the future of certified forest products worldwide.

NOTES

This chapter is adapted from "STORA: The Road To Certification," prepared by Rick Fletcher, James McAlexander, and Eric Hansen of Oregon State University for the Sustainable Forestry Working Group, 1997.

1. This section is based on the following publications: Sven Rydberg (1988), *The Great Copper Mountain,* Värnamo, Sweden; and Tommy Forss and Kurt Netzler, *Chronicles, The Annals of STORA.*

2. This revenue figure and all others in the paper were tabulated using 1 U.S. dollar equals 7.477 Swedish krona as the conversion rate.

3. This material was prepared by Eric Hansen and Stephen Lawton, both of Oregon State University, and Stefan Weinfurter, Universitat fur Bodenkulter, Vienna, Austria.

4. This material adapted from "J. Sainsbury Plc. and The Home Depot: Retailers' Impact on Sustainability," prepared by James Alexander and Eric Hansen for the Sustainable Forestry Working Group, 1997.

Chapter 12

The Wall of Wood

Born during the "cut and run" days of early twentieth century America, Weyerhaeuser Co. defied conventional industry logic by holding onto timberlands after they were cut rather than walking away. By the late 1930s, the company faced a decision: What to do with previously logged land on which natural regeneration had been ineffective? It decided to regenerate forests and grow timber as a crop, first by seeding harvested areas in the 1940s and then, starting in the 1950s, by planting seedlings. Beginning in the 1960s, Weyerhaeuser began producing seedlings in nurseries and integrated replanting into its plantation operations. By following this strategy, Federal Way, a Washington-based company, has become the world's largest private owner of standing softwood timber, North America's largest producer of softwood lumber, and the world's largest supplier of softwood pulp.

Weyerhaeuser initiated sustained-yield forestry to provide a guaranteed and consistent supply of wood, not out of direct concern for the environment. However, the company has come to realize that by investing in a long-term strategy, its decisions have positive ecological and economic consequences that will amplify into the future. Over the past 30 years, Weyerhaeuser has developed a form of sustainable forestry based on high-yield plantations that are among the most productive in the world. This high-yield model provides higher returns while simultaneously minimizing overall environmental impacts by producing high-quality wood and fiber on substantially fewer, continuously regenerated, acres. In this sense, the Weyerhaeuser forestry model may facilitate both environmental and economic sustainability.

A Business Based on Softwood Timber and Wood Products

Weyerhaeuser was an $11.1 billion company in 1996 with the forest products segments accounting for slightly more than $10 billion in sales. The company provides products and services through three different business sectors: Timberlands and Wood Products; Pulp, Paper, and Packaging; and the Weyerhaeuser Real Estate Company. Figure 12.1 shows a five-year trend of the approximate contributions to operating earnings made by each sector.

Weyerhaeuser's corporate strategy is based on the premise that the highest returns from timber growing are achieved through an emphasis on high-value softwood timber and wood products over fiber for pulp and paper. As a result, Timberlands is run as a profit center, rather than as a cost center (internal supplier) for the product sectors. This arrangement does on occasion create friction with the downstream manufacturing operations, since company lands are not solely managed to produce raw material targeted to the needs of the wood products, pulp, paper, and packaging businesses.

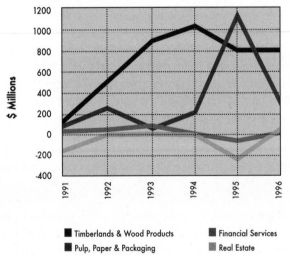

Figure 12.1.

Approximate Contributions to Operating Earnings, 1991–1996.

■ Timberlands & Wood Products   ■ Financial Services
■ Pulp, Paper & Packaging   ▨ Real Estate

Timberlands and Wood Products

In 1996, operating earnings for the Timberlands and Wood Products sectors together totaled $805 million on $5.2 billion in net sales. The Timberlands businesses produce primarily high-grade timber and sawnwood products—specifically, Douglas fir (*Pseudotsuga menziesii*) and western hemlock (*Tsuga heterophylla*) in the U.S. Northwest and loblolly pine (*Pinus taeda*) in the U.S. Southeast. In addition to operating company tree farms, Timberlands manages genetics and selective breeding programs, as well as seed and seedling operations at the company's orchards and nurseries. These nurseries grow 257 million seedlings per year, of which 50 million are used on the company's 5.3 million acres of timberland. The balance, 207 million, is sold to third parties. Finally, Timberlands provides timber to Weyerhaeuser's manufacturing facilities and to other forest products customers.

The Wood Products businesses manufacture structural and appearance-grade lumber, along with plywood and engineered products. In addition, Weyerhaeuser operates a wholesale building materials distribution business in the United States. Products consist of softwood lumber, plywood, veneer, composite panels, oriented strand board, doors, treated products, logs, and chips. In addition to Douglas fir and loblolly pine, the Wood Products business also extensively uses white spruce (*Picea glauca*) aspen (*Populus* spp.), sourced from Crown concessions in Canada, and other hardwood species from U.S. locations.

Most tree species used by the company, with the exception of aspen and other hardwoods, are evergreens, or softwoods.[1] Despite the nomenclature, softwood species are favored over hardwoods in structural applications because they are stronger, more durable, and more flexible. All softwood species, however, are not created equal: Douglas fir has strength and quality characteristics that are unmatched by other species. Douglas fir is particularly competitive in Japanese post and beam housing construction where customers are willing to pay premium prices for high structural performance and quality. Monterey or radiata pine (*Pinus radiata*), grown in the temperate tropics, and loblolly pine are less expensive to grow but are of lower structural quality.

Western Timberlands manages approximately 2.1 million acres located in Oregon and Washington. Southern Timberlands manages another 3.2 million acres in the states of Arkansas, Oklahoma, Louisiana, Mississippi, Alabama, Georgia, and North Carolina. In addition, Weyerhaeuser Canada oversees the company's long-term Crown license arrangements in the provinces of British Columbia, Alberta, and Saskatchewan, which cover approximately 23 million acres. Weyerhaeuser's combined 1996 timber inventory on U.S. and Canadian lands was approximately 26.6 billion cubic feet, of which 75 percent was softwood. To increase the yield and quality of the timber supply from its U.S. timberlands, the company engages in the active suppression of nonmerchantable species, precommercial and commercial thinning, fertilization, and operational pruning.

Pulp, Paper, and Packaging

In 1996, the Pulp, Paper, and Packaging (PPP) sector had sales of $4.6 billion on $6.7 billion in assets, representing approximately half of the company's asset base. While earnings soared in 1995 due to unusually strong pulp and paper prices, low PPP prices in the balance of the 1990s depressed PPP earnings relative to Timberlands and Wood Products (Figure 12.1). PPP products include market pulp, newsprint, fine paper, containerboard packaging, and paperboard. The PPP business uses wood chips and other residuals from the timber and wood products businesses but still sources raw material from non-Weyerhaeuser suppliers. PPP also secures a growing percentage of its raw material needs from recycled fiber.

Weyerhaeuser Real Estate Company

The final business, Weyerhaeuser Real Estate, is a remnant of the company's diversification efforts from the 1970s to the early 1980s. With the aim of countering the cyclical nature of the core businesses, Weyerhaeuser invested in ventures ranging from hydroponic farming to disposable diapers. Efforts to refocus the company begun in the late 1980s have led Weyerhaeuser to shed most of these operations. Weyerhaeuser Real Estate develops single-family housing and residential lots, as well as master-planned communities. Operations are concentrated in selected metropolitan areas in Southern California, Nevada, Washington, Texas, Maryland, and Virginia. In 1996, the real estate business made $35 million in operating earnings on sales of $804 million. Over the past several years, a number of operating units within Weyerhaeuser Real Estate have been sold.

The Evolution of Weyerhaeuser Forestry

Weyerhaeuser was one of the first U.S. forest products companies to move to the West Coast from the mid-western lake states at the turn of the twentieth century. As the mid-western forests of Michigan, Wisconsin, and Minnesota were depleted, it seemed clear that America's westward expansion would require additional timber. The great northwestern forests in California, Oregon, and Washington held the most promise. In 1900, Frederick Weyerhaeuser and a number of partners founded the Weyerhaeuser Timber Company with a purchase of 900,000 acres of forestland in Washington belonging to Northern Pacific Railroad.

In the early years, timber was harvested to provide returns to the original investors and to pay for the survey of the land holdings. By the late teens and early 1920s, the company decided to invest in milling facilities rather than just sell logs on the open market. In the 1920s and early 1930s harvest levels rose as the company acquired additional lands and mills. During this period, logged-over land was left to natural regeneration. However, the success of such regeneration, especially with the more desirable Douglas fir, was mixed at best.

Tree Farming

By the late 1930s, the company faced a decision: Should the land that had been cut be sold or could it be retained and continue to earn a profit?[2] Contrary to prevailing industry wisdom, Weyerhaeuser believed that timber could be grown like a crop on a sustainable and profitable basis. Management decided to retain and reforest the land holdings. In 1941, the company established the nation's first privately owned tree farm to demonstrate that intensive management of previously harvested forestlands could be profitable. The Clemen's Tree Farm became a laboratory where the company undertook experiments in forest regeneration, aerial seeding, fire protection, and brush control techniques. This research indicated that an economically viable tree farming operation would require more than merely allowing secondary forests to regenerate where virgin forests had once stood.

During the 1940s, the company's forestry focused on how to better regenerate cut-over land to Douglas fir through seeding methods. At the same time, the company also worked to change tax policies that discouraged the long-term holding of productive forestlands and helped to shape fire suppression policies that reduced the risk for long-term investment. By the end of the 1940s, Weyerhaeuser had begun the transition to a "tree growing" company. This was in sharp contrast to much of the rest of the industry, which was still largely focused on liquidating old-growth stands or cutting timber from public lands to feed downstream mill operations.

High-Yield Forestry

By the early 1950s, it was evident that even though seedling methods were superior to natural regeneration, a more systematic, research-based approach would be required if tree farming were to be profitable. Research was concentrated first on how to make the dense second-growth stands of Douglas fir, established after logging the old growth, saleable in a shorter time. By the late 1950s, however, it became increasingly clear that planting seedlings following harvesting was a better and more reliable method of forest regeneration.

As a consequence, the company began, by the mid-1960s, to develop a new forestry model that would come to be known as high-yield forestry (HYF). Research initially concentrated on management practices such as planting and stocking control to regenerate the forest rapidly. The company established forest nurseries to produce the seedlings required for replanting. Biological and financial variables were linked to develop practices that provided a continuous cycle of planting, growth, harvesting, and replanting.

Improving a Good Idea

To that end, soil surveys were conducted across the company's entire land base. The resulting data were used to construct simulations of forest growth and yield. These simulations helped to improve the technical and scientific aspects of HYF, which include:

- Planting techniques to optimize siting, spacing, and rooting
- Pest and noncommercial vegetation control to minimize competition
- Fertilizing regimes and thinning practices to maximize growth
- Road-building techniques to minimize erosion and compaction
- Harvest timing to maximize yield while maintaining wildlife habitat
- Protecting soil productivity
- Ensuring rapid reforestation

Once established, forests managed under these practices grew at unprecedented rates, compared to unmanaged natural stands. The company then began to work to decrease mortality rates of seedlings in the nurseries that up to that time had a dismal 50-percent survival rate. Additional applied sciences aided in the development of seed orchard techniques for efficient cone and seed production and collection, as well as nursery techniques for establishing reliable germination, culturing, and fertilizing practices that led to healthy seedlings with strong root systems. Seedling mortality has dropped to less than 5 percent today.

Breeding Better Trees

In the early 1970s, the company began to make collections of healthy, vigorous trees from natural stands of Douglas fir and loblolly pine. These "plus trees" formed the basis of the tree-improvement program for each species. Through selective breeding, trees that were inferior for growth, form, and adaptability were eliminated from the population. Beginning in the 1980s, the tree-improvement program began to focus on breeding varieties for specific end-use characteristics such as increased volume (fast growth), strength (specific gravity), and stem quality (straightness, branching habit).

Although HYF seeks to maximize growth of the primary commercial species at each site while suppressing other noncommercial species, the resulting forest is not a pure monoculture. Complete repression of undesired species is impossible because seeds are dispersed by wind, water, and animals. The result, while significantly different from a natural forest, is far from sterile either visually or ecologically.[3] Nor does HYF imply that the wrong species are introduced onto unsuitable sites. In addition to Douglas fir and loblolly pine, Weyerhaeuser has active regeneration programs in hemlock, noble fir, longleaf pine, and some species of hardwood.

A cycle of HYF begins when harvest plans are drawn up and information on physiographic and soil features of the next round of planting sites is analyzed to indicate which of the available genetic varieties is optimum for a site. A forester will place an order for seed of that variety that will be germinated and grown at the nurseries for 1 to 3 years before it is planted on those sites and managed according to the optimum silvicultural practices (planting density, thinning, pruning, and so on), which are designed to lead to a particular desired forest community and final harvest product.[4]

Although large quantities of seedlings are sold to other forest landowners, the most advanced seedlings are used exclusively on Weyerhaeuser lands. Within a year of harvesting any given region, those seedlings are planted by hand following predetermined spacing and patterns that will maximize the desired qualities. Over the forest growth cycle, most sites are thinned to remove less desirable trees, thus providing more growing space to ensure that the remaining trees are larger and straighter. Precommercial thinning allows the company to leave organic matter in the form of trunks and branches on the forest floor where their slow decay adds to the soil's organic matter and nutrient base, promoting future growth. Commercial thinning initially yields trees large enough for use in the pulp mills and, in subsequent thinnings, for lumber. Each tree may also be pruned once or more during its life. Lower branches are removed to reduce the presence of knots in future lumber products and to increase the production of clear wood. Aerial fertilization may be used to promote growth spurts throughout the rotation.

Patience Is Key

The goal of HYF is to produce wood that is, ultimately, more valuable to the end user. After 30 years in the Southeast (loblolly pine) and 45 years in the Pacific Northwest (Douglas fir), specific sites are harvested consistent with the needs of the company's downstream businesses. To ensure a sustainable flow of product, on average only 2 percent of Weyerhaeuser's forest sites in the West and 3 percent in the South are harvested in a given year. None of the harvest is wasted. Crowns and branches are left on site to return nutrients to the soil through decomposition. Bark may be burned for energy or sold as mulch for landscaping. Depending on the region where it is processed, the bulk of a log becomes either lumber and other solidwood products or ends up as chips for pulp production. The rest, sawdust, is used for particleboard or mulch, or is burned for energy.

HYF has resulted in significant increases in forest productivity on Weyerhaeuser's lands since the 1960s. Yields per acre have doubled in the Northwest and have increased fourfold in the Southeast, compared to a natural forest—making the company's lands some of the most productive in the world. However, HYF has also required great patience: Investments made in the 1960s will begin to yield economic returns early in the twenty-first century, as genetically improved plantations begin to reach maturity.

Since the decision to pursue tree farming in the late 1930s, the company's strategy has depended on a long-term outlook. Investment in lands that will not realize significant returns for 30 to 50 years requires planning and commitment. Weyerhaeuser believes that measured, deliberative actions that do not overreact to short-term changes in the marketplace produce a more stable operating environment. As one manager stated, " . . . deals are available all the time, companies are always making mistakes." This long-term commitment has made it possible for the company to take the time and make the investments necessary to effectively implement HYF management practices.

Weyerhaeuser Forestry

With its heritage as a pioneer in forest management, Weyerhaeuser developed a sense of commitment to environmental stewardship. However, during the 1970s and 1980s, the public's concern over forest practices escalated and regulatory pressures on the industry increased. Objections to clear-cutting and concern about dwindling old-growth forests resulted in increasingly strict regulations on the use of public and private lands for timber harvesting, particularly in the Pacific Northwest.

Since Weyerhaeuser owned much of its own timberland and did not source significant amounts of wood from public lands in the United States, the company focused its attention on its own specific issues and tended to disregard these concerns. Weyerhaeuser countered criticisms by reemphasizing its commitment to growing trees and providing scientific data to justify its practices and positions. But environmentalists were not persuaded by scientific and technical defenses of the company's practices. By the end of the 1980s, Weyerhaeuser's reputation began to suffer as it was judged by the industry's lowest common denominator.

Private Forest as Public Trust

Dogged by poor financial performance relative to competitors, Weyerhaeuser underwent a series of changes beginning in 1989 when incoming president and CEO John Creighton, Jr. started to refocus the organization on core forest products operations. In the early 1990s, the company set about regaining the public trust by seeking to understand the changing social forces that had led to the gap between public perception and Weyerhaeuser's practices. Management came to understand that the public viewed all forests as public trust resources, regardless of ownership. As one Weyerhaeuser manager noted, "We were not acting as stewards of the resource in the eyes of the people." Weyerhaeuser's attempts to justify its practices through mounds of data had failed to address fundamental public fears over the health and future of forests and ecosystems.

This understanding allowed Weyerhaeuser to move beyond the sustained yield practices of HYF into a broader definition of sustainability: HYF evolved into Weyerhaeuser Forestry as management moved to incorporate the concerns of external stakeholders into operational decisions. Regional Forest Councils and town hall style meetings in the communities directly affected the company's operations, and senior management began to enter into a dialogue that allowed them to better understand where public values dovetailed or diverged from those of the company. The outgrowth of these actions was the creation of a corporate stewardship statement and a set of resource strategies that provided a guideline for the company's more holistic model of sustainable forestry.

Weyerhaeuser's forestry now includes commitments and strategies to enhance water quality, fish and wildlife habitat, soil productivity, biodiversity, and the aesthetic, cultural, and historical values of forests, as well as to continually improve its performance as a responsible steward of the environmental quality and economic value of its forests. The company has pledged to accomplish its goals through sustainable forestry; standards set for all forest operations; management processes and practices based on scientific research and technology; and cooperative efforts with public agencies and other interest groups to develop balanced, cost-effective forest practices and regulations.

Economic and Environmental Sustainability

Weyerhaeuser Forestry challenges the company to assume a more proactive role in the protection of endangered species habitats and has facilitated the development of new tools and data for better management of the natural environment. The company's Habitat Conservation Plans (HCPs), for example, are designed to reduce impacts of forest management on threatened and endangered species through multispecies management. Watershed analyses that allow the company to manage forests as systems rather than addressing individually the scores of regulations that apply at one time allow the company to operate more flexibly.

Weyerhaeuser Forestry was challenged in 1989 when the spotted owl was listed as a threatened species in the Pacific Northwest. The management plan proposed by the Federal Fish and Wildlife Service would have defined a "no cut" zone 1.2 miles in radius around each identified spotted owl nest. Through its HCP approach, however, Weyerhaeuser demonstrated a conservation plan for the owl that also allowed more flexible use of its forest lands, opening up timber assets that had previously been defined as off limits under the "owl circle" management plan.

With five steelhead trout populations listed as endangered or threatened by the National Marine Fisheries Service in the mid-1990s, Weyerhaeuser Forestry will again be put to the test: Proposed management strategies by regulatory agencies include leaving 300-foot wide riparian zones on each side of fish-bearing streams. The company will have to again define an environmentally appropriate and socially acceptable management plan while retaining flexibility in the use of its lands: Reasonable set-asides of harvestable timber to achieve environmental goals can be offset by the higher yields associated with Weyerhaeuser's forestry practices.

The company believes Weyerhaeuser Forestry can achieve both environmental and economic goals, the latter through higher yields that reduce the need to cut or manage additional forestland, and the former through initiatives like HCPs and watershed analyses. Weyerhaeuser's high-yield plantation model also facilitates cleaner air and offsets the production of greenhouse gases through the higher oxygen production/carbon-dioxide fixing associated with forests composed of young fast-growing trees.

If the company successfully meets both economic and environmental goals through its Weyerhaeuser forestry, it should gain greater support or recognition from environmental groups and other stakeholders whose priority is the environmental values of forests. In 1998, though, Weyerhaeuser had still not convinced many leading environmental interests in the Pacific Northwest that its revised forest management plans and strategies would, in the long run, keep the forests robust and protect watersheds and wildlife (Box 12.1).

In the future, the company's high-yield model may help fulfill society's growing demand for wood products into the next century without shifting that demand onto the world's remaining old-growth and native forests. By harvesting more wood on less acreage, potentially more forestland can be reserved on public lands for other nonwood purposes, including the development of nontimber forest resources, wilderness, wildlife habitat, and recreation. Indeed, the Weyerhaeuser Forestry model raises interesting possibilities for the preservation of public lands, parks, and reserves around the world as population growth and economic development increase the

Box 12.1. Mixed Reviews from Environmentalists[5]

Has Weyerhaeuser lived up to its claims of practicing a more holistic approach to forestry that includes involving the community and protecting watersheds and wildlife? One group of stakeholders contacted in 1998, environmental groups in the region, were ambivalent about the company's environmental performance under its recently adopted forestry strategy. Although she gave the company mixed reviews, Laurie Wayburn, president of the Pacific Forest Trust, echoed the assessment most often expressed about the company as a plantation manager: "Weyerhaeuser is one of the best."

But other environmentalists' assessments reflected what appear to be irreconcilable differences with the company concerning issues of its plantation-style forest management. Since 1900, according to its own records, Weyerhaeuser had clear-cut concerning 4 million acres. As Charles Willer of the Coast Range Association in Oregon pointed out, the biomass that stands on land under industrial forestry is about 1/6 to 1/10 of that which stands in native forests, which "produces an impoverished condition of the landscape," even under the best plantation practices.

Environmentalists continue to argue that Weyerhaeuser's fundamental management practices are detrimental to long-term health. The company's 45-year rotations are well below the 120 to 200 years that ecologists argue are optimum for the region. The company continues to use clear-cuts and typically has less than the 15-percent accumulation of the coarse debris on the forest floor that is needed for full function of the forest in first-growth plantations, according to foresters and ecologists. In second- and third-growth plantations, debris can get as low as 3 percent. This ensures that the forest will remain impoverished.

Some environmentalists do consider Weyerhaeuser's Habitat Conservation Plans (HCPs) an improvement. Weyerhaeuser implemented a watershed protection plan on some of its land in The Tolt and Griffin/Tokul areas. George Pess, a stream ecologist with The Tulalip Tribes in Marysville, Washington, said that under that plan, once the company agreed to changes, it followed through. On one critical issue for watershed protection—culverts that block the passage of fish—the company agreed to fix all blockages on its land within five years. In 1998 Pess reported that the company was on schedule with the changes.

Another HCP for two nearly contiguous blocks of land that total about 400,000 acres in the Western Cascades of Oregon represented a "substantial improvement over past actions and practices, and was conceptually moving in the right direction," according to David Bayles, Conservation Director of Pacific Rivers Council. He and others credit the company with correctly understanding that implementing riparian zones where logging is prohibited within a proscribed distance from streams is insufficient to

(continued)

protect water systems: Other protective measures are essential further up the slopes. "Weyerhaeuser is ahead of most of the industry on that issue," according to Bayles. However, critics faulted the plan for weaknesses in the measures it proposed to protect unstable slopes, headwater streams, and species other than salmon.

Although Weyerhaeuser's management may believe that Weyerhaeuser Forestry can meet its economic objectives while protecting habitats, watersheds, and endangered species, many environmentalists will need more evidence before they will accept the company's claim. If, however, Weyerhaeuser does demonstrate that its HCPs and revised forestry practices deliver long-term ecological benefits, the company should enhance its credibility even more with the public, environmentalists, and regulators in the future.

demand for wood products worldwide while a variety of forces make wood supplies tighter.

By the year 2010, demand for paper and paperboard is expected to rise from approximately 250 million tons to over 450 million tons. Global demand for solidwood products is expected to increase almost 60 percent from 630 million cubic meters annually in 1989 to over 1 billion cubic meters annually in 2010. Softwood production overall is estimated to increase from 939 million cubic meters to a total of 1,085 million cubic meters, for growth of about 15 percent over the next 25 years. Given that demand growth is conservatively estimated at 1.3 percent per year, the softwood market would appear to be on a collision course, with demand outstripping supply for the foreseeable future.[6] As discussed in Chapter 1, a number of regional developments, many of them driven by environmental concerns, will hinder softwood supply expansion over the next 20 years. These trends indicate a market where softwood timber demand will outpace supply, creating a softwood timber deficit. As softwood timber products become more expensive due to restricted supply, substitute products—steel, concrete, and engineered wood products, such as wood I-beams, laminated veneer lumber, and oriented strand board—may grow in popularity.

Competitive Position

Weyerhaeuser's strategy is centered on the production of the highest value softwood timber in one of the most developed, environmentally conscious regions in the world. The company has focused its principal business on softwood species with structural and performance characteristics that yield superior quality lumber and wood products. For Douglas fir, this means virtually 100 percent of the harvest goes directly to the sawmill; for loblolly pine, more than half the harvest goes to the sawmill or plywood plant. Even with this sawlog mentality, however, over half of the cubic volume of wood ends up as residual materials, since only the highest-grade portions of each tree can be used for grade lumber. Most of the remaining material becomes pulping fiber that can be used in Weyerhaeuser's pulp and paper mills.

HYF as Core Competence

Weyerhaeuser has spent almost 50 years developing a knowledge of its forestlands. This enables the company to evaluate site conditions that influence breeding programs so that tree varieties are matched to sites, ensuring high levels of productivity. Stocking control in the form of spacing, thinning, and fertilizing helps to produce straight fast-growing trees with a maximum amount of clear wood for added value. These traits fulfill customers' needs while sustainable yields guarantee a reliable supply of product, and an advanced infrastructure guarantees the product reaches the customer in a timely manner.

Forestry initiatives that yield consistent, dependable results garner strong customer relationships and allow the company to enjoy premium prices on certain product lines where demand often exceeds supply. "The Weyerhaeuser name gets us in a lot of doors," noted one manager. This is particularly important to customers in the company's most important export market of Japan, where reputation and relationships are a prerequisite for continued success, and the company enjoys a large market share (10 percent of company sales are to Japan alone). Thus, a cycle exists where Weyerhaeuser Forestry helps to retain existing customers and attract new ones, which in turn legitimizes and drives the continuation of the forestry operations.

Because the company's asset base is in North America, it is beholden to longer investment and growth cycles than competitors in the warmer climates of Latin America and the Pacific-Asian region. Furthermore, Weyerhaeuser's 30 years of experience in softwood genetics make emulation of its productivity gains more difficult for most other companies. Three decades of capability building in forestry have enabled the company to establish a substantial first-mover advantage over its competitors, especially in the Pacific Northwest.[7]

Wall of Wood

Weyerhaeuser's early investment in the land base enabled it to build a portfolio comprising some of the best forest sites in the world. The majority of the company's lands in the Northwest are more moderately sloped, lower-elevation lands that are almost impossible to acquire today. Indeed, through almost 100 years of carefully planned acquisitions, the company has been able to accumulate a superior portfolio of forest lands.[8]

Through superior site quality and forest management methods, Weyerhaeuser has boosted forest productivity. The company estimates that its lands currently generate twice the wood per acre in the Pacific Northwest and four times as much in the Southeast compared to unmanaged (natural) forest land. For example, U.S. net annual tree growth in the 1990s is about 54 cubic feet per acre, twice the level of the 1950s. Average growth on Weyerhaeuser land, however, is more than 108 cubic feet per acre, with the company's prime, intensively managed lands producing an annual tree growth of up to 240 cubic feet per acre.

Furthermore, by 2010–2020, as the genetically improved plantations reach maturity, the company estimates there will be an additional increase in yield on its lands in both the Pacific Northwest and the U.S. Southeast. These yield improvements should translate into an increased wood harvest per acre of 50 percent in the Northwest and 100 percent in the Southeast. The result is a reliable, high quality, and growing volume of softwood timber (referred to internally as a "Wall of Wood"), which is expected to reach maximum yield in a time of constricting supply, beginning around the year 2005 (Figure 12.2).

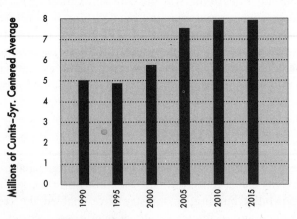

Figure 12.2.

Wall of Wood.

Environment as Competitive Advantage

Through its initiatives to establish a dialogue with communities, indigenous peoples, and environmental groups, the company has developed skill at cultivating stakeholder relationships. The company's experience with town hall style meetings to solicit stakeholder opinions has allowed it to develop an openness and willingness to accept outside views and an understanding of how best to engage in that kind of dialogue. Such constructive engagement has resulted in better experiences and relationships with both regulating agencies and the public through its open implementation of alternative regulatory initiatives like HCPs and Watershed Analyses.

Weyerhaeuser's willingness to cultivate stakeholder relationships should position it to capture the environmental high ground in the forest products industry. Indeed, environmentally driven constraints on timber harvests, particularly on public lands, provide a competitive advantage. Given its substantial reliance on well-managed, company-owned lands, Weyerhaeuser is advantageously positioned to provide a growing supply of high-quality softwood in a world of tightening environmental regulation and constricting supply. In fact, if the competitive game shifts away from sustainable forestry and toward timber mining—if the Russian Far East were to be logged at an accelerating pace to generate hard currency—then the Weyerhaeuser Forestry model would lose much of its basis for competitive advantage. In a very real sense, the environment can be a competitive ally for Weyerhaeuser. It is possible that a restricted supply of quality wood along with increased demand for softwood could amplify this cash flow trend over time.

Financial Payoff

Each stage of forest operations has financial implications: Investment decisions must be discounted over 30 to 50 years, placing great importance on the discount rate chosen. Because of HYF management, however, investments of $300 to $600 per acre can be recouped through doubled, tripled, or even quadrupled wood yields.

While Weyerhaeuser underperformed the industry during the 1980s in the equity markets, the company has performed significantly better over the past few years, due to refocusing efforts, reaching and maintaining its top quartile performance goal since 1993 (see Figure 12.3). Increased yields from Weyerhaeuser Forestry at today's market prices would significantly improve the company's cash flow position over the next 25 years (see Figure 12.4).

1985-90 Percent Change		1990-95 Percent Change	
MacMillan Bloedel	153%	Willamette	210%
International Paper	146	Louisiana-Pacific	202
S&P 500	80	**Weyerhaeuser**	**136**
Temple-Inland	71	Georgia-Pacific	112
Willamette Ind.	63	S&P 500	111
Westvaco	62	Bowater	96
Georgia-Pacific	49	Champion Int.	72
Union Camp	46	Westvaco	70
Louisiana-Pacific	39	Potlatch	67
Weyerhaeuser	**28**	Temple-Inland	59
Champion Int.	22	International Paper	59
Potlatch	13	Union Camp	56
Boise Cascade	-9	Boise Cascade	45
Bowater	-18	Stone Container	32
Stone Container	-19	MacMillan Bloedel	-1

Figure 12.3.

Total Shareholder Return.

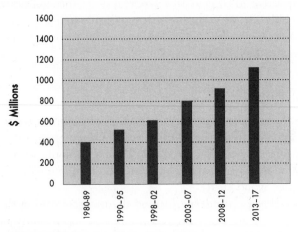

Figure 12.4.

Pre-Tax Free Cash Flow from Timberlands.
Note: At 1993–1995 Average Prices

Future Challenges[9]

Weyerhaeuser appears to have successfully integrated environmental concerns into its core strategy and value proposition. Significantly, environmental concerns have been incorporated within a broader model of product quality and competitive advantage, offering the potential for long-term sustainability, both economically and environmentally. However, a number of strategic challenges face the company as it looks toward the twenty-first century.

Strategic Questions

Should the company seek certification for its company-managed timberlands or its products?

Scientific Certification Systems (SCS), SmartWood, and the Forest Stewardship Council (FSC) certification programs all set forth principles and criteria for site-specific forest management, and all require third-party evaluation. A few niche players have used third-party certification to advantage. Virtually all of the large companies, however, have resisted the certification programs advocated by NGOs. Companies have preferred instead to identify outcome-based goals and design management systems that foster measurable progress toward those goals: The U.S. American Forest and Paper Association's (AFPA) Sustainable Forestry Initiative (SFI) represents such a "management system" approach, as does the ISO-14001 Environmental Management Standard.

So far, demand for certified wood products has been limited to niche markets in western Europe and North America: Worldwide, buyers have consistently shown that they will not pay a premium for environmental wood products. However, if a large industry player should break from the pack and seek third-party certification, the competitive dynamics of the industry may change as other large competitors feel compelled to follow. Such a move might have special consequence in the environmentally sensitive European market and in developing countries where certification could establish barriers to entry in an otherwise underregulated environment.

Third-party certification might make competitive sense for a large player that was able to integrate sustainable forestry and environmental performance into its core business strategy—that is, a firm that could offer a quality product at competitive prices plus satisfy the environmental criteria. Weyerhaeuser may be positioned to gain such an advantage. It may therefore make sense for the company to consider third-party certification, at least for several of its forest sites if not for its products in general.

How should the company handle chain-of-custody issues for the wood it buys from outside suppliers?

Despite high yields from its own well-managed forest lands, Weyerhaeuser is not self-sufficient in wood. Both the Wood Products and the Pulp, Paper, and Packaging businesses rely on non-Weyerhaeuser timber for their wood and fiber needs. Reliance on outside, usually small, private sources of wood and fiber suggests that it may become increasingly important to manage the impacts of these suppliers' practices, particularly if third-party certification is deemed useful or necessary in the future. Indeed, working with small private suppliers to improve its forest management practices could become a critical element in the company's management system should product certification ever be seriously considered as an option. Chain-of-custody tracking might also become important should environmental concerns force large consumer products companies such as Procter & Gamble into requiring first-tier suppliers to provide such data. Moving the supplier base toward sustainable forestry will not happen overnight. There may be first-mover advantage for the high-purchasing forest products player that can gain preferred access to the best, meaning sustainable, outside wood suppliers in specific geographic regions.

Is the company's current position in Canada consistent with the objectives of Weyerhaeuser Forestry?

With 20 percent of its assets residing north of the border, Weyerhaeuser is a substantial player in the Canadian forest products industry. In contrast to the United States, however, where only 39 percent of the forest lands are publicly owned, 94 percent of Canadian Industrial Forestlands are owned by the provincial governments. Crown ownership prevents private ownership of that country's forestlands and undermines the company's ability to apply Weyerhaeuser Forestry practices for the lands on which it operates. Given the opportunity to follow its best management practices, the company could derive substantially higher yields from less land in Canada. However, lack of land ownership makes it risky to establish high-yield plantations on the Crown concessions granted to the company, since there is no guarantee that the company will be able to maintain long-term control over these sites.

As a result, the company continues to log native forests on concessions covering more than 18 million acres granted to it by the provincial governments of British Columbia, Alberta, and Saskatchewan. And while the company avoids the controversial "ancient forests" of western British Columbia, the inconsistency between its U.S. and Canadian forest operations is evident. As environmental pressure mounts on the Canadian and provincial governments to alter their current concession systems, which favor job creation through harvesting at the expense of long-term management, Weyerhaeuser's position will be improved. If sustainable forestry is to be a centerpiece of the company's competitive strategy, it is in Weyerhaeuser's self-interest to push for policy changes consistent with the principles of Weyerhaeuser Forestry in Canada and gradually move away from its practice of harvesting native forest in favor of long-term plantation forestry.

What should Weyerhaeuser's posture be with regard to the Russian Far East?

In Weyerhaeuser's initial venture into Siberia, it became a partner with a local company in an effort to gain a timber concession. Following an experiment involving the planting of 1.2 million seedlings, the company withdrew from the venture, citing infrastructure and corruption problems that made the practice of Weyerhaeuser Forestry impossible.

Some in the environmental community saw Weyerhaeuser's entry into Russia as a preliminary step intended only to gain preferred access to Siberia's vast old-growth softwood forests. As in Canada, if Weyerhaeuser is serious about centering its strategy on sustainable forestry, it is in the company's self-interest to collaborate with environmental groups, NGOs, and governments to preserve as much of the Russian Far East old growth as possible, while working to identify and secure access to the few highest potential sites for Weyerhaeuser Forestry.

Does the Weyerhaeuser Forestry model transfer easily as the company looks to expand its operations internationally?

As Weyerhaeuser currently produces exclusively in the United States and Canada, it has learned a great deal about producing timber under conditions of intensive regulation and environmental scrutiny.[10] This should serve the company well as it seeks to expand its forest and production base offshore, where the growth markets of the future reside. Both Weyerhaeuser's forestry competence and its skills in engaging stakeholders should be transferable internationally. But extensions and adjustments for each new location will be required. Expansion into less-developed and underregulated countries will require the company to expand its stakeholder involvement model to address more difficult tests than it has dealt with so far, including issues of indigenous peoples, poverty, rural migration, and others related to sustainable development.

Weyerhaeuser will have to apply its core capabilities in selective breeding, tree improvement, and plantation management to species other than Douglas fir and loblolly pine. This should be possible to achieve so long as the company sticks with its focus on softwoods. Particularly attractive are overseas lands that have already been converted to plantations or marginal farming/grazing uses with the soil potential for the company to implement Weyerhaeuser Forestry practices. Opportunities in the temperate tropics of Argentina, Chile, Brazil, Australia, and New Zealand, where it is possible to grow softwoods on a fast-rotation basis, are especially desirable. However, should Weyerhaeuser decide to move into fast-rotation hardwoods, a great deal of new learning and perhaps finding a partner would be required to counter the already strong positions held by substantial competitors such as Aracruz, Mondi, and others.

The company appears to have ruled out investment opportunities that preclude full use of Weyerhaeuser Forestry. For example, they company has steered clear of tropical rainforests partly because of the controversial nature of these ecosystems but also because of its comparative lack of competence with tropical hardwood forests and soils. Similarly, the company has avoided entering politically unstable regions where long-term land ownership and stewardship become high-risk propositions. As a case in point, political instability caused the company to back away from its venture in the Russian Far East.

Does Weyerhaeuser's high-yield plantation forestry constitute a model of sustainable forestry?

Perhaps the biggest challenge facing the company in the future will be that of convincingly demonstrating that Weyerhaeuser Forestry constitutes a model of sustainable forestry. Given the lack of operational definitions concerning "sustainable development" and "sustainable forestry," it will be important that the company proactively make the case that its plantation-based model of high-yield, environmentally sensitive forest management constitutes a viable path toward sustainability.

Recognizing that many believe plantation-based models are unsustainable by definition (since they imply alteration of the composition and character of native forests), Weyerhaeuser will need to engage a broad range of interests if it is to gain general acceptance for its methods. This will probably require the development of a strong argument as to how such intensive plantation-based models fit into the big picture regarding the state of the world's native forests, biodiversity, and the need for sustainable development, especially in developing and emerging economies. A vision of how plantations, managed native forests, and preserved old-growth forests might sustainably coexist is probably essential if such intensive management practices are to gain broad acceptance.

NOTES

This chapter was adapted from "Weyerhaeuser Forestry: The Wall of Wood," prepared by Stuart Hart and Mark Milstein (Corporate Environmental Management Program, University of Michigan), with assistance from Rob Day (Management Institute for Environment and Business, World Resources Institute) for the Sustainable Forestry Working Group, 1997.

1. In contrast to hardwoods, which are deciduous (i.e., leaf-bearing).
2. Much of the motivation for "cutting and running" had to do with the tax consequences of holding on to cut-over land: The conventional wisdom suggested that it would take too long for the next generation of trees to mature.
3. For example, in coastal Washington, if 100 percent Douglas fir is planted, the forest matures into a mix of approximately 50 percent Douglas fir and 50 percent hemlock and other canopy species.
4. The variation of 1 to 3 years in the nursery is due to site-specific issues such as elevation, frost susceptibility, damage, and so on.
5. This section was prepared by Joseph Strathman.
6. The figures come from T. Lent, D. Propper de Callejon, M. Skelly, and C. Webster (1997), *Sustainable Forestry within an Industry Context.* New York: Environmental Advantage, prepared for the Sustainable Forestry Working Group, 1997.
7. For example, Weyerhaeuser has now entered a second generation in its Douglas fir genetics program and a third generation in its loblolly pine genetics program. The closest competitors appear to be a full generation behind in each of these species.
8. Site quality and slope are particularly important in the Pacific Northwest where mountainous terrain is the norm. Moderately sloped lands increase harvesting ease, reduce risks of landslides and accidents, make pruning easier, and so on. In the Southeast, much less emphasis has been placed on forest land holdings, since site quality varies less significantly in the dominantly flat terrain.
9. This section consists of the reflections of the case writers and does not represent Weyerhaeuser policy on these issues.
10. Weyerhaeuser has recently reached agreement to purchase timberland and related assets in New Zealand.

Chapter 13

Meshing Operations with Strategic Purpose

In the past several decades in the late twentieth century environmental pressures have begun to impinge on a variety of business sectors, including oil, chemicals, and transportation. The escalating level of activity in sustainable forestry in the forest products sector makes it a wonderful laboratory for the study of transformation toward more ecologically sustainable business practices. At the societal level, sustainable forest management (SFM) means preserving the environmental, economic, and social/cultural value of the forest base while simultaneously meeting the growing demand for fiber-based products. Efforts to implement SFM at the firm level, however, have proven more challenging, with companies initiating a wide range of new and innovative practices such as low-impact forestry methods, plantation forestry, and sustainable forestry certification.

To distill the key findings that can be gleaned from the 21 different business entities that were studied as part of the work of the Sustainable Forestry Working Group, the authors, Matthew Arnold, Rob Day, and Stuart Hart, used a conceptual framework derived from the literature on environmental strategy. The exercise yielded one predominant conclusion: SFM must reinforce the core business strategy to be successful. When companies mesh SFM operations with their core strategy, rather than append them at the margins, they can realize competitive and business advantages as well as environmental and social gains. As an extension of strategic thinking about SFM, some firms have begun to rethink the very definition of their businesses, reinventing themselves as "fiber service" companies as opposed to purely extractive forestry operations.

These companies are beginning to separate the services they provide—structure, smoothness, texture, and beauty or writing media—from the physical product they sell—lumber or paper. As the nontimber value of forests increases, pioneering companies are beginning to develop adjacent businesses on the forest lands, providing ecosystem services such as watershed management, carbon sequestration, and ecotourism. Although none of the companies studied can yet be categorized as the "fiber service company of the future," the cases demonstrate both the potential value of and the first movements toward this vision within the forest products industry.

Traditionally, most forest products—pulp, paper products, and construction materials—have been commodities, with very little differentiation between competitors in terms of product performance. Except for a few niche market products, producers have been price-takers who compete on the basis of operational efficiency while maintaining certain baseline standards of minimum quality. The transformation toward SFM, however, may offer the potential to change the "rules of the game" in the industry. As a consequence, players may be able to improve their competitive position through SFM. In this sense, SFM is related to the emerging field of corporate environmental management and strategy.

Perspective from the Environmental Strategy Literature

A number of authors have recently noted the distinction between a reactive versus a proactive approach to environmental management (e.g., Hunt and Auster, 1990; Smart, 1992; Porter and van der Linde, 1995). The former implies that companies adopt a stance of waiting for regulation or public pressure to define what is required of them with regard to the environment. By allowing regulation to dictate strategy, reactive firms have traditionally depended on defensive tactics, pollution control, and compliance as their primary vehicles for environmental management. Unfortunately, these approaches are costly and add little value since they are add-ons to existing production processes and product designs. Reactive approaches therefore differentiate environmental management practices from business practices.

Proaction, however, suggests that the company take the initiative to identify and implement solutions to environmental problems. Proactive firms seek to preempt regulation and public pressure by resolving issues before they become crises. By leading regulation and setting an example for other companies, proactive strategies may help companies protect the franchise and ensure their right to operate by building a reputation as responsible environmental actors (Bonifant, Arnold, and Long, 1995). More important, a proactive approach offers the potential to actually benefit the firm's core business processes by intervening farther "upstream" during strategic planning, product development, and process design. In an effort to capture these potential benefits, Hart (1995, 1997) developed a typology consisting of four proactive environmental management strategies:

- Pollution prevention
- Product stewardship
- Clean technology
- Sustainability vision

Through such efforts as waste reduction, energy efficiency, and materials management, a pollution prevention strategy focuses on minimizing or eliminating waste before it is created. Much like total quality management, pollution prevention strategies depend on continuous improvement efforts to reduce waste and energy use. The use of environmental management systems (EMS) is an important part of such an approach. Unlike pollution control, prevention strategies are driven by a compelling logic: pollution prevention pays. 3M, for example, has saved over $500 million through its aggressive initiatives in eco-efficiency.

Product stewardship focuses on minimizing not only pollution from manufacturing but also all environmental impacts associated with the full life cycle of a product. As a company moves closer to zero emissions, reducing the use of materials and production of waste requires more fundamental changes in underlying product and process design through such tools as design for environment (DFE), life cycle analysis, and supply-chain management. Xerox Corp., for example, through its Asset Recycle Management Program (ARM), uses leased Xerox copiers in the field as sources of high-quality, low-cost parts for their new machines, which saves the company more than $300 million annually. Even more important, Xerox's DfE machines demonstrate higher levels of reliability than do the newly built machines. Thus, through product stewardship, it is possible, in some cases, to improve the quality or functionality of the product.

Pollution prevention and product stewardship deal with "today's" products and processes. In this sense they are strategies that seek to integrate environmental management with business processes. Companies looking to the future, however, can also begin to anticipate and invest in tomorrow's technologies. Clean technology is characterized by a fundamental change in core competence that dramatically reduces the use of harmful materials or processes. Clean technology programs involve resource allocation decisions in which environmental factors are incorporated as part of the R&D and technology development processes of firms. This strategy requires the company to develop a new way to design or manufacture a product and, as such, offers the potential to leapfrog the competition, especially in emerging markets where large, new capital investments are required.

Finally, a sustainability vision is needed to give purpose and direction to proactive environmental strategies. A vision of sustainability is like a road map to the future, showing the way products and services must evolve and what new competencies will be needed to get there. Through stakeholder communication, management commitment, and articulation of future opportunities, top management recognizes that sustainable development can become a core part of the company's intent or long-term vision. Taken together, clean technology and sustainability vision foster anticipation of tomorrow's strategic opportunities by positioning the firm to capture the products and markets of the future.

The "Four Horizons" Framework for Sustainable Forestry

Building on the environmental strategy literature, Arnold and Day (1998) developed a framework to examine the business drivers for environmental investments. Their model synthesized and integrated the strategies described earlier into a model comprised of "four horizons"—franchise protection, impact reduction, product enhancement, and business redefinition. Figure 13.1 summarizes the core concepts from the environmental strategy literature and integrates them into the Arnold and Day model. Each of the

Four Horizons Categories

	Franchise Protection	Impact Reduction	Product Enhancement	Business Redefinition
Concepts from the Environmental Strategy Literature	Right to Operate	Pollution Prevention	Product Stewardship	Clean Technology
	Protect the Franchise	Waste Reduction	Life Cycle Management	Leapfrog
	Reputation	Eco-Efficiency	Design for Environment	Step Change
		EMS		Innovation
			Supply Chain Management	Sustainable Vision

Figure 13.1.

The Four Horizons Framework Can Be Used to Synthesize and Integrate Concepts from the Environmental Strategy Literature.

four horizons is described in more detail below and on the following page, with reference to the forest products industry.

Franchise Protection

Companies are obligated to perform certain duties—such as paying taxes, cleaning up waste, and protecting workers—to maintain their basic right to operate. Even in regions with weak regulatory systems, companies need to comply with basic ethical standards to avoid public censure by watchdog organizations or extra-regulatory actions by local authorities. In the forest products sector, pollution control, protection of workers in the forest, protection of endangered species, changes in forest management practices, and dialogue with stakeholders are often required of companies to retain the right to operate.

Impact Reduction

One of the greatest challenges in the forest products industry is the maximization of resource productivity (per unit of land or per tree). For companies with a secure right to operate, impact reduction—in the form of pollution prevention, eco-efficiency, or simply risk avoidance—may be adopted to enhance productivity. The 21 case studies indicate that SFM adds operating costs in natural forests but reduces operating costs in plantation forests. In natural forests, greater control and lower harvesting rates preserve native ecosystems but lower yield. On plantations, genetic manipulation and silviculture techniques increase yield. It is important to note, however, that not all efficiency-driven forestry is SFM, in that many efforts to reduce harvesting and silvicultural costs violate the principles of sustainable forestry outlined previously. Downstream, impact reduction can lead to greater efficiency in the use of the tree: More of each tree is used in product, and wood is matched to the quality and market value of the application.

Product Enhancement

Some companies take another step and link SFM to product quality. Providing certified "sustainable" wood to customers can provide benefits ranging from peace of mind to fashion. For instance, Jay Leno's desk on the *Tonight Show* is made of certified sustainably harvested wood, specially requested. In some markets, certification may yield price premiums over noncertified wood. These opportunities are rare, however, especially for products that are less visible in their end use. There can be quality benefits from sustainable practices that go beyond consumer preference for "green" products. Sustainably grown and harvested trees may be better cared for and protected than trees grown using standard forestry practices, resulting in more consistent or higher quality. The value-chain control necessary under certification protocols can also be used to ensure proper treatment of the log through the entire manufacturing process, helping to preserve the value of the wood.

Business Redefinition

A few pioneering companies are moving toward new definitions of the forest products enterprise. These companies define themselves as sustainable in process, product, and vision. They pursue low-impact forestry, carbon sequestration, ecotourism, social and community development, or nontimber product markets. Such business redefinition is leading toward the "fiber service" company of the future, where firms begin to separate the service they provide from the physical product they sell. Innovation is at the heart of the advantage to be gained by viewing SFM in this way, which holds the most promise for future growth in the industry.

The Difference Between Success and Failure

The 21 cases in the MacArthur Foundation project could easily be classified along one or more of the horizons in Figure 13.1. This confirmed that the framework successfully captured the full range of content associated with the emerging SFM construct. This classification, however, did little to distinguish the successful growing strategies for SFM from those that were contracting or failing. Several of the cases that appeared similar on the surface with regard to SFM, in fact, differed in subtle but important ways regarding the business aspect. By iterating between the data and the framework, an important missing dimension emerged that began to explain why some of the SFM cases succeeded while others failed: the degree to which SFM practices were integrated or "embedded" within the core strategy.

It appears that the success or failure of SFM is determined largely by whether or not the activity is undertaken as part of a larger competitive strategy or business vision. When SFM is differentiated from the business's overall strategy—appended rather than embedded—it does not contribute to the overall fitness of the business and fails to contribute to the development of a competitive advantage. In contrast, a company that engages in SFM with an eye toward lowering costs, improving quality, or establishing new markets can find "win-win" solutions to sustainability challenges.

Figure 13.2 maps selected cases from the Sustainable Forestry Working Group onto the four horizons framework, separating those whose SFM activities were embedded (integrated) from those that appeared to be appended (differentiated). A single case has been chosen to illustrate each of the cells in the table.

	Franchise Protection	Impact Reduction	Product Enhancement	Business Redefinition
Embedded	Woodmaster	Parsons Pine	Portico	Aracruz Celulose
Appended	Riocell	Brent Property	Home Depot	Collins Pine

Figure 13.2.

Successful Sustainable Forestry Initiatives Appear to Be Embedded in
the Firm's Core Strategy Rather Than Appended as an Afterthought.

Franchise Protection

Woodmaster

When a cash-rich multinational timber company decided to invest in a
tropical forest region where management had no previous experience, it
knew that the precedent set by its first efforts would be important. This
was particularly true in light of the company's poor reputation around the
world for perceived environmental and social transgressions. The company's first operation in the new region, described in the disguised case
study Woodmaster Co., was therefore designed with franchise protection
in mind. The company provided free health care to the local community,
restored a historical site, developed an entire village with all the necessary
infrastructure for its workers, and committed to sustainable forestry practices. Initially, the Woodmaster operation proved to be less financially
successful than management hoped. Nevertheless, the Woodmaster management team stuck to its commitment to responsible, sustainable
forestry. A new management team brought in by the parent company was
able to improve performance significantly, achieving an operational profit in 1997.

Because Woodmaster operates in a region with a weak regulatory system,
it is tempting to local forestry companies to violate their commitments in
order to reduce costs. Woodmaster's parent company, however, is using the
venture to demonstrate its visible commitment to responsible practices as

a way to build credibility for its environmental performance and to open doors to future investments in the region. These franchise protection efforts appear to have been successful, since the parent company has been invited by several other countries in the region to make investments.

Riocell S.A.

In the 1970s, while under foreign ownership, the Brazilian pulp- and paper-maker's franchise to operate was suspended by the government because of the company's poor environmental performance. Despite new ownership under Klabin, Riocell's management is still influenced by this episode, which has convinced this team of the importance of environmental protection. The company consequently spends more than is legally required on environmental protection both in the forest (with minimized impact techniques) and at the mill (with major investments in pollution control technology). Riocell management concludes that its profits are reduced by an average of 15 percent per year by these expenses.

While the company's end-of-pipe pollution abatement orientation does indeed ensure good relations with the surrounding communities, these franchise protection efforts are not designed or implemented with improved operational performance in mind. Instead, they are separated from the company's overall strategy of quality and customer service. The company seeks no significant advantage through these investments, and therefore it does not use SFM to provide benefit to the performance of the business. In fact, Riocell's poor financial performance (return on equity of 4.7 percent in the pulp boom year of 1995, then a net income loss of US$3.7 million in 1996) has made it the subject of rumors that it is up for sale by Klabin.

Impact Reduction

Parsons Pine

Parsons Pine, an Oregon company with annual sales of $7 million, has a business strategy to purchase wastewood at low prices and convert it to value-added products. The company enjoyed early success with mouse-trap blanks and later moved into sushi serving pallets and Shinto prayer boards for the Japanese market, all based around the same basic wastewood blank. Its cost advantage helped the company gain a significant share of these specialized markets. Parsons Pine also makes a wide variety of component parts and finished consumer products including knife blocks, wine racks, roll-top counter organizers, CD rack components, shoe organizers, toy blocks, and other applications. Parsons Pine has made an art of finding niche markets for its wastewood technology—80 percent of all of these products come from wood materials that were purchased as defect or scrap material from manufacturers in the United States and Canada—and focusing on flexible manufacturing methods that enable it to supply a wide variety of products from wood blocks that are less than 24 inches in length.

Parsons Pine's operations and strategy exemplify the potential for impact reduction in the manufacturing end of the forest products value chain. The company eases pressure on existing forests by satisfying demand for certain products, through the use of wastewood, and realizes a cost advantage in the market. A limitation of this approach, however, is that it is relatively easy for competitors to duplicate. Anyone can buy wastewood. So on the manufacturing side, impact reduction efforts often do not lead to a long-term competitive advantage—as is apparently the case with Parsons Pine, which reportedly has encountered price pressure both in its markets and wastewood input as competitors begin to copy the company's methods. Impact reduction may actually hold more promise in the forest, where such efforts preserve the value of the land base.

Brent Property

As a nonindustrial private forest, the Brent property is not a profit-making company. Nevertheless, the Brent family depends on 171 acres of farm and forest in Oregon's Willamette Valley for income. The family engages in impact-reducing SFM techniques such as selective felling and natural regeneration because the Brent's feel it is necessary in order to preserve the aesthetic value and health of the forest. The property, according to the case study, has the potential to provide over $66,000 in annual income. Yet the family has only generated $47,000 in net stumpage income since 1985 because SFM is not pursued with financial optimization in mind. This example, while small in scale, demonstrates how impact reduction undertaken without a guiding business strategy can serve to destroy rather than create financial value in the forest.

Product Enhancement

Portico

As a small Costa Rican producer of Royal mahogany (*Carapa guianensis*) doors in the early 1980s, Portico had difficulty achieving the high-quality standards necessary for export. Yet the small size of the Costa Rican market would not allow any significant growth for the company. By taking advantage of a debt-for-nature swap with a U.S. financial institution, Portico acquired its own forest holdings and, as part of the agreement, began low-impact, SFM practices. Soon management found that the higher costs in the forest were more than compensated by the higher quality standards it could achieve as a result of greater control over the entire value chain. For example, while Portico's previous practice of buying mahogany logs from independent loggers had cost much less per log, now management encountered less defects and waste.

Portico further enhanced the quality of the product and reduced costs by switching to a less-wasteful manufacturing technique, producing engineered doors from blocks of wood rather than solidwood doors, which warped and cracked. Portico also used certification of its forests to gain access to the lucrative U.S. market through Home Depot. While certification itself gave the company no obvious market premium, the company's high-quality and ecofriendly reputation helped it capture the southeastern U.S. market for its product. In 1997 Portico planned not only to expand its marketing to other regions of the world but also to begin selling its forestry expertise as a service to other forest products companies. Portico's product enhancement approach to SFM reinforces its business strategy to sell high-quality premium products in the United States and Europe and is helping the company develop a competitive advantage.

Home Depot

This nationwide do-it-yourself chain holds a strong corporate-responsibility ethic, which has led management to emphasize environmental protection in both its operations and products. Home Depot has publicly supported environmental education programs and social causes such as Habitat for Humanity and forest certification efforts. Home Depot has also given preference to forest products that are sustainably produced, including Portico doors and—at one time—Collins Pine shelving.

The company is using its commitment to SFM to enhance its product mix. Home Depot believes that by supplying these products it is providing more value for its customers, even if the customers are not willing to pay a price premium for sustainability at this time. As their 1994 Social Responsibility Report states, "Improving the environment is the ultimate home improvement we can make." Nevertheless, the number of sustainably produced products stocked—fewer than 10—is tiny for a chain whose average store carries about 45,000 items on a floor space just over 12,000 square meters. Home Depot has found it costly and difficult to supply such products in any volume, because its policy on SFM is in conflict with its business strategy of maintaining high sales volume and the related economies of scale that it brings. For instance, in late 1996 the company quit stocking Collins Pine's shelving, claiming that quality had declined. Collins Pine's interpretation is that Home Depot had been frustrated by the dual accounting system required under the chain-of-custody requirements for the use of the certification label. Either way, the incident illustrates how Home Depot has found it difficult to create a market advantage by simply adding on a commitment to SFM.

Business Reinvention

Aracruz Celulose S.A.

Aracruz Celulose, S.A., the world's producer of bleached eucalyptus pulp, exports its product around the world for use in fine writing papers, tissues, and other consumer products. From the very start, founder Erling Lorenzen has been driven to reinvent the pulp business. First, the company took advantage of a tax incentive for tree planting to buy and reforest cutover, degraded land in the impoverished state of Espirito Santo. By 1992, the company had acquired 200,000 hectares and planted 130,000 hectares with managed eucalyptus; the rest was restored as conservation land. Next, Aracruz tackled the problem of poverty head-on. In the early years, Aracruz was able to hire local people for very low wages because of its desperate situation. But instead of simply exploiting the abundant supply of cheap labor, the company embarked on an aggressive social-investment strategy, spending more on its social investments (schools, hospitals, housing) than it did on wages. Since that time, the standard of living in surrounding communities has improved. Aracruz's skilled labor force, which is a result of its social investment strategy, and its highly productive and easily accessible forests, provide the company with an enduring source of competitive advantage in the world market: It enjoys a low-cost and consistently high-quality source of fiber.

Aracruz has begun to move toward the vision of a "fiber service" company. It maintains natural forest holdings, but its fiber base consists of uniform plantations of eucalyptus that are closer in form and function to agriculture than forestry. The natural forest holdings, while mandated as a percentage of the holdings by law, are used to protect the eucalyptus plantations and to offset any impacts on the land surrounding Aracruz's. The differences between the natural forest holdings and the eucalyptus tracts are striking, demonstrating the conceptual shift underway at Aracruz and throughout the fast-growth plantation industry: When viewed as forestry, Aracruz's practices are very intensive and unnatural, but when viewed as agriculture, Aracruz's fiber crop is not very intensive, is incredibly productive, yet also provides many ecosystem services not provided by other crops such as grains or potatoes, especially since Aracruz's land was mostly degraded before the company moved in. Aracruz's SFM strategy therefore relieves pressure on natural forests rather than encouraging their destruction. The company is now looking to use its practices in solidwood applications, providing another form of fiber service through high-yield forestry. Today, Aracruz remains a forest products company in outlook and practice; but the first signs of movement toward a fiber service company are evident in this business success story.

Collins Pine

Collins Pine, which produces lumber for industrial and construction applications, has defined itself as a sustainable enterprise, striving to be "in tune" with community needs and emphasizing SFM. Fifty percent or more of the total raw materials utilized by Collins Pine at its California, Pennsylvania, and Oregon manufacturing operations originate from its Green Cross-certified forestry operations. The strategy is based entirely around sustainable production, which the company hopes will yield price premiums for certified products. This strategy, however, has met with limited success. Collins Pine sells products that are undifferentiated from competitors' products within commodity-driven markets. Consequently, its SFM efforts are not integrated into a core strategy that can create an advantage to offset the higher SFM costs incurred. Even though forest health on Collins Pine's managed stands is often visibly superior to neighboring lands, the company's strategy is not well designed to yield financial benefit from these improvements. Company attempts to "reinvent" the forest products industry will thus fall short as long as its SFM strategy remains paired to an inferior business concept.

Integrated Sustainability

The four horizons, however, are not mutually exclusive. Full integration of SFM into the business means finding ways to create advantages in all four areas. Three of the case study companies are beginning to realize this level of integration: Weyerhaeuser, Aracruz Celulose, and Portico. Weyerhaeuser provides the example here of how the four horizons can be combined in powerful combination to yield environmental and social as well as business results.

Weyerhaeuser

Weyerhaeuser Co., one of the world's largest forest products companies with net earnings of $463 million on sales of $11.1 billion in 1996, has operations that encompass the entire forest products value chain, from forestry to pulp and solidwood production to some retail and recycling. The company owns over 5 million acres of forest land in the United States. Dogged by poor financial performance, Weyerhaeuser underwent a series of strategic changes in the late 1980s when incoming president John Creighton set about refocusing the company on core forest products operations. Returning to its roots, the company once again began to define itself as "The Tree Growing Company." Since 1989, the company's financial fortunes have reversed: Weyerhaeuser reached top quartile performance in its industry for shareholder returns in 1993 and is well positioned to reap significant financial rewards from its investments in forests and forest operations over the next 10 to 20 years. Over the past decade, Weyerhaeuser has evolved an SFM strategy that effectively integrates all four levels of the horizons model (see Figure 13.3).

When the spotted owl and other environmental controversies threatened its holdings, Weyerhaeuser moved to *protect its franchise* by engaging stakeholders and addressing its concerns through a series of town hall style meetings. The company identified and adopted best management practices in the forest, undertaking extensive habitat conservation efforts, and

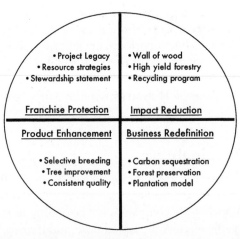

Figure 13.3.

Weyerhaeuser's Sustainable Forestry Strategy Combines Elements from All Four of the Horizons to Create an Integrated Whole.

formalized the practices in a stewardship statement and specific resource strategies that included goals for water quality, wildlife habitat, soil productivity, and cultural, historic, and aesthetic values. Through these actions, Weyerhaeuser not only minimized the impact of the continuing endangered species controversy on its operations but also reestablished its reputation in the community as a good corporate citizen. The franchise protection efforts allowed Weyerhaeuser to operate on land that would otherwise have been declared off-limits, a significant advantage for the company.

The Company also pursued *impact reduction* through SFM. It worked for decades to promote high-yield forestry (a form of plantation forestry) in its U.S. forest holdings through an ambitious program of genetics, selective breeding, and silvicultural innovation, all of which depend on soil preservation and minimizing impacts to its land base. These investments have already resulted in forests that are two to four times more productive than native forests and should produce further yield increases of 50 to 100 percent as the genetically improved forests finally mature early in the twenty-first century. This so-called "wall of wood" should produce substantial cash flow for decades to come, as softwood timber stocks around the world become increasingly scarce or expensive. The company's strength in paper recycling has also succeeded in reducing input costs in the pulp and paper business. The genetics and selective breeding initiatives have further served to *enhance the quality* of its product, by promoting the growth of higher quality "clear" wood that has the potential to fetch a premium in the market. Weyerhaeuser's forest practices also promote more consistency in wood quality—an important consideration for most lumber customers.

Finally, the company is beginning to think about sustainable development as a source of *business redefinition*. Weyerhaeuser's extensive forest holdings provide sequestration of carbon from fossil fuel combustion. Furthermore, by growing more wood on less ground, the company is in a position to advocate native forest preservation on a global basis. Indeed, it may be that the company's high-yield model can help fulfill society's growing demand for wood products without shifting that demand onto the world's remaining old-growth and native forests. The business angle to this approach may become evident in the near future, as the company's integrated value-chain approach gives it an advantage over competitors with regard to emerging forest certification schemes. Weyerhaeuser still has to wrestle with many issues and has not yet adopted fully the "fiber services" model of the future, but the company is finding business value from struggling with these issues.

The Weyerhaeuser case illustrates the power of fully integrating SFM into the business. Had Weyerhaeuser simply done the minimum necessary to comply with regulations, the spotted owl and subsequent endangered species controversies would have been a bigger cost for it than it actually was and might have threatened its license to operate. Had Weyerhaeuser done the minimum necessary regarding forest management, it may have maintained access to a wood and fiber source but would not have realized the preserved land value, high yield, and improvement in quality that has come from the Weyerhaeuser Forestry program. SFM has also enabled the company to explore new directions that could potentially generate significant future growth in the international market. Thus, it appears that forest product companies that integrate SFM into their strategy and use it to protect their franchise, reduce their impact, enhance their products, and redefine their business not only enhance their profitability but also generate growth and create important future opportunities.

The "Fiber Service" Company of the Future

Rather than constituting a threat to industry survival, this analysis suggests that the transition to sustainability represents an enormous business opportunity for forest product companies. As a reinforcement to business strategy, ecological and social sustainability becomes a core platform for competitiveness. SFM can be an important part of the effort to build, leverage, and extend competitive advantage. Given the apparent inevitability of industry transformation, it appears that forward-thinking companies would be well served to see how SFM might help them mesh their operations with a new strategic vision of sustainability.

Over the next 20 years as constraints on producers increase and the nontimber value of forests rises, the cost of forest loss will increase, which will escalate the costs of unsustainable timber production. In the future, it will not make financial sense to make phone books and fiber board from old-growth forests. To supply the world aggregate wood demand, which is projected to increase by 1 to 2 percent per year for at least the next decade, the fiber can come from only two major sources: increased efficiency in utilization of native forests or new production from plantations and nonwood sources. Companies that produce wood fiber will find ways to use more of every tree they cut. In many tropical forest operations, as much as 80 percent of every tree is still wasted.

Cutting operations and sawmills suffer from outdated equipment and techniques. There is an enormous opportunity for new investment and technology exchange. On the use side, companies will learn to use less wood in their products. For instance, homes and furniture will use wood more efficiently, and packing and shipping products will be reused. More substitutes for natural forest-derived fiber will emerge from the expansion of fast-growing plantations. Nonwood fibers from agricultural waste, hemp, and other sources will also account for a greater share of total fiber consumption.

These trends will force the forest products industry to reconsider the purpose of its business. Increasingly companies will shift from the current business definition of transforming trees into pulp and solidwood and toward a more service-oriented concept of "fiber service." All of the products currently supplied by the forest products industry are fiber based, whether as solidwood or pulp. In supplying the world's growing demand for fiber and the services it provides, in the face of resource scarcity, the forest products industry will increasingly turn to alternative sources of fiber: plantations, recovered fiber, certified-sustainable natural forests, and even nonwood fibers such as agricultural crops and polymers. In marketing, the fiber service company will look for new markets and customers for the services it provides—wood-based fibers are already used in textiles, and engineered-wood construction products are increasingly competing with plastics and other materials.

The forest service company of the future will also look to derive revenues and profits from the nontimber value that resides in forests through its operations. For companies operating sustainably in natural forests, this could mean initiating carbon sequestration programs, which will derive revenues from customers—in this case, other companies—that will pay them to keep their forests standing to offset the customers' carbon emissions. Commercially profitable ecotourism and habitat preservation efforts could also increasingly become part of the operations of forest-owning companies. Eventually, even agriculturally driven companies that relieve pressure on natural forests may look to own natural forests and thereby capture some of the value they preserve. Aracruz, for example, is already considering marketing the carbon sequestration services of the natural forest areas it is required by law to set aside as part of its plantation operations.

The idea of the "fiber service" company is one vision of industry evolution in light of dwindling natural forests, increased demand for fiber, and increased pressure from stakeholder groups. While none of the companies in this study evidenced this vision in its full form, a few pioneering companies were beginning to reinvent their business along these lines. Only time will tell whether or not the progression in SFM takes the direction that we foresee.

NOTE

This chapter was adapted from an article prepared by Matthew Arnold, Rob Day, and Stuart Hart, "The Business of Sustainable Forestry: Meshing Operations with Strategic Puipose," submitted to *Interfaces,* special issue on Ecologically Sustainable Business Practices, February 1998.

REFERENCES

Arnold, M., and R. Day (1998). *The Road to Sustainability: A Business Framework for the Millennium.* Washington, D.C.: World Resources Institute.

Bailey, K. (1982). *Methods of Social Research.* New York: The Free Press.

Bonifant, B., M. Arnold, and F. Long (1995). "Gaining competitive advantage through environmental investments," *Business Horizons* July-August, 37–47.

Bryant, D., D. Nielsen, and L. Tangley (1997). *The Last Frontier Forests.* Washington, D.C.: World Resources Institute.

Costanza, R., et al. (1997). "The value of the world's ecosystem services and natural capital," *Nature* vol. 387 (15 May), 253–260.

Food and Agriculture Organization (1995). *Forest Products Forecasts to 2010.* Rome, Italy: FAO.

Glaser, B., and A. Strauss (1967). *The Discovery of Grounded Theory: Strategies for Qualitative Research.* Chicago: Aldine.

Hart, S. (1995). "A natural-resource-based view of the firm," *Academy of Management Review* 20, 986–1014.

Hart, S. (1997). "Beyond greening: Strategies for a sustainable world," *Harvard Business Review* January-February, 66–76.

Hawken, P. (1997). "Natural Capitalism," *Mother Jones* March/April, 40–54.

Hunt, C., and E. Auster (1990). "Proactive environmental management: Avoiding the toxic trap," *Sloan Management Review* 31, 7–18.

Environmental Advantage (1997). *Sustainable Forestry Within an Industry Context.* Chicago: MacArthur Foundation.

Johnson, N., and B. Cabarle (1993). *Surviving the Cut: Natural Forest Management in the Humid Tropics.* Washington, D.C.: World Resources Institute.

Nilsson, S. (1996). *Do We Have Enough Forests?* International Institute for Applied Systems Analysis, Laxenburg, Austria.

Porter, M., and C. van der Linde (1995). "Green and competitive: Ending the stalemate," *Harvard Business Review* September-October, 120–134.

Sedjo, R., and D. Botkin (1997). "Forest plantations and natural forests," *Environment* December, 15–30.

Smart, B., ed. (1992). *Beyond Compliance: A New Industry View of the Environment.* Washington, D.C.: World Resources Institute.

Sustainable Forestry Working Group (1997). *Sustaining Profits and Forests: The Business of Sustainable Forestry.* Chicago: The John D. and Catherine T. MacArthur Foundation.

United Nations Environment Programme (1997). *World Urbanization Prospects.* New York: United Nations.

World Wildlife Fund (1997). *WWF's Global Annual Forest Report 97: Forests for Life*. Surrey, U.K.: World Wide Fund for Nature.

Yin, R. (1989). *Case Study Research*. Newbury Park: Sage.

Chapter 14

Lessons Learned

The cases discussed here, as well as the escalating number of other examples around the world and in all corners of the industry, demonstrate that sustainable forest management (SFM) is rapidly being transformed from a niche activity to significant market opportunities. The spread of SFM is a clear signal of a global readjustment in relations between the world's forests, the people and industries who grow and use them to make products, the markets through which these products flow, and the means the public uses to affect these relations.

Although these cases represent an early stage in this process of transformation, and each company exists in a unique set of circumstances, the aggregate experience of these businesses suggests several useful generalizations and observations about the contours, depth, and velocity of the evolution for more sustainable forest management.

1. The industry is rapidly evolving toward more sustainable production.

Sustainability is emerging as a major driver throughout the industry, affecting firms large and small from the floor of the forest to the retailer of forest products. Although the velocity of this transformation is accelerating, it varies in intensity and scope among different sectors of the industry and among regions of the world. Retailers, producers, and consumers in Europe are leading the change, with the United States following more slowly. Latin America and Asia are notable laggards. Markets for certified products are developing most rapidly where a few buyers or sellers can influence market performance, as in the pulp and paper industry, in high-end retail chains, and in specialty manufacturers.

Within each region, a handful of pioneers are leading the transformation. In Europe, the UK Buyers' Group is fundamentally changing the conditions for market access with major retailers, such as B&Q and J. Sainsbury Plc., by requiring their forest products suppliers to use wood from forests that are certified to be well-managed. The pressure from such large retailers has provided an opportunity for industry leaders such as STORA and Assi-Doman to seek certification and to become the first European producers to market certified products. Dwindling supplies of wood worldwide and environmental pressures are among the most powerful forces behind the transformation toward sustainability. Multinational firms seeking new markets and new forest resources are accelerating the industry's globalization and will exert greater pressures on natural forests as they shift large amounts of capital and technology to remote regions. These same firms, however, may help transfer sustainable practices to the Southern Hemisphere as they expand their forest base and manufacturing in those regions.

The continued growth in wood consumption worldwide will increase the importance of using raw materials efficiently. Latin American plantation forests, such as those operated by Aracruz and Riocel, are achieving high levels of eco-efficiency in their production of pulp and paper. They are destined to become leaders in supplying sustainable forest products to European markets and will exert greater competitive pressures on producers in the Northern Hemisphere. As fiber production increases in the Southern Hemisphere, primary and secondary processing in the region will expand dramatically.

2. THE MARKET FOR SUSTAINABLE FOREST PRODUCTS, THOUGH CURRENTLY TINY, IS SHIFTING AT AN ACCELERATING RATE FROM NARROW NICHES TO SIGNIFICANT MARKET OPPORTUNITIES.

Pressures from both the demand and supply sides of the market will lead to rising volumes of forest products that come from sustainably managed forests. Consumers are placing greater emphasis on the life cycle of forest products, from the forest floor, to emissions, and disposal, while retailers and intermediate customers are pressuring producers to supply sustainable forest markets. Buyers' groups made up of commercial and industrial wood product customers are generating demand significant enough to encourage some major producers to try and gain a competitive advantage by coming early to market with differentiated sustainable forest products to either capture early price premiums, gain marketshare, and protect customer good-will. The emphasis on certified products, which initially concentrated on solid wood products, is shifting to paper products in response to pressure from European publishers.

It is still unclear, however, whether the market for certified forest products will continue to grow and consolidate as a differentiated market segment or become a standard in the market. Before sustainability can take hold in the forest and the industry, all of the links of the "value chain"—from the forest through processors, manufacturers, retailers, and whole-salers to the end consumer—need to develop in a more synchronized fashion. As the market demand grows, the capacity to provide reliable supply must grow correspondingly. In the late 1990s, only 1 percent of the world's woodlands were certified, much of it in community-based projects in developing countries with little or no access to international, large-scale markets. Some of the currently certified resource does not reach the marketplace as certified product. The rising demand for certified wood products expressed through the development of buyers' groups threatens to outstrip the supply of certified, sustainably managed wood. In many regions of the world, it is the lack of certified processing and manufacturing capacity that is restricting the flow of product from sustainably managed and certified forests to retailers searching for certified products. The lack of a verifiable "chain of custody" that follows the product from forest to consumer is a major stumbling block for the creation of a robust sustainable forestry industry.

3. CERTIFICATION IS RAPIDLY BECOMING SYNONYMOUS WITH SUSTAINABLE FOREST MANAGEMENT AND WILL INCREASINGLY SET A HIGHER STANDARD FOR FOREST MANAGEMENT PRACTICES.

The rise of nongovernmental certification indicates that the process is able to serve as a credible third-party evaluation tool for forest management and practice with the public. In some tropical countries, certification has become essential for companies to get permission to log those forests. Certification, so far, however, has had a far greater political than market impact. Confusion over and conflict among multiple certification schemes/systems within the industry is a major hurdle for the expansion of certification. In the near future, though, market forces will contribute to the convergence toward a single certification scheme.

4. CERTIFICATION CAN OPEN UP BUSINESS OPPORTUNITIES.

As the experience in these case studies demonstrates, well-managed businesses that have committed to certified products can reap business advantages. In most instances, certification has not produced profits for companies through consumer price differentials or premiums, nor have consumers yet demonstrated a willingness to pay higher prices for certified products. But companies that produce certified material (e.g., MTE, Collins Pine, and Colonial Craft) have profited through preferential market access, increased influence in the distribution channels with retailers, favorable public relations, and in some cases initial price premiums. And Sweden's STORA expects to achieve preferential market access and premium prices in the European pulp and paper markets as the first European producer of certified product. Processors' commitments to certified forest products have also improved their access to harvest on nonindustrial forests and to opportunities for new products and specialized markets that they would otherwise have missed. Under the pressure of international scrutiny, governments have displayed greater flexibility toward forestry operations that are committed to sustainability than toward those that are less so. The evolution toward certified forest products will continue to create openings that well-managed businesses will be able to exploit to profitable advantage.

5. SUSTAINABLE FORESTRY IN TANDEM WITH GOOD BUSINESS PRACTICES CAN CONTRIBUTE TO COMPETITIVENESS THROUGH LOWER COSTS AND THE ABILITY TO HELP COMPANIES ENSURE LONG-TERM, RELIABLE, HIGH-QUALITY SOURCES OF WOOD.

As the experience of companies presented here suggests, although the initial costs for sustainable forestry may be higher in the short term, the long-term benefits from greater yields in productivity and quality can reduce costs. Aracruz Celulose made significant investments in both forest management and its publish and paper facilities to ensure that the operations are sustainable and eco-efficient. The strategy enabled Aracruz to establish a low-cost position in the plantation pulp market and gain market share by providing a type of pulp produced by less-polluting technology. The company has gained a strategic competitive advantage and built up technological leadership in eco-efficient, vertically integrated plantation forestry that supplies a consistent, high-quality source of fiber.

As the consumption of wood products grows and available forest land declines, the ability of companies to guarantee a stable, high-quality supply of wood fiber will become a pressing strategic consideration in the forest products industry. Those firms that demonstrate that they can manage these finite resources in a sustainable manner will continue to gain access to forest lands, and they will ensure a future steady supply of consistent, high-quality fiber—both competitive advantages in a world where wood supplies are tight. The continued shortage of wood fiber will also provide a competitive opportunity for those firms that invest in technology to make use of wood waste and engineered wood products that reduce the dependency on virgin fiber. In fact, more efficient consumption patterns of wood will be a by-product of the transition to greater sustainability. Parsons Pine, for instance, already produces a variety of products with flexible manufacturing systems that use scrap material. And Portico S. C. in Costa Rica has captured a leading share of the high-end mahogany door market through sustainable forestry, which ensures its supply of tropical hardwood, and the production of engineered doors, which use less virgin material.

6. THE TRANSITION OF THE FOREST INDUSTRY TOWARD SUSTAINABILITY WILL REQUIRE A CULTURAL TRANSFORMATION.

The forest products industry is the historical artifact of a series of decisions about products, people, and expectations that developed by moving to replacement, whether in technology or resource. Forestry, like other natural resources industries, has followed the pattern of a mobile industry drawn to forest frontiers or new resources. With the disappearance of frontiers, an era of resource abundance has given way to an era of limits in forest resources, both physical and political.

Sustainability and sustainable development are part of a long process that involves broad participation and continuous adaptation. Sustainable development for the industry constitutes a new contract with society to preserve the environmental, economic, and socio-cultural value of the forest base while also meeting the burgeoning demand for wood products. The process of sustainable forestry is the antithesis to the conventional conduct of business, which rewards short-term performance and immediate needs. It challenges the way the industry measures growth and development, builds and runs mills, and trains and manages employees. Implementing sustainable forest management at the company level, which calls for innovation at every step of the industry's value chain and at all levels of a company, fundamentally requires a change in the way the industry—and society—is organized.

Significant cultural change is typically driven by a perceived crisis. The forestry industry's transition to greater sustainability is no exception. Worldwide, forest activists and conservation organizations have successfully communicated the crisis of the world's loss of forests by revealing the scars left on the landscape and by showing the destroyed lives of forest dwellers. They have also successfully connected forest loss to the widespread extinction of species and undesirable shifts in global climate patterns. The industry itself has contributed to the sense of urgency by confronting the reality of a dwindling forest base through a drive for efficiencies, consolidation, retooling, and other changes. These factors, and many more, that have created an awareness in the industry among regulators and the public that forests are under siege are forcing the industry to adopt more sustainable forestry. The commitment to continued improvement in the management and use of forests that defines sustainable forest management offers the best prospect to help the industry successfully tackle the issues raised in the future about the wise, but also profitable, management of the world's forest resources.

NOTE

This chapter was prepared by Michael B. Jenkins.

Glossary of Terms

agroforestry The management of forest and food crops together in one system.

biological diversity The variability among living organisms from all sources, including inter alia, terrestrial, marine, and other aquatic ecosystems, and the ecological complexes of which they are a part; this includes diversity within species, between species, and of ecosystems.

bleached hardwood kraft pulp (BHK) Cellulose produced from short-fiber trees in the kraft (sulfate) process.

board foot A common measurement of timber transactions (mbf = 1000 board feet). A board foot is one foot by one foot by one inch.

boltwood Small logs or sections of larger logs that have been split. A bolt is usually less than four and a half feet long.

chain of custody The monitoring process of the production and distribution channels of forest products from forest floor to end product.

character wood Visual and/or structural defects in wood that constitute criteria for lowering the grade of the lumber in traditional lumber production but that consumers may find attractive in the products they purchase. Mineral stains and large knots are criteria for devaluing lumber, but they are preferred characteristics for wood used in some lines of furniture manufacturing.

clear-cut A method of logging that removes all merchantable trees in a designated area in one cut.

clear-cut with reserve A variation of clear-cutting that leaves some stand structure usually in the form of scattered trees or patches of trees throughout the cut block. Generally no more than 20 percent of the original stand is left. The reserves are generally left uncut after the block is regenerated.

community forestry Management and control of forests by local people.

conservation brokerages Organizations that facilitate the conservation of land areas by arranging and/or managing agreements between land owners and tax agents. Landowners cede specified development rights in perpetuity in exchange for lower property taxes.

conservation trusts Diverse public and private arrangements for committing land and water property to environmental purposes in perpetuity.

conversion The outright loss of forest land to other uses, which often creates permanent fragmentation. Conversion may also mean the changing of one forest type to another—for example, a stand containing several species may be converted to a single species plantation.

custom-grade lumber Lumber graded to the specific needs of individual customers as opposed to that graded only according to national standards.

cut-off blocks Pieces of wood larger than trim ends such as cants, left over from cutting wood to specified lengths. These pieces are typically chipped and converted to pulp or burned for energy.

debt-for-nature exchanges Agreements whereby generally first world organizations or governments redeem third-world debt to channel money toward conservation.

directional felling A logging technique that allows the logger to control the falling path and landing of a cut tree.

ecoculture Scientific manipulation of ecosystems to achieve sustainable flows of desired products, services, and qualities.

eco-efficiency Focuses on creating additional value by better meeting customers' needs while maintaining or reducing environmental impact. It includes reducing material and energy intensity of goods, reducing toxic dispersion, recycling more material, maximizing sustainable use of renewable resources, extending product durability, and increasing the service intensity of products.

ecolabel A labeling system intended to distinguish one product or service from another on the basis of its environmental impact.

Eco-Management Auditing System (EMAS) A voluntary management scheme of the European Commission established in 1993 to promote continuous environmental performance improvement of industrial activities by committing sites to evaluate and improve their environmental performance and provide relevant information to the public.

ecosystem A community of all plants and animals and its physical environment, functioning together as an interdependent unit; the complex interactions between a community of living organisms and its environment.

ecosystem management Forest management based on the goal of maintaining a healthy ecosystem.

edgings Waste wood produced by a machine called an "edger" used in saw milling operations when cutting and squaring lumber from a slab or cant.

effluent-free pulp Pulp produced in a mill whose liquid effluents are totally recycled, a technology still under development.

Elemental Chlorine Free technology (ECF) Pulp bleached with chlorine dioxide instead of chlorine gas.

engineered wood products Those wood products that are not made of solid wood but wood products made from multiple wood product components that may be glued together to make a structurally sound product, such as laminated veneer lumber.

environmental certification The process by which independent, third-party groups evaluate and monitor forest management practices and production for their environmental impact against a set of standards.

environmental services The nontimber services that a forest provides, such as watershed and soil protection, wildlife habitat, carbon sequestration, and recreation.

even-aged stand A stand of timber of a single age created through treatment such as a clear-cut, or a natural event such as a hurricane, that creates conditions that eliminate trees of varying ages and replaces them with trees that all begin to grow at the same time.

fingerjoint technology Technology that allows the joining of two or more pieces of wood together to form one structurally sound piece. Typically, the ends to be joined are cut so that the finger-like notches of each piece of wood join together and can be glued with an adhesive.

flexible manufacturing The ability to deviate the process of producing a product in stream or to modify production processes to adapt to new designs as may be demanded by the market. Small to midsized companies are often better positioned to do this than are larger companies.

forest An aggregation of trees and the natural processes that sustain it.

forest fragmentation The reduction of overall forest cover and the isolation of forest patches.

The Forest Habitat Classification System A system originating in Europe of classifying site potential for growing trees on the basis of identifying the ecological characteristics of particular plant associations.

The Forest Incentive Program A program that offers federal money to cost-share activities fostering healthy forests. FIP, as it is called, was created along with the Forest Stewardship Program as part of the 1990 Farm Bill.

forest integrity The composition, dynamics, functions, and structural attributes of a natural forest.

forest management All of the activities performed on tree stands over a period of time while timber matures to meet specific objectives. These objectives might include creating a stand of a given species, encouraging the growth of high-value species, creating a forest similar to a natural one, or enhancing aesthetic quality.

forest products value chain The value added to a unit of wood as it progresses through the steps of manufacturing and sales distribution (from forest floor to livingroom floor).

forestry Regimes of actions that protect and modify forests to achieve desired products, services, and ecological conditions.

Forest Stewardship Council (FSC) An independent, nonprofit, nongovernmental organization founded in 1993 as an international accrediting organization to ensure public credibility and rigorous standards of forest product labels in the marketplace.

Forest Stewardship Program A federal program that offers NIPF owners technical support and cost sharing for conducting forestry practices and, in some states, for preparing forest management plans.

full-time family-wage jobs Competitive employment wages for the geographic area and job description, including a full benefits package.

"green" mutual funds Marketed portfolios of investments in environmentally sound and sustainable enterprises.

Green Seal Founded in 1989, Green Seal is a nonprofit organization that attempts to reduce the adverse environmental impacts associated with the production, use, and disposal of consumer products through the award of an environmental "seal of approval."

group selection A system that removes trees in defined groups to create stand openings with a width less than two times the height of adjacent mature trees and that manages the area as an uneven-aged stand.

group shelterwood A system in which small openings are created in the stand such that the adjacent trees shelter the new generation. The size or density of leave-tree groups will decrease throughout more future stand harvests, until the mature overstory has been completely removed.

hardwood Generally, one of the botanical group of trees that has broad leaves, in contrast to the needle-bearing conifers; also, wood produced by broadleaf trees, regardless of texture or density.

high grading The process of removing the commercially valuable species of trees in a stand and leaving those of less commercial value. The trees that regenerate as a result of this process are predominantly the species with less commercial value.

high-yield forestry Maximized volume of harvest on a per acre basis.

household enterprise Family-based systems of earning a livelihood.

international environmental management systems under ISO Environmental management systems that have been developed by the ISO to create an internationally recognized benchmark for environmental management.

International Tropical Timber Organization (ITTO) A membership organization composed of 53 member countries established in 1987 to promote conservation and sustainable development of tropical forests through international cooperation between producing and consuming countries of tropical timber.

ISO A federation of national standards bodies united to promote standardization worldwide. The ISO develops and publishes international standards to facilitate exchange of goods and services and to foster mutual cooperation in intellectual, scientific, technological, and economic spheres.

ISO 9000 A series of international quality standards developed by the International Standards Organization, the guiding principle of which is the prevention of defects through the planning and applications of best practices at every stage of business—from design to installation and service.

ISO-9000 Certification The assessment of a particular quality system against the requirements of one of the ISO-9000 standards, which results in the issuance of a certificate which confirms that a particular quality system conforms with the standards' requirements.

ISO 14000 A series of standards for environmental management developed by the International Standards Organization. The standards include an environmental management system (EMS) that allows a company to manage, measure, and improve the environmental aspects of its operations. Improvement of the EMS is achieved through system audits.

landscape (in forestry terms) A geographical mosaic composed of interacting ecosystems resulting from the influence of geographical, topographical, soil, climatic, biotic, and human interactions in a given area.

live-edge square Edging waste where one side has been cleanly edged but the other side has not (the live edge is the bark side of the edge versus the edged side). For example, if lumber boards are cut from a flitch (a log sawn on two sides from which veneer is sliced), the boards are run through an edger to cut off the barked log sides. This produces a liveedge square.

medium density fiberboard (MDF) Like particleboard, MDF is a panel product made by compressing very small particles of wood with a binding agent. The difference between the two is the higher density of MDF. There are also LDF (low density) and HDF (high density) fiberboard products, each of which is technically designed for different uses.

minimal cultivation A forest management method that is aimed at generating the least possible intervention in the environment.

monoculture plantation A plantation that consists of tree crops of a single species.

natural capital assets Natural processes that generate income and well-being at lower cost than by equivalent human provision.

natural forest Forest areas where most of the principal characteristics and key elements of native ecosystems such as complexity, structure, and diversity are present.

natural regeneration Forest regeneration that is not promoted by people—in other words, without any planting.

nonindustrial private landowners Owners of forest land who do not manage their forest lands for continuous commercial harvesting purposes. Can be owners of either small or large forestland acreage.

old-growth forests Forests composed of timber stands in which no cutting has been done; also called first-growth forests, virgin forests.

oriented strand board (OSB) Panels made of narrow strands of wood fiber oriented both lengthwise and crosswise in alternating layers combined with a binding agent, similar to MDF.

particleboard A sheet material manufactured from small pieces of wood or other ligno-cellulose materials (e.g., chips, flakes, splinters, strands, shreds, shaves) agglomerated by use of a binder together with one or more agents such as the following: heat, pressure, humidity, a catalyst.

peeler log A log used in the manufacture of rotary-cut veneer.

plant associations Combinations of plants that either only grow or grow best depending on the existence of the other plants in the group.

plantation Forest areas lacking most of the principal characteristics and key elements of native ecosystems as defined by FSC-approved national and regional standards of forest stewardship, which result from the human activities of planting, sowing, or intensive silviculture treatments.

plantation forestry The management of an area planted with trees of one or more species, usually but not exclusively for wood production.

riparian forests Streamside systems of trees, patterned by water flows, and their sustaining processes.

riparian set-asides Reserves of trees that are not cut along water corridors. Leaving them prevents sedimentation of streams and maintains stream temperatures by providing shade, which is important for protecting aquatic life.

rotation Period of time between commercial harvests, which varies depending on growth conditions and is different for various regions, species, and timber management objectives.

seed tree systems A silviculture system in which selected trees or tree groups are left standing after initial harvest, providing a seed source for natural regeneration. After natural regeneration is achieved, the seed trees may or may not be removed.

selection systems (single tree, group, and strip) An uneven-aged system that develops or maintains a mixture of three of more distinct, well-represented age classes. The management goal is to create and maintain the stand and age structure over time.

shelterwood cut A silvicultural treatment in which some mature overstory (the higher, older trees) trees are standing to provide shelter for a regenerating understory. This regenerating younger strand is essentially even-aged. Once regeneration is successful, this overstory shelterwood may be removed or left. An uneven-aged stand exists while the shelterwood is still standing.

shelterwood tree systems (uniform and group) A silviculture system in which mature trees are removed in a series of cuts to achieve a new even-aged stand under the shelter of remaining trees.

uniform shelterwood systems Individual leave-trees distributed relatively uniformly throughout the stand unit.

silviculture Scientific manipulation of forest ecosystems to produce sustainable flows of desired wood products. The science and art of cultivating tree crops to yield a harvestable resource or other forest values and benefits. It includes any mechanical and chemical treatments that may be involved in the process.

single-tree selection An uneven-aged system in which new age classes are created by the removal of individual trees of all size and classes, more or less uniformly throughout the stand.

site The area (environment) in which stands grow, or even a particular tree grows. Site resources are factors such as light, heat, water, available space, and other nutrients that influence the growth of trees.

site index A measure of the capacity for a given site to grow trees. It usually is a reflection of height growth that can be expected on a given site at some base age. Most commonly, a 50-year base age is used. So, for example, a site index of 100 at base age 50 means that the dominant (tallest) and co-dominant trees of that species will average 100 feet in height after 50 years of growth.

skid trails Paths through the forest floor through which felled trees are skidded or dragged out of the forest.

SmartWood A program of the Rainforest Alliance and the world's first timber certification program in existence. SmartWood is accredited by the Forest Stewardship Council and has certified forestry operations in 10 countries worldwide.

softwood One of the botanical group of trees that generally have needle or scalelike leaves—conifers, such as gingko, larch, spruce; also, the wood produced by such trees, regardless of texture or density.

stand A community of trees that grow together at a particular place, which foresters can effectively manage as a unit. Transitions between stands in a natural forest are often gradual and can be the result of changes in site. For example, a slope may have varying water availability, thus creating a line between two forest types. Silviculture, however, can create "hard lines" between stands.

stand structure The variation in size, age, spacing, and height of trees in a stand.

stocking The density of trees on-site as compared to the maximum possible for that site. Optimal stocking refers to a density at which trees are achieving the best possible annual growth. An understocked stand is one that has not yet fulfilled its potential.

strip selection A variation of group selection where trees are removed in strips rather than in groups.

stumpage fees or stumpage cost Money paid for standing timber (i.e., trees on the stump) or the value of standing timber at a particular point in time.

sustainable forest A forest that people are able to attain and preserve for the particular long-term ecological processes, services, and products they want.

sustainable forest management Adaptive capacities for timely and sufficient response to economic, ecological, and social changes that otherwise would undermine the desired long-term processes, services, and products of a forest.

sustainable forestry The pursuit of innovation, investment, and institutional reform for regimes of actions that improve long-term ecological processes and productivity while satisfying human wants.

sustained yield As applied to a policy, method, or plan of forest management, the term implies continuous production, with the aim of achieving, at the earliest practicable time, an approximate balance between net growth and harvest, either by annual or somewhat longer periods.

temperate forests Forests grown in milder climate regions throughout the world, which contrast with tropical forests (hotter climates) and boreal forests (cold climates).

The Tree Farm Program A program created by the forest product industry in the 1940s to encourage investment in forestland to help provide a continuing source of wood and fiber. The program continues today and is the oldest of the national programs designed to encourage NIPF forestry.

thinning The removal of selected trees to achieve the management objectives for a particular stand of trees.

Timber Stand Improvement (TSI) The collective term for thinnings and other management techniques that improve the condition of a stand of trees. A commercial thinning is one in which trees are actually sold. A pre-commercial thinning is a treatment in which no sale occurs, but rather a stand is being treated in anticipation of a later sale.

"A Tragedy of the Commons" Garret Hardin's term for the consequences of disparity between motives to use and to invest in a renewable common property resource. Individuals have reason to use the resource but none to invest in its renewal, with consequent decline of resource productivity and benefits for all users.

trim end Small pieces of wood left over from cutting wood to specified lengths. These pieces are typically chipped and converted to pulp or burned for energy.

uneven-aged stand A stand of trees of various ages. It is produced from smaller scale disturbances ranging from a single tree felling to groups of trees that are selected for harvest.

valued manufacturing The process of adding value to the same unit of wood material in the production and/or sales scheme.

veneer A surface covering made of thin sheets of wood of uniform thickness—rotary cut, sliced, or sawn—for use in plywood, laminated construction, furniture, veneer containers, and so on.

The Sustainable Forestry Working Group

Individuals from the following institutions participated in the preparation of this publication.

Environmental Advantage, Inc.
Forest Stewardship Council
The John D. and Catherine T. MacArthur Foundation
Management Institute for Environment and Business
Mater Engineering, Ltd.
Oregon State University, Colleges of Business and Forestry
Pennsylvania State University, School of Forest Resources
University of California at Berkeley, College of Natural Resources
University of Michigan, Corporate Environmental Management
    Program
Weyerhaeuser Company
The World Bank, Environment Department
World Resources Institute

Members

Project Director: Michael B. Jenkins, John D. and Catherine T.
    MacArthur Foundation Program on Global Security and
    Sustainability

337

Project Coordination and Support: Greg Lanier, John D. and
    Catherine T. MacArthur Foundation

Project Contributors

Matthew Arnold, World Resources Institute, Management Institute
    for Environment and Business

Tasso Rezende de Azevedo, Instituto de Manejo e Certificacao
    Florestal e Agricola

John Begley, Weyerhaeuser Company

Bruce Cabarle, World Resources Institute, Management Institute for
    Environment and Business

David S. Cassells, The World Bank, Environment Department

Rachel Crossley, Environmental Advantage, Inc.

Robert Day, World Resources Institute, Management Institute for
    Environment and Business

Betty J. Diener, University of Massachusetts, Boston

Jamison Ervin, Forest Stewardship Council

Richard A. Fletcher, Oregon State University Extension Service

Eric Hansen, Oregon State University, Department of Forest
    Products

Stuart L. Hart, University of Michigan, Corporate Environmental
    Management Program

Stephen B. Jones, Auburn University

Isak Kruglianskas, Universidade de Sao Paulo, Faculdade de
    Economia, Administacao e Contabilidade

Keville Larson, Larson and McGowin, Inc.

Stephen Lawton, Oregon State University, College of Business

Tony Lent, Environmental Advantage, Inc.

Catherine L. Mater, Mater Engineering, Ltd.

Scott M. Mater, Mater Engineering, Ltd.

James McAlexander, Oregon State University, College of Business

Bill McCalpin, John D. and Catherine T. MacArthur Foundation

Mark Miller, Two Trees Forestry

Mark Milstein, University of Michigan, Corporate Environmental
    Management Program,

Larry A. Nielsen, Pennsylvania State University, School of Forest
    Resources

Diana Propper de Callejon, Environmental Advantage, Inc.

John Punches, Oregon State University, Department of Forest Products

Richard Recker, Oregon State University, Sustainable Forestry Partnership

A. Scott Reed, Oregon State University, College of Forestry

Jeffrey Romm, University of California at Berkeley, College of Natural Resources

Nigel Sizer, World Resources Institute, Biological Resources and Institutions/Forestry Frontiers Initiative

Michael Skelly, Environmental Advantage, Inc.

Thomas Vandervoort, Vandervoort Public Affairs & Communications

Court Washburn, Hancock Timber Resource Group

Michael P. Washburn, Pennsylvania State University, School of Forest Resources

Mark Webb, Mark Webb & Co.

Charles A. Webster, Environmental Advantage, Inc.

Peter Zollinger, FUNDES/AVINA Group

About the Authors

Michael B. Jenkins is the executive director of Forest Trends, a new organization dedicated to the conservation and sustainable management of the world's forest resources. He has been the associate director for the World Environment and Resources Program of the John D. and Catherine T. MacArthur Foundation since 1989. His responsibilities with the program include all grant making in Latin America and the Caribbean, as well as overarching program management. He holds a master's of forest science from Yale University. Before joining the foundation, he worked for three years as agroforester in Haiti with the USAID Agroforestry Outreach Program. Prior to that he worked with a Washington-based development program, Appropriate Technology International, as a technical advisor. In the late 1970s, Jenkins was a Peace Corps volunteer in Paraguay, working in agriculture, apiculture, and forestry projects. He has traveled and worked throughout Latin America, Asia, and parts of Africa, and speaks Spanish, French, Portuguese, Creole, and Guarani.

Emily T. Smith, an author and journalist, has been writing books on the environment and business since 1995. Prior to 1995, during a fourteen-year tenure at *Business Week* magazine, she held several positions, including that of bureau chief in the Boston office and science editor. As science editor, she covered technology, science, and environmental issues, as well as the pharmaceutical, biotechnology, computer, and chemical industries extensively. She has won two National Magazine Awards and in 1993 won the Overseas Press Club prize for environmental coverage for her writing on the 1992 United Nations Conference on Environment and Development and the integration of environmental concerns with economic development. She holds a master's degree in mass communications from the University of Minnesota and a bachelor's degree in art history from the University of Washington.

Index

Abbot of Trappist Abbey, Inc., 172
Acacia plantations, 31, 249
Addison Corp., 211
Adhesives
  in fingerjointing, 102
  soy-based, 95, 103–104
Aesthetics. *See* Nonmarket benefits
Africa
  eucalyptus plantations, 238
  pressure to harvest remote forests, 38
  tropical hardwood exports, 55
Air quality
  cost of emissions control, 252–253
  emissions from pulp mills, 245, 248, 251
  fires in Indonesia and Malaysia,
    258–259
  greenhouse effect, 4–5, 290
Alabama, private ownership case history,
    161, 162, 180–182
Alien species, 244, 245
Almanor Forest, 118
American Forest and Paper Association,
    43, 63, 64, 79, 297. *See also*
    Sustainable Forestry Initiative
Andersen Corp., 200
Andersen Windows Company, xv, 200
Aracruz Celulose S.A., 44, 236–248,
    254–257, 308, 312–313
Argentina, restrictive log export policy, 72

Asia
  attitude toward certified products,
    83–84
  economic crisis, 8
  engineered products, 47
  plantation forestry, 25
  pressure to harvest remote forests, 38
  tropical hardwood forests, 54, 231 n.4
Asian Timber, 83
Asia-Pacific, transition from exporter to
    importer, 34, 38, 56
Assi-Doman, certification, 60, 71, 78, 273
Atterbury Consultants, Inc., 97
Auburn Machinery, Inc., 100
Audits
  chain of custody, 64, 73
  forest management, 64, 120
Austin Green Builder Program, 127
Authority. *See* Land ownership; Political
    aspects

Banking, 2, 112, 113. *See also* Economic
    aspects; World Bank
Barama, Ltd., 24
Biodiversity, 4, 11, 40, 133–134, 168, 244,
    259, 327
Biotechnology, 2
  in plantation forests, 35, 235, 255
  in softwood forestry, 283, 293, 301 n.7

Bloomquist, Eric. *See* Colonial Craft, Inc.
Body Shop, The, 81
BOLFOR project, 109–110
Bolivia
    BOLFOR project, 109–110
    IMR/CIMAL, 75
Boltwood, 99, 327
B&Q Plc., 74, 75, 80, 321
Brazil
    Champion International Corp., 26
    cost of fiber production, 33
    CPTI Technology and Development,
        108
    deforestation of, 216, 218
    eucalyptus hardwood plantations, 31, 32
        Aracruz Celulose S.A., 44,
            236–248, 254–257, 308, 312–313
        Riocell S.A., 246, 248–257, 308, 309
    hardwood exports, 54, 223–225, 247
    indigenous peoples, 243
    mahogany harvest moratorium, 72
    national economics, 232 n.6–7
    Precious Woods, Ltd., 8–9, 66–67,
        221–231
Breeding, 238–239, 255, 286. *See also*
        Biotechnology
Brent Tract, 161, 176–178, 308, 310
British Columbia
    area required for fiber production, *vs.*
        plantations, 32, 33
    clear-cutting in, 62, 148, 149
    Forest Practices Code, 153
    harvest from Crown lands, 50, 148, 283
    Vernon Project, 148–156
Brokers, 205, 328
Brown Trucking and Logging, 99
Builders' Supply, 115. *See also* Collins Pine
        Co.
Business aspects. *See also* Certification;
        Demand; Supply
    capital
        intensity, 41–42, 231 n.2
        of Precious Woods in Brazil, 221,
            233 n.14
        from product to, 17–18, 112–113
    downsizing and outsourcing, 241, 243
    financials
        Brent Tract, 177–178
        Collins Pine Co., 128
        Colonial Crafts, Inc., 199

Frederick Property, 182
Freeman Farm, 169
    Lyons Family Tract, 171–172
    Moniminee Tribal Enterprises,
        141–142
    Parson's Pine Products, 196–197
    Precious Woods, Ltd., 226–228
    Riocell S.A., 251–254, 309
    STORA, 264, 276–277
    Trappist Abbey Forest, 174
    VanNatta Tree Farm, 166, 167
    Vernon Project, 153–154
    Weyerhaeuser Co., 295–296, 314
financing, 18–19, 22, 66, 93, 128
franchise protection, 305, 306, 308–309,
        314–315
harvest bidding process, 137–138
integration of sustainability into
        operations, 302–318, 320–321,
        323–324
mail-order, 197–198
new opportunities, 39–40, 190
public relations, 72, 117, 120–121,
        146–147
pulp and paper industry, 40–42
redefinition toward "fiber service,"
        303, 305, 307, 312–313, 314,
        315–318
role of science and technology, 1, 22, 93
strategic planning, 304
tropical timber industry, 217–220
vertical integration, 22, 207–208, 215
Buyers' groups, 2, 60, 70–71, 270, 321

Cabinetry, 2, 191
California. *See also* Collins Pine Co.
    silviculture in, 118
    state policy, 72
Canada. *See also* British Columbia
    cost of fiber production, 33
    ISO-14000 standards for, 83
    sale of recycled wood waste, 102–103
    SFM standards, 60
    share of U.S. softwood market,
        1991–1995, 195
    softwood production, 30
    Weyerhaeuser Co. in, 283, 298–299
    Windhorse Farm, Nova Scotia, 185–186
Canada, Ministry of Forestry, 148,
        149–150, 151–153, 155

Canada–U.S. Softwood Lumber
    Agreement, 195
Canadian Pulp and Paper Association, 71,
    83
Canadian Standards Association, 64, 79, 81
Canon, 84
Carter-Sprague, Inc., 100–101
Cary Property, 161, 179–180
Case histories
    certified wood products marketing. *See*
        Collins Pine Co.
    eucalyptus hardwood plantations,
        236–257, 308–313
    log sorting, 148–156
    mahogany door production, 207–215,
        308, 311
    private land ownership, 162–182
    sustainable forest management, 66, 131,
        132–148, 281–301
    use of recycled wood waste. *See* Parsons
        Pine Products
Celos management system, 222
Certification, 25, 329. *See also* Forest
        Stewardship Council-certification
    advantages of, 66, 202–204, 275–276,
        306–307, 322–323
    by Associated Oregon Loggers, 165
    by Canada, 60, 64, 79, 81
    competitive risks of, 277–278
    cooperative, 185–186
    by Greenpeace, 149, 153
    historical background, 60, 63
    for loggers, 95, 96, 97
    for manufacturers, 199–204, 206,
        322–323
    marketing, 59–85, 123–128, 200–201,
        323
    of panels, 47–48
    of Portico S.A., 210
    of Precious Woods, Ltd., 225
    for private landowners, 54, 182–186
    by Rainforest Alliance. *See* SmartWood
        program
    sham, 63
    of softwood, 51–52
    from solid wood to paper, 76–79
    of STORA. *See* STORA, Forest
        Stewardship Council-certification
    by Sweden, 60
    of U.S. hardwoods, 53

Weyerhaeuser Co. and, 297
Certified Forest Products Council, 71
Certified Logger's Program, 96
Chain of custody audit, 64, 73
Chain of custody certification (COC), 144,
    199–200, 203
Chain of custody issues
    disadvantages for small landowners, 183
    documentation, 272
    outside suppliers, 298
    synchronization, 322
    tracking and inventory, 94, 129
Champion International Corp., 25, 26–27,
    79, 81, 226, 229
Chapman, Chip, 184–185
"Character wood." *See* Recycling, of wood
    waste
Chile
    cost of fiber production, 33
    eucalyptus plantations, 32, 238
    Rio Condor project, 1
    Rio Iata Group, 34
    Savia International, Ltd., 1–3, 39
Chlorine free products, 61, 240–241, 247,
    251
Clean technology, 304–305
Clear-cutting, 327–328
    alternatives to, 149, 151–153, 166–167,
        168, 174, 176
    in British Columbia, 62, 148, 149
    as strategy for disease, 165
    by Weyerhaeuser Co., 291
Coast Range Association, 291
Collins Pine Co., 6, 60, 66, 67, 80, 114–130,
    308, 313
Collins Resources International, Ltd., 115
Colonial Craft, Inc., 66, 185, 189–190,
    199–206
Color scanning, 105
Commission on Sustainable Development,
    37–38
Community involvement, 17, 120, 152,
    155, 256, 312
Company image. *See* Public relations
Competitive strategy, 7–8, 91, 117,
    197–198, 277–278, 302–303
Composite products. *See* Engineered wood
    products
Computers. *See* Software
Conflict resolution, 17

Connecticut/Vermont Project, 186
Connor AGA, 144
Construction lumber. See Mills; Sawnwood
Consumers
  attitudes and perceptions of, 67–70, 77,
    79, 124, 125
  education of, 19, 75, 147
  environmental activists and, 62
  surrogates for, 69–70
Conventional forestry management, 4–5,
    12, 16
  in combination with progressive
    management, 164–165
  Forest Stewardship Council audits, 64,
    120
  number of trees damaged per acre, 137
  objectives and results, 90
  problems with, 148
  remedies for, 119–120
  in the tropics, 106
  by U.S. Forest Service, 114
Cooperative certification, 185–186
Cost. See Economic aspects
Costa Rica, 207–215. See also Portico S.A.;
    Precious Woods, Ltd.
  access to resources in, 66
  regulations, 212–213
CPTI Technology and Development, 108
Crops, as fiber, 46, 317
CSA (Canadian Standards Association),
    64, 79, 81
Curry-Miller Veneers, Inc., 49

Deforestation. See also Harvest
  in Brazil, 216
  in developing countries, 4, 207
  indirect effects of, 216–217
  in Latin America, 216
  in Mexico, 27
  in South America, 4
  in the U.S., 4
Demand
  buyers' groups effect on, 70–71
  convergence with supply, 73–76,
    217–218
  global, 41, 45, 292
  gulf between supply and, 25, 27–29,
    73–76, 124–125, 140–141, 204
  motivation behind industrial, 71–72
Design for environment (DfE), 304, 305

Developing countries, 3, 4, 27, 207
Dioxins, transition to chlorine free
    products, 61, 240–241, 247, 251
Disease and pest control, 165, 244, 245, 250
Distribution channels, 117, 125, 127, 141,
    150, 197, 205–206
Do-it-yourself chains (DIY). See also
    Home Depot
  Collins Pine Co. marketing, 126–127
  environmental activists and, 71, 72
  Homebase, 272, 273
  1995+ Group and, 71, 80
  in the United Kingdom, 71, 72, 80, 272
Donghia Furniture and Textile, 71
Doors, 311
  affiliation to produce, 200
  grilles for, 199, 200
  mahogany. See Portico S.A.
  by Weyerhaeuser Co., 283

"Ecolabels." See Certification
Ecological aspects, 12. See also Nonmarket
    benefits
  of deforestation, 4, 216–217
  global warming, 4–5
  of plantation forests, 235–236, 244–245,
    250, 266
  of private land ownership, 163
  transition from silviculture to
    ecoculture, 5, 165, 268–269
Ecological Landscape Planning, 267–268
Economic aspects
  advantages of plantation forests, 32, 33
  consumer willingness to pay for
    certification, 67–68
  of conventional forestry, 18, 152
  costs
    environmental protection, 251–254,
      255
    Forest Stewardship Council-
      certification, 63, 183
    Indurite process, 107
    labor, 32
    multiple-sensor machine, 106
    Ponsse harvesting system, 99
    Sorbilite composite molding
      system, 89
    sustainable forestry management,
      8, 122, 123
    wood, 116, 145

from product to capital, 17–18, 112–113
profit *vs.* sustainability, for private
  landowners, 162–163
reforestation funding, 66
research funding, 91, 93
of SFM technology development,
  91–93, 108–110
three contextual planes of, 22
of transition to SFM, 132, 139, 152,
  264–268
value of standing forest, per hectare, 5
Education and training
  computer literacy, 92
  of consumers, 19, 75, 147
  of personnel, 6, 26, 93, 94, 95, 108,
    136–137, 209
  by private landowners, 168
  in sustainable forestry strategy
    development, 21–22
  in wood waste recycling, 194
Eisen, Mark, 80
Emissions. *See* Air quality
Employment. *See* Job opportunities
Endangered species, 26, 29, 245, 289–290,
  315
Engineered wood products, 43–48, 88–93,
  329
  adhesives used in, 95, 103–104
  butcher block construction, 212
  consistency of, 35
  European market for, 56
  Indurite technology, 106–107
  medium density fiberboard, 35, 45, 46,
    88, 332
  oriented strand board, 35, 44–45, 46–47,
    88, 283, 332
  with polyvinyl chloride, 201
  Sorbilite composite molding system, 88,
    89, 95
Enso Oyj, 62
Environmental activists
  on do-it-yourself chains, 71, 72
  in Germany, 271
  influence of, 61–62
  in Japan, 84
  in Sweden, 266, 267–268, 276
  on third-party certification, 64–65, 79
  on Weyerhaeuser Co., 291–292
Environmental Advantage, Inc., 228
Environmental aspects, 55–56

cost of environmental protection,
  251–254, 255
of engineered wood products, 46–48
opportunities for U.S. hardwoods,
  53–55
of plantation forests in Brazil, 244–246
restrictions, 29, 39, 42–43
Environmental Enterprises Assistance
  Fund, 214
Environmental management systems
  (EMS), 304, 305
Estate planning, 165, 168, 170, 179–182
Eucalyptus hardwood plantations, 31, 32,
  236–248, 254–257
Europe
  government regulation in, 73
  market for certified products, 66, 74–76,
    85, 124, 185–186, 225, 270–273, 275
  market for chlorine free products, 247
  softwood roundwood supply, 30
  tropical hardwood imports, 54–55
Evergreen Indus-Trees Limited, 107
Exotic species, 244, 245

Federal Paper Board Co., Inc., 24, 41
Fiber
  cost of production, 33
  crops as, 46, 317
  redefinition toward "fiber service," 303,
    305, 307, 312–313, 314, 315–318
  use of recycled fiber content, 82–83
Fibrex, 201
Fingerjoint technology, 95, 101–102,
  103–104, 190, 205, 329
Finland, 41, 62, 77
Finnish Paper Mills Association, 68
Fire, 4–5, 258–259, 285
FLIPS 96 software, 98
Flooring, 127, 144, 193
Florida, private ownership case history,
  161, 179–180
Forest fires, 4–5, 258–259
Forest Habitat Classification System, 133,
  136
Forest plantations. *See* Plantation forests
Forestry management. *See* Conventional
  forestry management; Sustainable
  forestry management
Forest Stewardship Council certification,
  36–37, 63–65, 330

Forest Stewardship Council certification
    (*continued*)
    advantages of, 2, 19
    of BOLFOR in Bolivia, 110
    of Collins Pine Co., 124
    companies that applied for, xvi, 1
    cost of, 63, 183
    hectares approved for, 60
    historical background, 63
    of STORA, 6, 8, 24, 44, 60, 71, 262–280
    of U.S. public lands, 81
ForestView software, 97, 98
Fort Howard Corp., 44
Franchise protection, 305, 306, 308–309,
    314–315
Frederick Property, 161, 162, 180–182
Freeman Corporation, 127
Freeman Farm, 161, 162–163, 167–170
Fuelwood, 28, 192
FUNDECOR, 66
Furniture. *See also* Do-it-yourself chains
    Collins Pine Co. marketing, 126–127
    Eco-furniture, 81
    garden, 75
    of medium density fiberboard, 47
    office, 47
    ready-to-assemble, 45, 47
    recycled wood waste in, 190, 197
    species used in global market, 2
    volume of hardwood used, 1995, 53

Gabon, 38
Gap, Ltd., The, 81
General Agreement on Trades and Tariff
    (GATT), 34, 72
Genetic engineering. *See* Biotechnology
Geographic information system (GIS), 95,
    97–98, 267, 269
German Magazine Publishers Association,
    271
German Paper Producers Association, 272
Germany, market, 270, 271, 272
GIS. *See* Geographic information system
Global market, xiv, 84–85
    barriers to, 124–125
    convergence of supply and demand,
        73–76
    creation of new niches, 3, 6, 15, 16, 40,
        66–67, 193, 310
    for doors, 214

emerging, 24–25, 26–56, 59–85, 320–322
for furniture and cabinet wood, 2
mail-order, 197–198
off-shore competition, 197–198
for paper and paperboard, 292
percent of global gross national
    product, 10
percent of total world trade, 10
public relations in the, 72, 117, 120–121,
    146–147
for pulp, 28–29, 237–238, 254
quality of wood supply, 5–6, 116
for solidwood products, 292
for tropical timber, 54, 217–218,
    223–225
for veneer, 48–49
in wood waste, 195
Global positioning system (GPS), 97
Global warming, 4–5, 290
Glossary, 327–336
Golden Hope Plantations Bhd, 47
Goodwill, 128–129
Grading of wood, 143, 146
"Green" certification. *See* Certification;
    Forest Stewardship Council
    certification
Green Cross, 210, 313. *See also* Scientific
    Certification Systems
"Green" financing, 18–19, 128
Greenhouse effect, 4–5, 290
Greenpeace, 61, 62, 149, 152, 236, 266, 271
"Green" political parties, 74
"Green Purchasing Network," 84
Green Seal, 330–331
Greenweld Process, 95, 102–103
Gross domestic product (GDP), pulp and
    paper industry, 40
Guchess International Group, 247
Guyana
Barama, Ltd., 24
pressure to harvest remote forests, 38

Habitat, 133–134, 168, 266, 269, 289–291,
    317. *See also* Nonmarket benefits
Habitat for Humanity, 71
Hancock Timber Resource Group, xv
Hardwoods, 28, 45, 331. *See also* Veneer
    global production, 30–32, 52–53
    manufacturing, in mix of species,
        139–140

tropical, 53, 54–55, 207–215
Harvest
    in Costa Rica, 208, 209
    ecology and economics of, 131–156
    high-thinning, 176–177
    impact of planned *vs.* unplanned, 108
    impact reduction, 98–99, 108, 305, 306,
        309–310, 314, 315
    of mahogany, Brazil moratorium, 72
    in plantation forests, 240
    of remote forests, 38
    strategies and scheduling, 165, 168, 175,
        249, 268, 287–288
    in the tropics, 38, 54, 108–109
Hemp, 317
Herman Miller, 47
High-thinning, 176–177
High-yield forestry, 285–287, 293, 300–301,
        315, 331
Historical background
    certification, 60, 63, 76–78
    conventional logging practices, 4–5
    pulp and paper industry in Brazil,
        248–249
    STORA, 263–264
    sustainable forest management, 3
    Sustainable Forestry Working Group,
        xiii–xvi
    Weyerhaeuser Co., 281, 284–289
Homebase, 272, 273
Home Depot, The, 6, 68, 79, 80, 308,
        311–312
    Collins Pine Co. marketing, 125, 126,
        311–312
    Colonial Crafts, Inc. marketing, 203
    Portico S.A. marketing, 211, 212, 214,
        311
Hunting. *See* Nonmarket benefits

IBAMA, 220, 222
IKEA, 47, 56, 78
IKPC Group, 248
Impact reduction, 98–99, 108, 305, 306,
        309–310, 314, 315
IMR/CIMAL, 75
Indigenous peoples, 243
Indonesia
    hardwood production, 31, 33
    plantation forests, 236, 257–259
Indurite process, 106–107

Industrias Klabin, 248
Information and telecommunication, global,
        39, 40
Infrastructure, lack of, 29, 38, 222
Injury rate, 96
Innovations, 15–19, 25, 307. *See also* Labor
        issues, education of personnel
    in efficiency, 36
    in lumber mills, 51, 53
    retooling, 6, 26, 144
    standardization and, 35
    in sustainable forestry management,
        15–19, 25
Insects. *See* Disease and pest control
Intergovernmental Panel on Forests, 9
International Paper Co., 24, 41
International Standards Organization (ISO),
        36, 60, 64, 331
    ISO-9000 standards, 331
    ISO-14000 standards, 83, 251, 331
International Tropical Timber Agreement,
        37
International Tropic Timber Organization
        (ITTO), 54, 65, 331
Inventory of timber. *See also* Chain-of-
        custody issues
    increase in, 134
    prelogging, 209, 222–223
ISO. *See* International Standards
        Organization

J. Sainsbury Plc., xvi, 44, 71, 74, 272–273, 321
Jann, Roman, 221. *See also* Precious Woods,
        Ltd.
Japan
    attitude toward certified products, 84
    imports, 34, 55, 293
    softwood sawnwood market, 50–51
Job opportunities, 16, 147–148, 153
    linked with social investment, 312
    in secondary wood products
        manufacturing, 190–191
Jurisdiction. *See* Land ownership; Political
        aspects

Kane Hardwoods, 53, 115
Keweenaw Land Association, 66
Key Bank, 112, 113
Kimberly-Clark, 41
Knoll Group, The, 47, 138–139

Koppensky Kombinant, 29
Korea, Sunkyong Group, 24
Kornas, certification, 60
Kreibich System, 102, 103–104
Kreibich and Associates, 103

Labor issues. *See also* Job opportunities
    education of personnel, 6, 26, 93, 94, 95,
        108, 136–137, 209
    workers' compensation, 96
    working conditions, 243–244
Lake Superior Land Company, 26
Landowners
    bottlenecks and solutions, 95
    key needs and constraints of, 94, 121
Land ownership
    governmental, 50, 52, 148, 283
    private. *See* Private land ownership
    from private to councils and
        communities, 17
Landscape perspective, 16, 331–332
Laser-based ranging systems, 105
Latin American Region Caterpillar, Inc.,
        108
Legal aspects, Rio Bravo project in Chile, 2
Life-cycle analysis, 304, 305
Loft Bed Store, 81
Loggers
    bottlenecks and solutions, 95
    certification for, 95, 96, 97
    key needs and constraints, 94
    training, 6, 26, 93, 94, 95
Logging. *See* Harvest
Logging contractors, 94, 95
    Canadian bidding policy, 149–150
    incentives and penalties for, 136–137
    "lottery" contract awards, 137–138
Log homes, 150, 151
Log sorting
    average log size, 5
    for profit, the Vernon Project, 148–156
Lorentzen, Erling. *See* Aracruz Celulose
        S.A.
Lowe's, 211
Lumber. *See* Mills; Sawnwood
Lumber brokers, 205
Lyons Family Tract, 161, 163, 170–172

MacMillan Bloedel, 60
Mahogany doors. *See* Portico S.A.

Maine, Certified Logger's Program, 96, 97
Maine Tree Foundation, The, 96
Malaysia
    attitude toward certified products, 83
    Golden Hope Plantations Bhd, 47
    hardwood production, 31, 33
    Samling Strategic Corp., 24
    tropical hardwood exports to Europe,
        54, 55
Manadnok Paper Mills, 84
Manne, Robert, 2
Manufacturers. *See* Wood products
        manufacturers
Marketing. *See* Global market
Mater Engineering, Ltd., 93, 100, 103, 205
Matsushita Electric, 84
McNulty, John W., 67
Medium density fiberboard (MDF), 35, 45,
        46, 88, 332
Menominee Tribal Enterprises (MTE), 53,
        66, 131, 132–148
Mercer, Larry, 80
Mexico, deforestation of, 27
Miami Corp., 98
Mills. *See also* Pulp and paper industry
    in Brazil, 218–219, 227
    certification of, 138–139
    computer literacy in, 92
    in Costa Rica, 208
    electricity costs of, 192, 219, 230
    emerging technology for, 99–107,
        110–112
    impact reduction, 305, 306
    in Indonesia, 257
    of Menominee Tribal Enterprises, 139,
        144
    sawnwood
        consolidation of, 53
        retooling and retraining in, 6, 26,
            144
    small, and engineered wood products,
        91–92
    wood waste recovery in, 192–193
Minnesota, Forest Stewardship Council
        certified public lands, 81
Models
    certification, 182–186
    sustainable forestry, 305–307
    sustainable wood products, xiv
Molding, 2, 102, 199, 200

Monoculture. *See* Plantation forests
MTE. *See* Menominee Tribal Enterprises
Multiple-sensor machine (MSM), 105–106
Musical instruments, 150, 151

New Zealand
    cost of fiber production, 33
    plantation forests in, 32
New Zealand Research Institute, Ltd., 102
Niche market, 3, 6, 15, 16, 40, 66–67, 193,
    310
1995+ Buyers' Group, 71, 80, 185, 270, 272
NIPF. *See* Nonindustrial private forest
    landowners
Nongovernmental organizations, 113, 267.
    See also Environmental activists
Nonindustrial private forest landowners
    (NIPF), 157–158, 182–187, 332.
    See also Private land ownership
Nonmarket benefits, 14, 15, 40, 158,
    162–163, 168, 267–268, 290, 315
North American Free Trade Agreement
    (NAFTA), 34
Northeast Ecologically Sustainable Timber
    Company, 184
Norwest Corp., 208
Nurseries, 283, 286, 287

Oceania, 30, 34
Oregon. *See also* Collins Pine Co.
    private ownership case histories. *See*
        Brent Tract; Trappist Abbey Forest;
        VanNatta Tree Farm
    recycled wood waste, case history. *See*
        Parsons Pine Products
    silviculture in, 118–119
    state policy, 72
Oregon Competitiveness Council, 91
Oregon Wood Products Competitiveness
    Corp., 191
Organizational products, 193, 194, 196, 197
Oriented strand board (OSB), 35, 44–45,
    46–47, 88, 283, 332
Owen, Carlton N., 25, 27
Ownership. *See* Land ownership; Political
    aspects
Oyama Forest Products, Inc., 151

Pacific Forest Trust, 291
Pacific Northwest, 29, 33, 62, 190. *See also*

British Columbia; Oregon;
    Washington
Pacific Rivers Council, 291
Pallets and crates, 43, 101
Panel industry, 27, 43, 46
    global demand, 28, 29, 47
    increased efficiency in, 36
Paraguay, eucalyptus plantations, 238, 246
Parsons Pine Products, 189, 190–199, 308,
    309–310
Particleboard, 34, 45–46, 332
Partnerships, 9–10, 17, 112–113, 117–118
Pennsylvania. *See also* Collins Pine Co.
    Freeman Farm, 161, 162–163, 167–170
    Forest Stewardship Council-certified
        public lands, 81
    Lyons Family Tract, 161, 163, 170–172
    silviculture in, 119
Pesticides. *See* Disease and pest control
Philippines, restrictive log export policy, 72
Plantation forests, 2, 5, 25, 333
    advantages and limitations, 9, 32, 39,
        235–236
    in Brazil, 235–259
    in Costa Rica, 66
    softwood. *See* Weyerhaeuser Co.
    in the Southern Hemisphere, 32–34
    in Sweden, 265
Plaza Hardwood, Inc., 65
Plywood
    alternatives to. *See* Oriented strand
        board
    market for, 33, 45, 257
    percent of tropical hardwood harvest, 54
    by Weyerhaeuser Co., 283
Political aspects, 22
    of conventional forestry, 18
    government oversight of management,
        37–38
    government regulation, 72–73
    government *vs.* institution authority,
        20–21
    Green parties, 74
    of indigenous communities, 12
    local and municipal policy, 73–74
    of Russia, 29
    of scale, 13, 20
Pollution prevention, 252–254, 304–305,
    309. *See also* Air quality; Water
    quality

Polymers, 317
Polyvinyl chloride, 201
Ponsse, Ltd., 99
Ponsse log harvesting system, 99
Portico S.A., 8, 66, 80, 190, 207–215, 308,
   310–311
Portugal, eucalyptus plantations in, 238
Poverty and deforestation, 3, 27, 28, 207
Precious Woods, Ltd., 8–9, 66–67, 221–231
Price. *See* Economic aspects, costs
Private land ownership. *See also* Taxes
   to councils and communities, 17
   highly variable intensity of, 175–176
   impact reduction with, 310
   joint, 249
   for long-term investment, 54
   models for certification, 182–186
   opportunities for, 157–187
   percent of commercial forest land in,
      157
   portrait of owners and land, 158–162
   raw materials for STORA from, 278
Procter & Gamble Co., 72, 81
Products matrix, xiv
Product stewardship, 304
Professional advice, 162
Protected areas, impact of surrounding
   areas on, xiii
Public relations, 72, 117, 120–121, 146–147
Publishers and printers, 77, 271, 272
Puertas y Ventanas de Costa Rica, 207
Pulp, Paper, and Packaging sector,
   Weyerhaeuser Co., 282, 284
Pulp and paper industry. *See also* Mills
   annual sales, 1997, 40
   in Brazil, 235–259
   business challenges, 40–42
   capital intensity of the, 41–42
   certification trend in, 76–79
   consolidation in the, 41
   drop in number of mills, 41
   eco-efficiency in, 240–241
   global market, 28–29, 237–238, 254
   increased efficiency in, 36
   in Indonesia, 257–259
   percent of GDP, 40
   standardization in, 35
   in Sweden. *See* STORA
   transition to chlorine free products, 61,
      240–241, 247, 251

trends in, 27, 41–42, 44, 250
use of recycled fiber content, 82–83

Quality of timber stands, 134–135
Quality of wood, 5–6, 116, 190–199,
   223–224

Rainforest Alliance, 184
Rain Forest Café, The, 81
Rainforests. *See* Deforestation; Tropical
   forests
Rasmussen, Michael J., 198
Recreation. *See* Nonmarket benefits
Recycling, 43, 44, 81–83
   of packing and shipping products, 317
   Sorbilite composite molding system, 88,
      95
   of wood waste, 93, 95, 99–100, 190–199
Reforestation, 38, 66, 221, 250, 285–286
Regulations
   Brazil, 220
   Canada–U.S. Softwood Lumber
      Agreement, 195
   compliance, 94
   Costa Rica, 212–213
   the power of government, 72–73
   reactive *vs.* proactive approach to,
      303–304
   Sweden, 265, 269
Research and development
   forestry and forest products, U.S., 111
   funding for, 91, 93
   Indurite process, 106–107
   politics of academic institutions
      and, 91
   Robo-Eye scanning system, 105–106
Resource managers, 183, 184–185
Rimu, 106
Riocell S.A., 246, 248–257, 308, 309
Rio Condor project, 1–3, 39
Rio Conference of 1992, 9–10, 37, 210
Rio Iata Group, The, 34
Riparian areas, 16, 26, 290, 333
Robert Berry and Sons, 97
Robo-Eye scanning system, 105–106
Rock-Tenn Co., 44
Russia, 34, 62
   percent of world's coniferous forests, 29
   political aspects of investment in, 29
   Weyerhaeuser Co. and, 299

Saddles, 151
Sainsbury's. *See* J. Sainsbury Plc.
Samling Strategic Corp., 24
Sand County Almanac, A, 187
Sanyo, 84
Savia International, Ltd., 1–3, 39
Sawnwood
    Collins Pine Co. marketing, 127
    conversion rate of natural logs to,
      228–229, 233 n.15
    "green" market for, 81–82
    kiln-dried, 145–146
    Menominee Forest data, 135, 136
    softwood, 49–52, 283
    tropical hardwood, 54, 224, 228–229,
      247
Scale, 13–14, 20, 22
Scandinavia
    certification in, 43, 44, 52, 76–78
    cost of fiber production, 33
Scanning technologies, 94, 95, 104–106
Scenic values. *See* Nonmarket benefits
Schrafl, Anton E., 221
Scientific Certification Systems (SCS)
    accreditation by the Forest Stewardship
      Council, 124, 132
    Collins Pine Co. and, 115, 118, 120
    Kane Hardwoods and, 115
    Menominee Tribal Enterprises and, 134
    STORA and, 263, 274
    Tecnoforest del Norte and, 210
Scientific data
    in development of the landscape
      perspective, 16
    to monitor global changes, 12
    role in sustainable management
      planning, 1, 22
Scott Paper, 41
Scrap wood, 2, 46. *See also* Recycling, of
    wood waste
Secondary forests, 5, 285
Seed and seedling nurseries, 283, 286, 287
Seven Islands Land Co., 53, 67
SFI. *See* Sustainable Forestry Initiative
Shared certification, 185
Shelterwood, 333–334
Shelving, 126. *See also* Do-it-yourself
    chains
Silva Forest Foundation, 149
SILVAH software, 171

Silvania Foresty, 109
Silviculture, 5, 118–119. *See also*
    Conventional forestry
    management
Sivin, Bernard Jerry. *See* Parson's Pine
    Products
Small Business Forest Enterprise
    Program, 148
Small Business Innovative Research
    program, 111
SmartWood program
    Abbey Forest and, 175
    Col ‚ıial Craft, Inc. and, 199, 202
    Menominee Tribal Enterprises and,
      132, 134, 144
    Precious Woods, Ltd. and, 225
    private land owners and, 184, 186
Social aspects. *See also* Job opportunities
    investment in, 312
    poverty and deforestation, 3, 27, 28,
      207
    of scale, 13, 20
    of sustainable forestry, 14, 15, 243–244,
      256, 302, 324–325
Software
    employee training on, 92
    for forest management, 93, 97, 98, 171
Softwood, 28, 45, 334. *See also*
    Weyerhaeuser Co.
    manufacturing, in mix of species,
      139–140
    production of, 29–30, 292
    as sawnwood, 50–52
    transformation of, Indurite process,
      106–107
Solidwood business, 27. *See also* Mills;
    Sawnwood
Sony, 84
Sorbilite, Inc., 89
Sorbilite composite molding system, 88, 89,
    95
South Africa, eucalyptus plantations in,
    238
South America
    deforestation of, 4
    plantation forestry, 25
    softwood production, 30
Soy-based adhesives, 95, 103–104
Spain, eucalyptus plantations in, 238
Spatial scale, 13, 20, 22

Stakeholder involvement, 2–3, 9, 17, 128, 295
States Industries, Inc., 81
Steelcase, 47
Ston Forestal, 66
STORA, Forest Stewardship Council certification, 6, 8, 24, 44, 60, 71, 262–280, 323
Streams and rivers, 333
  conservation, 26, 290
  restoration, 16
Sunkyong Group, The, 24
Supply. *See also* Global market, quality of wood supply
  convergence with demand, 73–76, 217–218
  fragmented, 41, 48
  gulf between demand and, 27–29, 73–76, 140–141, 204
  and sales chains, 22, 26
  stimulation of, 71–72, 84–85
  of tropical timber, 217–218
Suriname, 38, 222
Sustainable Forest products matrix, xiv
Sustainable Forestry Initiative (SFI), 64, 65, 79, 297
Sustainable forestry management (SFM).
      See also High-yield forestry
  adaptive learning in, 21–22
  challenges for, 19–21, 121–123, 277–278
  as core business strategy, 302–303, 307
  cost of, 8, 122, 123
  definitions, 6, 9, 12–14, 335
  emerging technologies, 93, 95, 110–112.
      See also Engineered wood products
  frameworks for, 305–307, 308
  globalization and, 9–10, 84–85. *See also* Global market
  innovations, 15–19, 25
  integration with business practices, 302–318, 320–321, 323–324
  motivations for, 7, 61–62
  new business opportunities, 39–40, 66–67
  objectives and results, 90
  by "principle and interest," 115–116
  by private owners, 157–187
  social aspects. *See* Social aspects, of sustainable forestry

  strategies, 116–118, 313–318
  in the tropics, 207–215, 216–231
Sustainable Forestry Working Group
  historical background, xiii–xvi
  members, 337–339
Sustained Development Institute, 147
Sweden. *See also* STORA
  certification system, 60
  IKEA, 47, 56, 78
  national economics, 270
  preservation ethic in, 265–267
Swedish Society for Nature Conservation, 266
Swedish Steel Co., 263
Sylvania Forestry, 75
Syre, David, 1

Taxes, 158, 166, 170, 174, 178–182
Techflor Industrial S.A., 247
Tecnoforest del Norte, 209, 210. *See also* Portico S.A.
Temporal aspects, of scale, 13, 20
3M, 304
Timberlands and Wood Products sector, Weyerhaeuser, 282
TimberTracker, 272, 273
Total quality management (TQM), 251, 256–257
Tourism, 18, 303, 317
"Tragedy of the commons," 14, 336
Trappist Abbey Forest, 161, 163, 172–176
Trappist Monks of Guadalupe, Inc., 172
Tree Farm Program, 168, 186, 335
Tree farms. *See* Plantation forests
Trillium Corp. U.S.A., 1
Trim Block Drying Rack system, 100–101
Trim ends. *See* Recycling, of wood waste
Tropical Forest Foundation, 108
Tropical Forestry Action Plan, 9
Tropical forests. *See also individual countries*
  difficulties for SFM in, 216–231
  hardwood, 53, 54–55, 207–215, 218–231
  less-known species, 224, 227
  rate of regeneration, 232 n.12
Trusts, 181, 288–289, 328
Tulalip Tribes, 291
Turner Construction Co., 82
Turner Corp., The, 71

Umbrella certification, 183, 184–186
United Nations Conference on
    Environmental Development
    (UNCED), 9–10, 37, 210
United Nations Environment Program
    (UNEP), 244
United Soybean Board, 103
United States. *See also individual states*
    annual revenue of public *vs.* private
        companies, 112
    cost of fiber production, 33
    hardwood export, 1988–1996, 53
    hardwood production, 31
    market trends, 79–82, 123–128,
        316–318
    mill operations, 24
    off-shore competition, 197–198
    Pacific Northwest, 29, 33, 62, 190. *See
        also* British Columbia; Oregon;
        Washington
    per capita paper consumption, 41
    veneer production, 1996, 48
Uruguay, eucalyptus plantations in, 238,
    246
U.S. Department of Agriculture, 168
U.S. Department of Defense, 73
U.S. Environmental Protection Agency,
    Cluster Rules, 43
U.S. Forest Service, review of certification,
    81

Value-added production, 47, 91, 94, 95,
    105, 112, 189–190
Value assessment, 5, 17–18
"Value chain," xiv, 73, 74, 76
VanNatta Tree Farm, 161, 162, 164–167,
    178
Veneer, 34, 336
    by Collins Pine Co., 127
    marketing, 48–49, 127
    by Moniminee Tribal Enterprises, 140,
        141
    by Weyerhaeuser Co., 283
Vernon Project, 148–156

Wallpaper, 75, 84
Walt Disney World, Inc., 144
Washington, Weyerhaeuser Co., 281–301
Waste recovery contractors, 192
Waste reduction, 305, 306

recycling of wood waste, 93, 95, 99–100,
    190–199
    scanning technologies for, 94, 95,
        104–106
Water quality
    cost of effluent control, 252–253
    plantations and pulp mills in Brazil,
        245, 248
    small woodlots, 183
Watersheds, 5, 11, 17, 168, 290
Water usage, by plantation forests, 245,
    246
Weldwood, 79
West Africa, hardwood production, 52
Western Wood Products Association, 69
Weyerhaeuser Co., xv, 6, 71, 81, 281–301
    in British Columbia, 155
    Collins Pine Co. and, 115
    historical background, 281, 284–289
    productivity of forests, 294
    Siberian joint venture, 29
    sustainable forestry strategy of, 314–316
    tree farming by, 285
Weyerhaeuser Real Estate Co., 282, 284
Wildlife. *See* Endangered species
Willamina Lumber Co., 104
Windhorse Farm, 185–186
Window and door grilles, 199, 200
Wisconsin, Menominee Tribal Enterprises,
    66, 131, 132–148
Wood. *See also* Hardwood; Sawnwood;
    Softwood
    amount converted to processed product,
        36, 90
    average log size, 5
    converted value of, from wood waste,
        194
    financial loss due to cup and warp, 102
    grading, 143, 146
    percent Forest Stewardship Council
        certified, 1995, 8, 60
    transformation with the Indurite
        process, 106–107
    from volume to quality, 15–16, 116
Wood drying technology, 100–101
Wood hardening technology, 95
Woodmaster Co., 308–309
Wood producers
    bottlenecks and solutions, 95
    key needs and constraints, 94, 121

Wood products manufacturers
  bottlenecks and solutions, 95
  conventional, barriers to SFM, 91–93
  marketing by, 124–125, 193, 321–322
  product enhancement by, 305, 306–307,
    310–312, 314
  of semi-finished goods, 224–225
  shared certification costs with, 185
Wood supply. *See* Global market, quality
    of wood supply
Wood waste. *See* Mills, wood waste
    recovery in; Recycling, of wood
    waste
Work. *See* Job opportunities

Workers' compensation, 96
Working conditions, in Brazil plantation,
    243–244
World Bank, 60, 113, 225, 258
World Commission on Forests and
    Sustainable Development, 9–10
World Resources Institute, 65
World Wildlife Fund, 60, 62, 63, 266, 272
WTD Industries, Inc., 103

Xerox Corp., 304
X-ray scanning systems, 105

Yield-Pro scrap recovery system, 100